Free Radica

The Practical Approach Series

SERIES EDITOR

B. D. HAMES
Department of Biochemistry and Molecular Biology
University of Leeds, Leeds LS2 9JT, UK

★ **indicates new and forthcoming titles**

Affinity Chromatography

Anaerobic Microbiology

Animal Cell Culture
  (2nd edition)

Animal Virus Pathogenesis

Antibodies I and II

★ Antibody Engineering

Basic Cell Culture

Behavioural Neuroscience

Biochemical Toxicology

Bioenergetics

Biological Data Analysis

Biological Membranes

Biomechanics—Materials

Biomechanics—Structures and
  Systems

Biosensors

Carbohydrate Analysis
  (2nd edition)

Cell–Cell Interactions

The Cell Cycle

Cell Growth and Apoptosis

Cellular Calcium

Cellular Interactions in
  Development

Cellular Neurobiology

Clinical Immunology

Crystallization of Nucleic
  Acids and Proteins

Cytokines (2nd edition)

The Cytoskeleton

Diagnostic Molecular Pathology
  I and II

Directed Mutagenesis

★ DNA and Protein Sequence
  Analysis

DNA Cloning 1: Core
  Techniques (2nd edition)

DNA Cloning 2: Expression
  Systems (2nd edition)

★ DNA Cloning 3: Complex
  Genomes (2nd edition)

★ DNA Cloning 4: Mammalian
  Systems (2nd edition)

Electron Microscopy in
  Biology

Electron Microscopy in
  Molecular Biology

Electrophysiology

Enzyme Assays

★ Epithelial Cell Culture

Free Radicals
A Practical Approach

Edited by

NEVILLE A. PUNCHARD

Department of Biology and Health Science,
University of Luton

and

FRANK J. KELLY

Cardiovascular Research, The Rayne Institute,
St Thomas' Hospital, London

OXFORD UNIVERSITY PRESS
Oxford New York Tokyo

Oxford University Press, Walton Street, Oxford OX2 6DP

Oxford   New York
Athens   Auckland   Bangkok   Bombay
Calcutta   Cape Town   Dar es Salaam   Delhi
Florence   Hong Kong   Istanbul   Karachi
Kuala Lumpur   Madras   Madrid   Melbourne
Mexico City   Nairobi   Paris   Singapore
Taipei   Tokyo   Toronto

and associated companies in
Berlin   Ibadan

Oxford is a trade mark of Oxford University Press

Published in the United States
by Oxford University Press Inc., New York

Users of books in the Practical Approach Series are advised that prudent
laboratory safety procedures should be followed at all times. Oxford
University Press makes no representation, express or implied, in respect of
the accuracy of the material set forth in books in this series and cannot
accept any legal responsibility or liability for any errors or omissions
that may be made.

A catalogue record for this book is available from the British Library

Library of Congress Cataloging in Publication Data

Free radicals : a practical approach / edited by Neville A. Punchard
and Frank J. Kelly.
(The practical approach series ; 168)
Includes bibliographical references and index.
1. Free radicals (Chemistry)—Analysis—Laboratory manuals.
2. Antioxidants—Analysis—Laboratory manuals.   I. Punchard,
Neville A.   II. Kelly, Frank J.   III. Series.
QP527.F725 1996   574.19'285—dc20   96-11221   CIP

ISBN 0 19 963560 9 (Hbk)
ISBN 0 19 963559 5 (Pbk)

Typeset by Footnote Graphics, Warminster, Wilts
Printed in Great Britain by Information Press, Ltd, Eynsham, Oxon.

Preface

Free radical research has changed dramatically over the last fifty years. In the early days, studies were conducted with high energy radiation and toxic chemicals as the sources of radicals. In the 1990s, however, these type of investigations represent only a small proportion of the enormous amount of work conducted in free radical research. With the realization that oxygen and nitrogen radicals are involved in normal cell metabolism, as well as many disease states, free radical research now features in most disciplines in the life sciences.

Along with the increasing acceptance of free radicals as commonplace and important biochemical intermediates, they have also been implicated in a large number of human diseases. Conclusive evidence of their involvement is, however, lacking in many instances and this has led not surprisingly to some confusion, especially with those researchers on the periphery of the field. This situation has arisen primarily as a consequence of the difficulties in detecting and measuring free radicals that generally have lifetimes measured in microseconds and which are extremely difficult to measure *per se*, not least in the clinical situation. Investigators, in most cases, have to rely on the measurement of products of free radical reactions that are often also transitory in nature.

Despite these difficulties, sensitive and specific methods to detect free radicals and their biological oxidation products at sites of tissue injury are available. These methods are subject to continual improvement. However, developers of these new, or improved techniques, and those who use them, have a responsibility to ensure that they are applied in an appropriate manner, and that the limitations of such methods are always carefully considered.

Although enormous strides forward have been made in understanding the role of free radicals in the pathogenesis of many diseases, there is still a considerable distance to go before we are in a position to use the knowledge to reduce the consequence of, or prevent, many of these diseases. Our primary aim in compiling this book has therefore been to provide detailed practical knowledge for those wishing to undertake free radical research and hopefully to make this exciting aspect of biological research more easily accessible to new investigators.

From the very nature and diversity of the subject, it would be impossible to cover all techniques which are currently employed in free radical research. We have, however, tried to include details of the fundamental techniques in the area, as well as including some promising new techniques which have recently become available.

Lastly, we would like to extend our thanks to the authors not only for the quality of their contributions, but also for their patience during the preparation of this text.

London   F. J. K.
Luton   N. A. P.
March 1996

Contents

Contents

Part II.   Free radical detection: biochemical methods

5. Intrinsic (low-level) chemiluminescence   65

Yury A. Vladimirov

6. Visual assessment of oxidative stress by multifunctional digital microfluorography   83

Makoto Suematsu, Geert W. Schmid-Schönbein, Yuzuru Isimura, and Masaharu Tsuchiya

Part III.   Measurement of free radical products

8. **Peroxides and other products**   119

R. K. Brown and F. J. Kelly

Contents

xiv

Contents

14. Measurement of products of free radical attack on nucleic acids

M. Lindsay Maidt and Robert A. Floyd

Part IV.   Measurement of antioxidants

15. Glutathione

Mary E. Anderson

Contents

Contents

Contributors

MARY E. ANDERSON
Department of Biochemistry, Cornell University Medical College, 1300 York Avenue, New York, NY 10021, USA.

SHARYN G. ARMSTRONG
Cell Biology Group, The Heart Research Institute, 145 Missenden Road, Camperdown, Sydney, NSW 2050, Australia.

COLIN D. BINGLE
DH Department of Toxicology, The Medical College of St Bartholomew's Hospital, Dominion House, 59 Bartholomew Close, London EC1 7ED, UK.

D. R. BLAKE
The Inflammation Research Group, ARC Building, The London Hospital Medical College, 25–29 Ashfield St, London E1 2AD, UK.

R. K. BROWN
Cardiovascular Research, The Rayne Institute, St Thomas' Hospital, London SE1 7EH, UK.

GRAHAM W. BURTON
Steacie Institute for Molecular Sciences, National Research Council of Canada, 100 Sussex Drive, Ottawa, Ontario, Canada K1A OR6.

J. BUTLER
Paterson Institute for Cancer Research, Christie Hospital NHS Trust, Manchester M20 9BX, UK.

CHING K. CHOW
Department of Nutrition and Food Science, University of Kentucky, Lexington KY 40506-0054, USA.

GIDON CZAPSKI
Department of Physical Chemistry, The Hebrew University of Jerusalem, Jerusalem, Israel.

MALGORZATA DAROSZEWSKA
Steacie Institute for Molecular Sciences, National Research Council of Canada, 100 Sussex Drive, Ottawa, Ontario, Canada K1A OR6.

ROGER T. DEAN
Cell Biology Group, The Heart Research Institute, 145 Missenden Road, Camperdown, Sydney, NSW 2050, Australia.

Contributors

ROBERT A. FLOYD
Oklahoma Medical Research Foundation, Department of Biochemistry and Molecular Biology, University of Oklahoma Health Science Center, Oklahoma City, OK 73104, USA.

SHANLIN FU
Cell Biology Group, The Heart Research Institute, 145 Missenden Road, Camperdown, Sydney, NSW 2050, Australia.

STEVEN GIESEG
Cell Biology Group, The Heart Research Institute, 145 Missenden Road, Camperdown, Sydney, NSW 2050, Australia.

SARA GOLDSTEIN
Department of Physical Chemistry, The Hebrew University of Jerusalem, Jerusalem, Israel.

H. R. GRIFFITHS
Division of Chemical Pathology, Centre for Mechanisms in Human Toxicity, University of Leicester, PO Box 138, UK.

M. GROOTVELD
The Inflammation Research Group, ARC Building, The London Hospital Medical College, 25–29 Ashfield St, London E1 2AD, UK.

BARRY HALLIWELL
Neurodegenerative Disease Research Centre, King's College, Manresa Road, London SW3 6LX, UK.

G. E. HAWKES
Department of Chemistry, Queen Mary and Westfield College, Mile End Road, London E1 4NS, UK.

J. HAWKES
Department of Chemistry, Queen Mary and Westfield College, Mile End Road, London E1 4NS, UK.

YUZURA ISHIMURA
Department of Biochemistry, School of Medicine, Keio University, Tokyo, 160, Japan.

HARPARKASH KAUR
Neurodegenerative Disease Research Centre, King's College, Manresa Road, London SW3 6LX, UK.

FRANK J. KELLY
Cardiovascular Research, The Rayne Institute, St Thomas' Hospital, London SE1 7EH, UK.

Contributors

MICHAEL KINTER
Departments of Microbiology and Pathology, University of Virginia Health Sciences Center, Charlottesville, VA 22908, USA.

E. J. LAND
Paterson Institute for Cancer Research, Christie Hospital NHS Trust, Manchester M20 9BX, UK.

RODNEY L. LEVINE
Laboratory of Biochemistry, National Heart, Lung, and Blood Institute, National Institutes of Health, Bethesda, MD 20892-0320, USA.

J. LUNEC
Division of Chemical Pathology, Centre for Mechanisms in Human Toxicity, University of Leicester, PO Box 138, UK.

E. LYNCH
Department of Conservative Dentistry, The Dental Institute, The London Hospital Medical College, Stepney Way, London E1 2AD, UK.

M. LINDSAY MAIDT
Oklahoma Medical Research Foundation, Department of Biochemistry and Molecular Biology, University of Oklahoma Health Sciences Center, Oklahoma City, OK 73104, USA.

RONALD P. MASON
Health Sciences, National Institutes of Health, Research Triangle Park, NC 27709, USA.

DETLEF MOHR
Biochemistry Group, The Heart Research Institute, 145 Missenden Road, Camperdown, Sydney, NSW 2050, Australia.

JASON D. MORROW
Department of Pharmacology and Medicine, Vanderbilt University School of Medicine, Nashville, TN 37232-6602, USA.

D. P. NAUGHTON
The Inflammation Research Group, ARC Building, The London Hospital Medical College, 25–29 Ashfield St, London E1 2AD, UK.

NEVILLE A. PUNCHARD
Department of Biology and Health Science, Faculty of Applied Science and Computing, University of Luton, Park Square, Luton, Beds. LU1 3JU, UK.

L. JACKSON ROBERTS
Department of Pharmacology and Medicine, Vanderbilt University School of Medicine, Nashville, TN 37232-6602, USA.

Contributors

DARET K. ST CLAIR
Graduate Center for Toxicology, University of Kentucky, Lexington, KY 40506-0054, USA.

GEERT W. SCHMID-SCHÖNBEIN
Institute for Biomedical Engineering, University of California, San Diego, La Jolla, CA 92093-0412, USA.

EMILY SHACTER
Division of Hematologic Products, Center for Biologics Evaluation and Research, Food and Drug Administration, Rockville, MD 20852-1448, USA.

EARL L. STADTMAN
Laboratory of Biochemistry, National Heart, Lung, and Blood Institute, National Institutes of Health, Bethesda, MD 20892, USA.

ROLAND STOCKER
Biochemistry Group, The Heart Research Institute, 145 Missenden Road, Camperdown, Sydney, NSW 2050, Australia.

MAKOTO SUEMATSU
Department of Biochemistry and Medicine, School of Medicine, Keio University, Tokyo 160, Japan.

MASAHARU TSUCHIYA
Department of Biochemistry and Medicine, School of Medicine, Keio University, Tokyo 160, Japan.

YURY A. VLADIMIROV
Department of Biophysics, State Medical University of Russia, Malaja Pirogovskaja 1a, Moscow, Russia.

JOY A. WILLIAMS
Division of Hematologic Products, Center for Biologics Evaluation and Research, Food and Drug Administration, Rockville, MD 20852-1448, USA.

Abbreviations

AMProp	2-amino-2-methyl-1-propanol
t-BH	*tert*-butylhydroperoxide
BHT	butylated hydroxytoluene
BSA	bovine serum albumin
BSTFA	*N,O*-bis(trimethylsilyl)trifluoroacetamide
CA	catecholamines
CD	circular dichroism
cDNA	complementary DNA
CFSE	carboxyfluorescein diacetate succinimidyl ester
CHPD	conjugated hydroperoxydiene
CIA	chloroform/isoamyl alcohol mixture
CL	chemiluminescence
cRNA	complementary RNA
cyt(III)	ferricytochrome *c*
cyt(II)	ferrocytochrome *c*
DCF	dichlorofluorescein
DCFH	dichlorofluorescin
DEPC	diethyl pyrocarbonate
DHB	dihydroxybenzoate
DIPE	*N,N′*-diisopropylethylamine
DMA	dimethylamine
DMF	dimethylformamide
DMPO	5,5-dimethyl-1-pyrroline *N*-oxide
DMSO	dimethyl sulphoxide
DNase	deoxyribonuclease
DNPH	2,4-dinitrophenylhydrazine
dNTPs	deoxynucleoside triphosphates
DOPA	3,4-dihydroxyphenylalanine
DTNB	5,5′-dithiobis-(2-nitrobenzoic acid)
DTPA	diethylenetriaminepenta-acetic acid
DTT	dithriothreitol
e_{aq}^-	aqueous electron
EDTA	ethylenediaminetetra-acetic acid
ELISA	enzyme-linked immunosorbent assay
E_m	emission wavelength
EPR	electron paramagnetic resonance
ESR	electron spin resonance
EuT	Eu^{3+}-tetracycline
E_x	excitation wavelength
FACS	fluorescence-activated cell sorting

FADU	fluorescence-activated DNA unwinding technique
GC–MS	gas chromatography–mass spectrometry
GI	gastrointestinal
GPX	glutathione peroxidase
GSH	reduced glutathione
GSSG	glutathione disulphide
HBSS	Hanks' balanced salts solution
HEPES	N-(2-hydroxyethyl)piperazine-N'-(ethanesulphonic acid)
HNA	hydroxynonanal
4HNE	4-hydroxy-2-nonenal
HPLC	high performance liquid chromatography
HPLC-EC	high performance liquid chromatography with electrochemical detection
IgG	immunoglobulin G
LDL	low density lipoprotein
LOO$^{\bullet}$	lipid peroxyl radical
LOOH	lipid hydroperoxides
LPO	lipid peroxidation
mBBr	monobromobimane
MDA	malondialdehyde
MOPS	3-(N-morpholino)propanesulphonic acid
MTBSTFA	N-methyl-N-$tert$-butyldimethylsilyl trifluoroacetamide
NaAc	sodium acetate
NaDOC	sodium desoxycholate
L-NAME	N^{ω}-methyl-L-arginine methyl ester
NBT	nitroblue tetrazolium
NEM	N-ethylmaleimide
NEMP	N-ethyl morpholine
NICI	negative ion chemical ionization
NMA	normal melting agarose
NMR	nuclear magnetic resonance
$^1O_2$	singlet oxygen
$O_2^{\bullet-}$	superoxide anion radical
OA	osteoarthritis
ODS	octadecylsilyl
oligo(dT)	oligonucleotide chain of deoxythymidine
OPA	o-phthaldialdehyde
OPD	o-phenylenediamine
OFR	oxygen free radical
PBN	phenyl-$tert$-butylnitrone
PBS	phosphate-buffered saline
PBST	PBS–Tween
PCR	polymerase chain reaction
PFBB	pentafluorobenzylbromide

PG	prostaglandin
PH	phospholipid hydroperoxides
PI	propidium iodide
PM	photomultiplier
PMA	phorbol myristate acetate
PNK	polynucleotide kinase
POBN	α-(4-pyridyl-1-oxide)-*N-tert*-butylnitrone
PUFA	polyunsaturated fatty acid
Q12	dodecyl-triethylammonium phosphate
RA	rheumatoid arthritis
RNase	ribonuclease
rNTPs	ribonucleoside triphosphates
RNS	reactive nitrogen species
ROS	reactive oxygen species
RS	resorcinol
SA	salicylic acid
SDS	sodium dodecyl sulphate
SIM	single ion monitoring mode
SOD	superoxide dismutase
SSA	5-sulphosalicylic acid
SSC	standard saline–citrate
SSPE	standard saline–phosphate
SU	salicylurate
α-T	α-tocopherol
γ-T	γ-tocopherol
α-TAc	α-tocopheryl acetate
TBA	thiobarbituric acid
TBARS	thiobarbituric acid reactive substance
TBS	Tris-buffered saline
TCA	trichloroacetic acid
TE	Tris–EDTA solution
TEA	triethanolamine
TEP	tetraethoxypropane
TMA	trimethylamine
TMP	tetramethoxypropane
TNB	5-thio-2-nitrobenzoic acid
TPP	triphenylphosphine
TSP	sodium 3-(trimethylsilyl)-[2,2,3,3-^{2}H$_4$]propionate
2VP	2-vinylpyridine
XOO$^{\cdot}$	undefined peroxy radical

1

Introduction

NEVILLE A. PUNCHARD and FRANK J. KELLY

1. Free radicals

A free radical is defined as any atom or molecule that possesses an unpaired electron. It can be anionic, cationic, or neutral. In biological and related fields, the major free radical species of interest have been those of oxygen, that is oxygen free radicals (OFRs). Ground state molecular oxygen (dioxygen) is unusual in that it has two unpaired electrons in its outer orbital and much of its chemical reactivity results from its bi-radical properties. However, although dioxygen is an oxidant, it is relatively unreactive. Step-wise single electron additions to (reduction of) molecular oxygen generates a unique spectrum of more reactive intermediates, the OFRs. The term OFR includes the superoxide anion free radical ($O_2^{\cdot-}$), the hydroxyl radical ($HO^{\cdot}$), and lipid (L) and other (X) peroxy radicals ($LOO^{\cdot}$ and $XOO^{\cdot}$). More recently, through research into nitrous oxide ($NO^{\cdot}$), the active moiety of endothelial derived relaxing factor, and into air-borne pollution, there has been growing interest in nitrogen-centred free radical species such as peroxynitrate ($^{-}OONO^{2}$) and peroxynitrite ($^{-}OONO$).

OFRs are part of a greater group of molecules often called reactive oxygen species (ROS) that are all more strongly oxidizing than molecular oxygen itself. These include hydrogen peroxide (H_2O_2), lipid peroxide (LOOH), singlet oxygen ($^1O_2$), hypochlorous acid (HOCl), and other N-chloramine compounds. Singlet oxygen often confuses newcomers to the area because, although it represents an excited state of oxygen, it is not a free radical. $^1O_2$ exists in two states. In one (delta), both electrons are paired with opposite spin and exist in one orbital leaving the other empty, and in the other state (sigma) the two electrons occupy different orbitals as in the ground state, but are of opposite spins.

OFRs are not the only free radical species. For example, carbonyl, thiyl, and nitroxyl radicals can all exist. Many other free radical species can be formed by biological reactions, for example phenolic and other aromatic species are often formed during xenobiotic metabolism as part of natural drug detoxification mechanisms. Free radicals are also generated by electron-transport chain reactions and play a part in many endogenous synthetic pathways. In addition, because free radical reactions are important in many organic synthetic reactions they are thus of great interest to the industrial chemist.

However, it is in the biological arena where the study of OFR species has attracted by far the greatest interest in recent years. In this field, the majority of studies have been in the areas of chemistry, biochemistry, and pathology. Although the majority of studies have been in human and animals there is increasing interest in the physiological and pathological role of free radicals in plants.

OFRs are potentially very toxic to cells. Due to their highly reactive nature they can readily combine with other molecules, such as enzymes, receptors, and ion pumps, causing oxidation directly, and inactivating or inhibiting their normal function. Some of the products of OFR attack of other molecules can interfere with nucleic acid function, generating alterations in the base sequence with the potential for mutations, leading in extreme pathological situations to cancers or germ-line mutations. Changes in normal proteins and other structures by free radical species can also generate novel immunogenic structures.

One of their most destructive effects is the initiation of lipid peroxidation as this can result in run-away chain reactions, leading to destruction of the cellular membrane. A radical can abstract a hydrogen from polyunsaturated fatty acid (PUFA) in the cell membrane to generate a conjugated diene which, after rearrangement, will readily combine with oxygen to give a lipid peroxyl radical (initiation). This, in turn, can abstract a hydrogen from another PUFA (propagation) to give a lipid hydroperoxide and a new lipid radical that can then repeat the chain of events. If not terminated this chain reaction will lead to destruction of the cell membranes, breakdown of compartmentalization and release of lysosomal enzymes and subsequent autolysis. Termination of this chain of events can only be achieved by the reaction going to completion or by chain-breaking antioxidants that destroy the free radicals produced.

2. Sources of free radicals

OFRs and other free radicals are generated by normal metabolic processes such as the reduction of oxygen to water by the mitochondrial electron transport chain. The ability to utilize oxygen as an energy source is the result of a long evolutionary process during which mitochondria have evolved from symbiotic intracellular organisms. Approximately 2 million years ago blue–green algae began the process of converting the atmosphere of the earth from a reducing into an oxidizing one. The product of their metabolism, oxygen, would have been toxic to much of life that then existed. The symbiotic relationship between primitive bacteria and a eukaryotic ancestor not only allowed this ancestor to survive in the increasingly oxygen-rich, yet toxic atmosphere, but also gave it the potential for generating sufficient amounts of energy to allow more sophisticated organisms, such as mammals, to evolve. However, we still live in a precarious balance with oxygen. Increases above the normal atmospheric 21%, such as seen with hyperbaric oxygen, or exposure to normal oxygen before adequate antioxidant defences are developed, such as with premature babies, still produce toxic reactions and tissue injury.

In the normal metabolic reduction of oxygen to water there is the potential for the generation of free radical species. The biological process is not 100% efficient, and it is considered that between 1 and 5% of all oxygen used in metabolism escapes as free radical intermediates. Oxidative reactions are very important in other biological reactions and many of these also have the potential for generating free radicals under physiological conditions. For example, cytochrome P450s and cyclooxygenase both generate free radical intermediates. Thus, the production of free radicals would appear to be a natural part of cellular metabolism.

One of the major areas of interest in free radicals is in the field of inflammation, where tissue damage produced by free radicals has long been considered to contribute to the injury process. One of the early events in the normal inflammatory reaction is increased activation and migration of leucocytes into the inflamed area. Leucocytes, namely macrophages and neutrophils, possess the enzyme NADPH oxidase, which catalyses the one-electron reduction of oxygen to $O_2^{\bullet-}$ which can, potentially, through a series of reactions catalysed by metal ions generate other OFRs. Leucocytes also possess other enzymes, e.g. myeloperoxidase, which generates HOCl from H_2O_2. Such ROS are normally used by leucocytes to kill ingested or extracellular bacteria. Children suffering from chronic granulomatous disease die from bacterial infections because they have a genetic defect which prevents their leucocytes from generating OFRs.

Unfortunately, free radicals and other oxidants produced at sites of inflammation will, if not controlled, attack the tissues of the host. In addition, metal-containing proteins, for example haemoglobin released from lysed erythrocytes, are released at sites of inflammation. The iron or other metal contained in these proteins can react with free radicals and ROS through a series of reactions called the modified Haber–Weiss or Fenton reaction, to generate other, more toxic, species, such as $HO^{\bullet}$. The importance of this secondary iron-generated ROS in bacterial killing is seen by the ability of some bacteria to escape the immune system simply by having very little iron in their cells.

Tissue damage itself will decouple electron transport chains, release compartmentalized reactions, and generate oxygen and other free radical species. The observation that antioxidants, which protect against OFRs and ROS, also possess anti-inflammatory activity supports the concept that free radicals contribute to tissue damage at sites of inflammation. Confusingly, individuals with chronic granulomatous disease are reported to have normal inflammatory reactions, suggesting perhaps that tissue damage itself or other mechanisms, such as ischaemia (see below), may be more important sources of the free radicals produced during normal tissue damage. However, this does not remove the possibility that leucocytes may act as a major source of ROS-generated tissue damage during abnormal inflammatory reactions.

Another potential source of free radicals is ischaemia. Ischaemia, although normally rare, occurs during myocardial infarctions and the transplantation of organs. As a result of oxygen deprivation ATP is metabolized to xanthine and hypoxanthine, while the enzyme xanthine dehydrogenase is proteolytically converted into xanthine oxidase. Upon reperfusion, and restoration of oxygen, xanthine oxidase metabolizes hypoxathine and xanthine to uric acid, but in doing so produces O_2^-. This production of OFRs plays a major part in generating the tissue damage seen in ischaemia/reperfusion injury, as seen by the ability of inhibitors of xanthine oxidase, i.e. allopurinol, and antioxidants, to protect against such damage in experimental models. However, xanthine oxidase may not be the only source of OFRs during reperfusion. Activation of complement by proteases and other mechanisms leads to activation of neutrophils and their subsequent recruitment into the reperfused tissue. The fact that depletion of neutrophils or prevention of their adhesion to endothelium, the first stage in their recruitment, is also protective in ischaemia/reperfusion injury suggests that these may be an equal, if not more important, source of ROS. Recently, there has been growing interest in the role of the very fine microvascular blood vessels in inflammation and the possibility that activation of leucocytes and other events during inflammation may itself lead to occlusion of the fine blood vessels and result in ischaemia.

In addition to those mentioned above, there are other possible sources of OFRs. For example, ionizing radiation is a very good generator of OFRs. Although, fortunately, exposure to radiation from radioisotopes is a relatively rare event, exposure to ultraviolet radiation is much more common. The effects of sunburn are all too familiar to most of us. The decrease in the protective ozone layer brings with it the increasing risk of exposure to high levels of ultraviolet radiation and the increased risk of skin cancer. In response to this danger there is growing interest in ultraviolet radiation and free radicals in the skin.

There is little doubt that activated leucocytes at sites of inflammation contribute to tissue damage, and conversely that generation of free radicals can damage tissue and generate inflammation. The main debate is about how important is inflammation-induced free radical production, and whether scavenging free radicals with antioxidants can significantly prevent tissue damage, reduce inflammation, and bring about remission of the condition. Chronic inflammatory conditions where production of OFRs has been suggested to be important and where antioxidants might be of therapeutic benefit include rheumatoid arthritis and inflammatory bowel disease, namely Crohn's disease and ulcerative colitis. However, although there is evidence that free radical production occurs in these conditions, it has still to be demonstrated that antioxidants have a therapeutic role.

Other processes of interest that generate OFRs and ROS are those that occur in food. Any food will eventually oxidize on storage, even at temperatures as low as $-70\,^\circ C$. Although most natural foods such as meat, fish, and vegetables contain antioxidants, these can gradually be consumed, or overwhelmed, by oxidation processes. Processed foods, because of antioxidant depletion during production, often have large amounts of synthetic antioxidants added to them, such as butylated hydroxytoluene (BHT). There is thus much interest in the food industry in OFRs. Much of the original chemistry used to study OFRs has been derived from the food industry, the classical example being the measurement of malondialdehyde, which was first designed to measure peroxidation in food.

3. Antioxidants

The importance of oxidative free radical process can be seen by the number and diversity of the biological antioxidants present in cells. Such antioxidants include vitamin C, which acts as a cytosolic antioxidant; vitamin E, which acts as a membrane antioxidant; and glutathione, which acts to protect both cytosol and membranes against free radical attack. Also present are the glutathione-dependent enzymes, glutathione peroxidase (GPX), glutathione reductase, and glutathione transferase, catalase (which breaks down H_2O_2 to oxygen and water), and the enzyme superoxide dismutase (SOD), which converts $O_2^{\cdot}$ into H_2O_2. Some of these enzymes exist in several forms. Membrane, cytosolic, and plasma forms of GPX have all been reported. Similarly, there are mitochondrial, cytosolic, and extracellular forms of SOD. Mitochondrial SOD is very similar to bacterial SOD, confirming the common ancestry of this organelle. Other important antioxidants include carotenoids and ubiquinones. This vast network of intracellular and extracellular anti-oxidant defences is convincing proof that OFRs are produced under normal physiological conditions and that their levels must be tightly regulated for cell survival.

In addition to these natural antioxidants, many other antioxidants are used in the study of OFRs and other ROS, such as dimethyl sulphoxide (DMSO) and BHT. Other chemicals, such as chelating agents, are used to prevent the generation of OFRs and ROS. For example, EDTA and desferrioxamine are employed to remove metal ions that would otherwise react with OFRs to produce more toxic species.

Many antioxidant defences either are, or are dependent on, essential micronutrients. Vitamins C and E, for example, are essential vitamins whereas the antioxidant enzymes SOD and GPX are dependent on the essential micronutrients zinc and selenium, respectively, for their activity. As a consequence there is considerable interest in the role of diet in protection against free radical attack. However, there is still much debate on whether a deficiency of micronutrients exists in the western world with its rich diet, and whether consuming increased amounts of antioxidants offers increased protection. There is also the concern that increased ingestion of antioxidants does not guarantee increased antioxidant protection at the site of potential tissue injury. The absorption and transport of nutrients are regulated very closely by homeostatic processes and, as such, increased availability of a micronutrient does not necessarily follow increased consumption.

The argument is more complex when one considers that the intake of certain substrates prone to free radical attack, such as PUFAs, must also be taken into consideration. There is a vast wealth of literature on the roles of essential, i.e. the ω3 and ω6, and non-essential, i.e. ω9, PUFAs in health and disease. In addition, ω3 fatty acids, such as arachidonic and linolenic acid, and ω6 fatty acids, such as eicosapentanoic acid, can act as substrates for the production of the pharmacological mediators prostaglandins and leukotrienes. The complexity of such relationships is only now beginning to be understood.

4. Measurement of free radicals

There are many ways of studying free radicals, but all have to deal with the major problems associated with their high reactivity, namely their relatively short half-lives and migration distances. These features make the measurement of free radicals very difficult. OFRs can only be studied directly *in vitro* by physico-chemical methods such as electron spin resonance (ESR: refer to Chapter 2 for more details). For practical reasons such methods are limited to *in vitro* studies as are some others, for example pulse radiolysis (Chapter 4).

OFRs can be measured *in vivo* by 'trapping' them using spin traps which can then be measured *ex vivo* by ESR (see Chapter 2 for more details). Similarly, they can be trapped *in vivo* and detected *ex vivo* by their reaction with other chemicals such as salicylic acid (Chapter 7). Utilizing the fact that reactions involving OFRs give rise to chemiluminescence allows their reaction to be monitored both *in vitro* and *in vivo* in exposed organs (Chapter 5).

By far the most common forms of measurement made in this area are those of the products of free radical attack on biological substrates. Virtually all possible targets of OFR attack can be measured, including the products of attack on lipids (Chapters 8–10), proteins (Chapters 11 and 12), carbohydrates (Chapter 13), and nucleic acids (Chapter 14). Methods used in these measurements can be relatively crude and dependent on single chemical derivatization, or sophisticated and dependent on the separation of different products by high performance liquid chromatography (HPLC) and gas chromatography–mass spectrometry (GC–MS). Nuclear magnetic resonance (NMR) offers the ability to study the changes in a range of products simultaneously (Chapter 3) in a wide range of samples with relatively little preparation of the sample.

Measurement of free radicals is not complete without the measurement of defence systems that protect against them. Measurement of antioxidants (Chapter 15), agents or enzymes, can range from simple enzymatic assays to quite complex chemistry. Caution should be exercised before reaching conclusions based on the change in the level of any one antioxidant. Given the nature of enzymatic action and the potential for interaction between the different systems, the overall effect of the antioxidant system is probably much greater than any individual component. Currently, there is much interest in the concept of free radical stress, where the overall balance between generation and production is more important than the measurement of any single component. However, the measurement of such stress using a single system can only be performed *ex vivo* or *in vitro* and those methods that are available use exposure of fluids or tissue to a single OFR and are thus artificial and severely limited.

One of the major problems in measuring free radicals, and their products *in vivo*, is the problem of sampling. Inflamed tissue contains large numbers of leucocytes that can potentially generate free radicals. The process of sampling by biopsy or physical handling and processing of tissues can stimulate these cells to produce free radicals. Free radical measurements can therefore reflect the number of leucocytes present, rather than prior free radical-induced events, hence there is a large potential for misinterpretation of any findings. As in the case of food, biological samples once removed from the body oxidize, even at $-70\,^\circ$C, resulting in changes in levels of products and antioxidants. Thus, in addition to the problems of analysis, there is also the problem of storage.

5. Conclusion

The problems involved in the study of free radicals have to a certain extent slowed down our understanding of the biology of OFRs. We know that free radical production occurs, but its importance is still open to debate. For example, as previously discussed, it is still under debate whether free radicals represent cause or effect in inflammation. To use an analogy, investigators are currently in the situation of detectives investigating a possible murder. They have some of the clues, but can neither recreate the crime scene nor find a body.

Undoubtedly, the effects of free radicals are more complex than simply damaging tissues. Free radicals are involved in the modulation of pharmacological mediators, such as prostaglandins and nitrous oxide. More recently they have been implicated in modulation of adhesion molecule expression, lymphocyte functions, transcription factors, and intracellular signalling. It is quite possible that disturbances in cell signals generated by free radicals are generally more important than cellular damage. Given that mammalian life has evolved with OFRs it would be strange if we had not evolved sensitive systems that respond to, and perhaps even utilize, OFRs and other ROS.

Fortunately, there is also considerable interest in applying the techniques of molecular biology to the study of free radical-related molecules and events (Chapters 16 and 20). Perhaps through the application of these sensitive techniques, in conjunction with the more classical approaches, we will finally be able to gather sufficient clues to end the debate on the role of OFRs. Eventually sufficient information may be available not only to recreate the crime and identify the victim, but also bring the case to trial with a successful prosecution. The case now rests with the detectives. Good luck.

Part I

Free radical detection: physico-chemical methods

2

In vitro and in vivo detection of free radical metabolites with electron spin resonance

RONALD P. MASON

1. Introduction

Electron spin (or paramagnetic) resonance (ESR or EPR) is a spectroscopic technique that detects the unpaired electron present in a free radical. As such, it is the only general approach that can provide direct evidence for the presence of a free radical. In addition, analysis of the ESR spectrum generally enables determination of the identity of the free radical.

A free radical in the magnetic field of a typical ESR spectrometer will result in the unpaired electron occupying one of two energy states (*Figure 1*). These states occur due to the interaction of the free radical's unpaired electron with the magnetic field, and thus exist only when the sample experiences a magnetic field. The energy level of these states varies with the magnetic field strength; at a field strength of 3400 Gauss, the energy difference between the two states corresponds to the energy of 9.4 GHz microwave radiation. The detection of an electron spin resonance spectrum involves increasing the magnetic field through the value at H_0 while simultaneously subjecting the sample to microwave frequency.

The ESR signal is due to the absorption of microwave energy at magnetic field strength H_0, which is detected by a diode. ESR spectra are, for instrumental reasons, recorded as the first derivative of the absorption peak. The concentration of free radicals is proportional to the integral of the absorption signal or, equivalently, to the double integral of the normal first-derivative spectrum. Although ESR is inherently quantum mechanical, an excellent text makes the subject accessible to anybody with even a modest background in mathematics and physics (1).

Early biological ESR studies of enzymes and their free radical metabolites have used one of two techniques, continuous rapid flow or freeze-quench. The freeze-quench method uses cold isopentane (*c.* −140°C) to freeze the free radical-containing mixture within a few milliseconds after initiation of

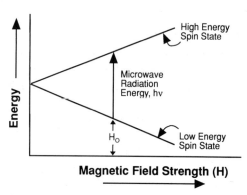

Figure 1. Energy of electron spin states as a function of magnetic field strength.

the enzymic reaction (2). This procedure will 'freeze' or stabilize many free radicals. Although this technique has been very successful in the study of metalloenzyme redox states, the study of free radical metabolites by freeze-quench has been limited almost exclusively to superoxide. Superoxide and many paramagnetic metal ions have linewidths which are so broad at room temperature that ESR signals cannot be detected, and although the use of low temperatures may affect the results, few alternatives are available.

2. Direct free radical detection with electron spin resonance

2.1 Freeze–quench technique

Unfortunately, the freeze–quench technique is of very limited utility in the study of organic free radical metabolites. When a free radical is immobilized in a frozen matrix, a multitude of superimposed spectra (one for each orientation of the free radical in the magnetic field) lead to broad, structureless composite line(s) at $g = 2.00$. Although such a signal proves the formation of a free radical, identification of the free radical metabolite will be ambiguous in a complex biological system. For this reason, this technique is not recommended for free radicals.

2.2 Quartz flat cells

If biological systems are to be studied at room or physiological temperatures, a serious instrumental constraint is imposed by the high dielectric constant of fluid water in the microwave region. If an ordinary capillary is used, the largest volume in the sensitive region of the instrument is about 50 μl. If a flat cell is used in the standard transverse electric (TE$_{102}$) cavity, then 100 μl samples can be used. The transverse magnetic (TM$_{110}$) cavity with a 17 mm wide flat cell, which holds about 200 μl in the sensitive region of the cavity, gives the highest molar sensitivity and resolution possible.

12

2.3 Steady-state condition

In the steady-state condition, the rate of radical formation is equal to the rate of radical decay. Any strategy that will increase the rate of free radical formation or decrease the rate of radical decay will help achieve the necessary 10^{-8}–10^{-7} M steady-state radical concentration.

Since in ESR the upper limit on the sample size is determined by the use of water as the solvent, high concentrations of cells, microsomes, or mitochondria should actually improve the sensitivity of the spectrometer to the extent that protein and membranes replace water. Since optical dispersion is not a limitation in achieving the optimum signal-to-noise ratio, the use of packed cells or microsomal protein concentrations as high as 40 mg/ml will primarily increase the rate of radical formation and, thus, the steady-state radical concentration. Any other approach that will increase the enzyme activity per unit volume should also be employed.

The rate of radical decay can be decreased by many approaches other than lowering the temperature of the enzymatic incubation as in the freeze–quench technique. Many free radical metabolites are aromatic radical cations and anions with decay constants that are pH-dependent. Radical cations, such as benzidine cation radical, are more stable at acidic pH values (3), whereas radical anions, such as metronidazole anion radical, are more stable at basic pH values (4). Orthosemiquinones, such as the catecholamine anion radicals, are stabilized by complex formation with Zn^{2+} (5).

2.4 Fast-flow technique

The rapid-flow technique enables the study of very short-lived free radicals with second-order decay constants near the diffusion limit of 10^{10}/M sec. Excellent reviews of this technique are available (6, 7). The major limitations of the rapid-flow technique are that flow rates as high as 8 ml/sec and a high concentration of enzyme with a fast turnover are needed to achieve the necessary 10^{-8}–10^{-7} M radical metabolite concentration. With few exceptions, these constraints have resulted in this technique being limited to the study of free radicals that are the one-electron oxidation products of horseradish peroxidase such as the acetaminophen phenoxyl radical (8). In this and other examples, the fast-flow technique is absolutely necessary to detect the primary free radical metabolites.

2.5 *In situ* technique

My interests have concentrated on the use of high-resolution ESR investigations to study relatively stable aromatic radical cations and anions with second-order decay constants on the order of 10^{5}/M sec. Many classes of free radicals can be detected under physiological steady-state conditions for periods of several minutes to over an hour at room temperature with the use of very simple procedures (9, 10). Most radical anions react with oxygen to form superoxide. Consequently, anaerobic conditions may be required to achieve the necessary 10^{-8}–10^{-7} M steady-state radical concentrations. The following protocol of a microsomal–xenobiotic incubation illustrates a simple anaerobic technique (*Protocol 1*).

Protocol 1. Detection of nitrobenzene anion radical in an anaerobic microsomal incubation (11)

Equipment and reagents

- NAPD$^+$
- Glucose-6-phosphate
- Glucose-6-phosphate dehydrogenase
- KCl–Tris–MgCl$_2$ buffer: 150 mM KCl, 20 mM Tris (pH 7.4), and 5 mM MgCl$_2$
- Nitrobenzene

- Fresh rat liver microsomes (40 mg protein/ml)
- Rubber stoppered serum bottle (*Figure 2*)
- Nitrogen tank (oxygen-free)
- ESR spectrometer

A. Preparation of incubation mixture

1. Mix nitrobenzene (2 mM) and an NADPH-generating system consisting of NADP$^+$ (0.8 mM), glucose-6-phosphate (11 mM), and 4 units of glucose-6-phosphate dehydrogenase in 3 ml of KCl–Tris–MgCl$_2$ buffer.

2. Add to rubber-stoppered serum bottle.

3. Bubble nitrogen gas into solutions for 5 min with the only exit being through the aqueous flat cell (see *Figure 2*).

4. Add 12 mg of rat hepatic microsomal protein (12) through the rubber stopper with a syringe.

5. Continue bubbling with nitrogen gas for 20 sec.

B. Sample handling

1. Lower the stainless-steel needle tubing below the surface of the solution.

2. Force solution into the aqueous flat cell with pressure of nitrogen gas until full.

3. Close ground glass cap (see *Figure 2*) and vent nitrogen pressure by inserting a second needle into the rubber stopper.

4. Remove needle tubing from the force-fitted septum in the bottom of the flat cell.

5. Mount the flat cell in the microwave cavity with aqueous cell holders.

6. Tune and operate ESR spectrometer to obtain spectrum of nitrobenzene anion radical (*Figure 3*).

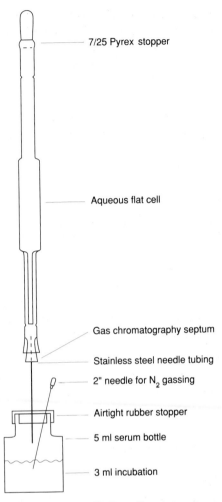

Figure 2. Apparatus for filling the ESR flat cell under a nitrogen atmosphere.

In unfavourable cases, the radical anion can be detected for only a few minutes before substrate depletion or enzyme inactivation occurs. Radical cations are, in general, less stable than radical anions; nonetheless, this *in situ* technique is useful in many cases where peroxidase-catalysed cation radical metabolites can be detected without using the fast-flow technique. If the per-oxidase forms a neutral radical, as in the case of the acetaminophen phenoxyl radical, the fast-flow technique must be used. Neutral radicals typically react with each other rapidly to form dimers due to the absence of charge repulsion, which decreases the decay rate of radical anions and cations.

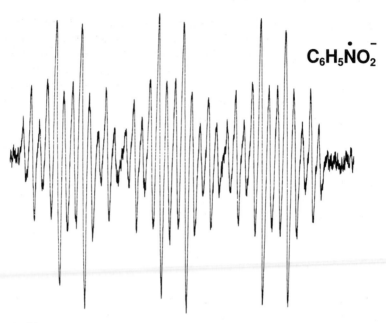

$C_6H_5\dot{N}O_2^-$

Figure 3. ESR spectrum of nitrobenzene anion radical. A total of 54 lines (some unresolved) results from hyperfine interaction of the unpaired electron with the nuclei that have spin, one nitrogen, and four hydrogens (modified from reference 11). Instrumental conditions were as follows: microwave power, 5 mW; and modulation amplitude, 0.63 Gauss.

2.6 Rapid sampling

A more convenient technique is to use aspiration (controlled by a stopcock) or a modified commercial rapid sampler to fill the flat cell. This approach has three major advantages. First, the position of the cell in the cavity is unchanged from one incubation to the next; hence, the tuning of the spectrometer is unchanged. Therefore, separate incubations can be examined with the spectrometer remaining in the operate mode so that only a small disturbance of the crystal current occurs. With this method, the biochemical controls can be examined under identical instrumental conditions. Care must be taken not to introduce air bubbles into the flat cell at any time, because the low dielectric constant of air causes a change in the cavity tuning. Air bubbles, as well as the old incubation, can be satisfactorily removed by the passage of quantities of water or buffer through the flat cell. Second, the tedious adjustment of the flat cell in the cavity is done only once. Third, kinetic information with dead times of a few seconds is obtained easily. The reaction can be initiated by the injection of a small volume of the substrate for the reaction into the serum bottle containing the biological material (purified enzyme, cells, or subcellular fractions). The incubation is then aspirated into the microwave cavity. The elapsed time from the initiation of the reaction in a stirred incubation until the sample is within the microwave cavity is less than 4 sec. The radical concentration time-dependence can be monitored by repetitively sweeping a small segment of the ESR spectrum. Alternatively, the magnetic field can be adjusted to the position of a particular line in the spectrum of an incubation at steady state, which is then replaced by a fresh incubation. This type of kinetic measurement is best done with the use of a field-frequency locking accessory, but satisfactory results can usually be obtained without one if the linewidths are over 0.2 Gauss (see Figure 5 of reference 11). Contamination of the new incubation by the old incubation is not usually a problem if large amounts of water are used to flush the system. However, in one unpublished case, polymerization products formed a precipitated, but active, horseradish peroxidase coating on the inside of the flat cell, which gave the appearance of a non-enzymic reaction when the controls were done. In such cases, the flat cell must be removed for washing with cleaning solution.

3. Spin-trapping detection of free radicals

With this ESR technique, a higher steady-state concentration of free radicals (as radical adducts) is achieved, which can sometimes overcome the sensitivity problem inherent in the detection of free radicals *in vitro* or *in vivo* (14–17). Spin trapping has been the most successful method for the detection of highly reactive free radicals *in vivo*. Since the concentration of endogenous radicals in biological tissues is generally near the sensitivity limit of ESR spectroscopy, the spin-trapping technique is not limited by natural background signals. The technique of spin trapping involves the addition of a primary free radical across the double bond of a diamagnetic compound (the spin trap) to form a radical adduct more stable than the primary free radical. This technique involves the indirect detection of primary free radicals that cannot be directly observed by conventional ESR due to low steady-state concentrations or to very short relaxation times, which lead to very broad lines.

3.1 *In vivo* spin trapping

All of the reported *in vivo* spin-trapping investigations have used the nitrone spin traps phenyl-*tert*-butylnitrone (PBN), α-2,4,6-trimethoxy-PBN ((CH$_3$O)$_3$PBN), α-(4-pyridyl-1-oxide)-*N*-*tert*-butylnitrone (POBN), and 5,5-dimethyl-1-pyrroline *N*-oxide (DMPO). Although the major difficulty of the spin-trapping technique *in vivo* is the mere detection of a radical adduct, other factors must be considered when spin traps are administered *in vivo*. For example, spin traps may affect the experimental system by inhibiting (or stimulating) enzymes or by producing toxicity. These possibilities have not seemed to affect *in vivo* work to date. There are several recent reviews which address the *in vivo* and *in vitro* applications of the spin-trapping technique (14–17). These reviews also discuss the effects of spin traps on enzymes and more general problems of spin trapping such as artefacts and ambiguities in the assignment of radical adduct structure.

Ideally, the radical adduct should be detected in the living animal (i.e. *in vivo* spectroscopy); sample handling would then be unnecessary. Since larger samples can be analysed using low-frequency ESR, this method could theoretically be used to study radical adduct production directly in small animals. Unfortunately, sensitivity is directly dependent on frequency, and low frequency instruments have, with three exceptions (18–20), been unable to achieve the sensitivity needed to detect the low concentrations of radical adducts generated *in vivo*.

3.2 Folch extraction of radical adducts

Results of the first *in vivo* spin trapping study were reported in 1979. In these experiments, the spin trap PBN and carbon tetrachloride were given to a rat through a stomach tube (21).

Protocol 2. *In vivo* spin trapping of the trichloromethyl radical metabolite of carbon tetrachloride (21)

Equipment and reagents

- Male, Sprague–Dawley rats: 250–300 g
- Phenyl-*tert*-butylnitrone (PBN): 1 ml of a 140 mM solution in 20 mM phosphate buffer, pH 7.4
- Carbon tetrachloride: 1.2 ml/kg body weight
- Corn oil
- Chloroform
- Methanol
- Anhydrous sodium sulphate
- Nitrogen tank
- No plasticware (will leach nitroxides into organic solvents)

A. Administration of spin trap and CCl₄

1. Fast the rats for 20 h.
2. Homogenize CCl_4, PBN, or both with corn oil.
3. Administer by stomach tube.

B. Folch extraction and sample handling

1. Kill treated rats after 2 h.
2. Immediately remove livers and homogenize in chloroform–methanol (2:1, v/v) in glass according to reference (22).
3. Dry sample with anhydrous sodium sulphate.
4. Remove chloroform layer and evaporate solvent under nitrogen gas until volume is reduced to 0.5 ml.
5. Transfer sample to 3 mm quartz tube and slowly bubble with nitrogen gas for 3 min using long needle or tubing.
6. Mount sample and tune and operate ESR spectrometer to obtain six-line spectrum of the PBN–trichloromethyl radical adduct (*Figure 4*).

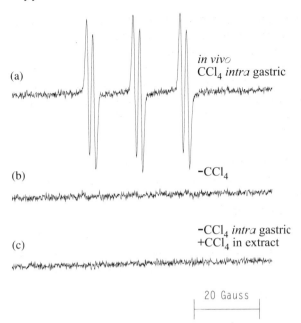

(a)

in vivo
CCl_4 *intra* gastric

(b)

$-CCl_4$

(c)

$-CCl_4$ *intra* gastric
$+CCl_4$ in extract

20 Gauss

Figure 4. Spin trapping *in vivo* of the trichloromethyl radical metabolite of CCl_4: (a) ESR spectrum of the chloroform/methanol (2:1) extract of liver from a rat 1 h post-administration of 0.8 ml/kg CCl_4 i.g. and 70 mg/kg PBN i.p.; (b) ESR spectrum of the extract of liver from a rat given only 70 mg/kg PBN i.p.; (c) as in (b), but liver was homogenized in a 2:1 chloroform/methanol mixture to which 25 μl of CCl_4/g of liver tissue had been added. Instrumental conditions were as follows: microwave power, 20 mW; modulation amplitude, 0.33 Gauss; time constant, 0.25 sec; scan rate, 10 Gauss/min (modified from reference 31).

The chloroform layer contains sufficient water, which has a high dielectric constant, that a conventional 3 mm quartz tube cannot be tuned in the microwave cavity unless the chloroform is first dried with anhydrous sodium sulphate. Alternatively the sample can be cooled to −70°C where the water is frozen, which lowers the dielectric constant of the sample (23). In any case, the sample must be bubbled with nitrogen to deoxygenate the solution because the oxygen present in air-saturated chloroform is high enough to broaden the lines of the ESR spectrum significantly. The greatest advantage of the extraction method is that the radical adduct is transferred from a sample with a high dielectric constant, the biological tissue, to a solvent with a lower dielectric constant, the chloroform solution. This enables the use of larger sample volumes, which increases sensitivity. In addition, the sample can be easily concentrated by evaporation of the chloroform layer. In favourable cases, the sample volume can be concentrated to a few microlitres where the loop-gap resonator can give greatly increased sensitivity (24). The greatest limitation of organic extraction is that only non-polar radical adducts, which are soluble in chloroform, such as the PBN adducts of the trichloromethyl and lipid-derived radicals can be detected. In addition, the radical adduct must be stable enough to survive not only the biological environment in which it was made, but the time required for the solvent extraction and evaporation. Of additional concern are at least two hypothetical problems. First, since the extraction technique requires homogenization of the tissue, this physical destruction may itself lead to free radical formation in some manner (i.e. the release of iron and subsequent hydroxyl radical generation). Second, nucleophilic attack by the radical precursor on a nitrone spin trap forms the corresponding hydroxylamine of the radical adduct. The oxidation of such hydroxylamines to radical adduct impostors is facile. Nucleophilic attack on the nitrone is greatly facilitated by non-aqueous solvents because water, with its high dielectric constant, stabilizes the charged and/or polar reactants in the nucleophilic reaction, and, in general, greatly inhibits this type of chemistry. For example, fluoride is a strong nucleophile in non-polar solvents, but not in water, where it is protonated.

3.3 Detection of radical adducts in biological fluids

A more recent approach to *in vivo* spin trapping has been to examine biological fluids such as urine, bile, and blood (or plasma) directly for spin adducts (16). Deoxygenation of samples is not usually necessary due to the low solubility of oxygen in water (approximately 250 μM), although deoxygenation will occasionally narrow the sharpest lines and is certainly necessary if deuterated PBN is being used to increase resolution (25). An elegant extension of this approach is the detection of the free radicals formed by brain global ischaemia/reperfusion in rat striatal perfusate samples obtained by intracerebral microdialysis (26).

In practice, both the extraction of tissues and the examination of biological fluids should be used. If the Folch extraction method is used, the aqueous/methanol layer should also be examined utilizing a flat cell. Regardless of how the samples are obtained for ESR analysis, the relatively short lifetime of many radical adducts, e.g. DMPO–superoxide radical adduct, will preclude their detection in *in vivo* experiments. Even the relatively stable DMPO–hydroxyl radical adduct, which appears commonly in *in vitro* investigations (as an impurity in the DMPO if nothing else), has never been detected *in vivo*. Clever experimental protocols can sometimes overcome the radical adduct instability problem. For example, in a perfused heart model, Arroyo *et al.* were able to detect DMPO–superoxide radical adduct after ischaemia/reperfusion by freezing the coronary effluent in liquid nitrogen to prevent radical adduct decay (27). The ESR spectrum of DMPO–superoxide was obtained immediately after thawing the effluent in the flat cell. In most cases, for matters of convenience, samples are collected at room temperature and stored at dry ice temperature.

In summary, radicals formed *in vivo* can be spin-trapped and detected by

ESR in Folch extracts of tissue or in biological fluids such as bile, blood, and urine. Only non-polar radical adducts such as those of trichloromethyl and lipid-derived radicals can be detected with organic extraction, whereas ionic free radical metabolites or macromolecules can be detected only in the predominantly aqueous biological fluids. In principle, ionic, polar, and non-polar radical adducts can be detected in bile, which contains biliary micelles.

Protocol 3. Biliary detection of radical adducts of CCl_4-derived free radical metabolites (28–30)

Equipment and reagents

- Male rats: 350–400 g
- CCl_4: 0.64 ml/kg
- PBN: 50 mg/kg dissolved in deionized water at 140 mM
- Oxygen and nitrogen tanks
- Eppendorf tubes

- Dry ice
- Potassium ferricyanide
- Polyethylene tubing (0.28 mm i.d. and 0.61 mm o.d.)
- ESR spectrometer

A. Administration of spin trap and CCl_4

1. Fast the rats for 20 h.
2. Anaesthetize rat with nembutal.
3. Cannulate bile duct with a segment of polyethylene tubing.
4. Inject PBN i.p. and CCl_4 i.g.

B. Collection and treatment of bile

1. Collect bile every 15 min into plastic Eppendorf tubes.
2. Freeze immediately on dry ice and store at $-70\,°C$ until ESR analysis (within a few days).
3. Thaw bile and transfer to quartz flat cell.
4. Bubble with oxygen to oxidize reduced radical adducts and then with nitrogen to narrow the spectral linewidth (or add 0.1–1 mM potassium ferricyanide).
5. Mount the flat cell in the microwave cavity with aqueous cell holders.
6. Tune and operate ESR spectrometer to obtain spectrum of three CCl_4-derived radical adducts.

3.4 *Ex vivo* radical adduct formation

Since the detection of radical adducts occurs *ex vivo*, the question of the possible *ex vivo* formation of these radical adducts needs to be addressed. Even in the original spin-trapping investigation of the *in vivo* formation of the trichloromethyl radical, Lai *et al.* checked for *ex vivo* radical formation by mixing Folch extract from fresh liver tissue, CCl_4, and PBN and then extracting this mixture with chloroform/methanol (21). After the extraction procedure was carried out, no radical adducts were detected.

Although it may not be logically possible to prove that *ex vivo* free radical formation has not occurred, it is possible to demonstrate that it is occurring and to design measures to prevent its occurrence. Trace transition metals catalyse many free radical reactions. Their presence either in the biological sample or in the collection vessel needs to be taken into account. In a study of α-hydroxyethyl radical formation by ethanol-treated alcohol dehydrogenase-deficient deermice, Desferal (a very strong Fe^{3+} chelator) was used to suppress Fenton chemistry during the collection and breaking of gall bladders (32).

In an investigation of *in vivo* hydroxyl radical generation in iron overload, biliary iron caused *ex vivo* hydroxyl radical formation, which could be totally prevented when bile samples were collected into 2,2′-dipyridyl, which stabilizes Fe^{2+} (33). *Ex vivo* radical adduct formation was undetectable even when Fe^{2+} was added to the 2,2′-dipyridyl in the collection tube. This chelator also suppressed a very weak radical adduct signal obtained when animals were administered PBN alone. In a related study of acute copper toxicity (34), the addition of both bathocuproinedisulphonic acid (a Cu(I)-stabilizing chelator) and 2,2′-dipyridyl to the bile collection tube was found to be necessary to inhibit *ex vivo* hydroxyl radical formation.

Ex vivo lipid peroxidation is clearly a concern in all *in vivo* experiments where sample handling and treatment occur and needs to be addressed no matter what technique is used to assess free radical formation. In an investigation of ozone-initiated free radical formation, α-tocopherol was added to the chloroform/methanol solution prior to extraction in an attempt to suppress *ex vivo* lipid peroxidation (35).

In the case of free radical formation in humans, the *ex vivo* experiment is the only one possible because nitrone spin traps have not been administered to humans. When blood taken from the coronary sinus during angioplasty was added to PBN, radical adducts were detected in up to 50% of the samples taken during reperfusion (36). Using this *ex vivo* technique, post-cardioplegia free radical production was detected in coronary sinus blood (37). In such an experiment, the possibility that the free radical was actually formed *in vivo* but trapped *ex vivo* is precluded by the short lifetime (less than 100 msec and, in most cases, much less) characteristic of the highly reactive free radicals which can be spin-trapped. Presumably these radical adducts are formed by the trapping of free radicals formed in the blood *ex vivo*, e.g. through lipid peroxidation.

3.5 Bioreduction of radical adducts

One of the major problems of the *in vivo* detection of radical adducts is the rapid bioreduction of the nitroxide moiety of radical adducts to ESR-silent hydroxylamines. In fact, all four of the radical adducts present initially in the bile of CCl_4-poisoned rats are totally reduced to their respective hydroxylamines and are undetectable (30). The radical adducts can be detected only after the oxidation of their respective hydroxylamines by air, pure oxygen, or ferricyanide. In fact, this *ex vivo* radical chemistry is necessary for the detection of many, if not most, radical adducts formed *in vivo*. The facile bioreduction of radical adducts will be a major problem for *in vivo* spectroscopy, even if the instrumental sensitivity problems can be overcome.

References

1. Weil, J. A., Bolton, J. R., and Wertz, J. E. (ed.) (1994). *Electron paramagnetic resonance. Elementary theory and practical applications.* Wiley, New York.
2. Ballou, D. P. (1978). In *Methods in enzymology* (ed. S. Fleischer and L. Packer), Vol. 54, pp. 85–93. Academic Press, London.
3. Josephy, P. D., Eling, T. E., and Mason, R. P. (1983). *Mol. Pharmacol.*, **23**, 766.
4. Mason, R. P. and Josephy, P. D. (1985). In *Toxicity of nitroaromatic compounds* (ed. D. E. Rickert), pp. 121–140. Hemisphere, New York.
5. Kalyanaraman, B. and Sealy, R. C. (1982). *Biochem. Biophys. Res. Commun.*, **106**, 1119.
6. Borg, D. C. (1972). In *Biological applications of electron spin resonance* (ed. H. M. Swartz, J. R. Bolton, and D. C. Borg), pp. 265–350. Wiley, New York.
7. Yamazaki, I. (1977). In *Free radicals in biology* (ed. W. A. Pryor), Vol. III, pp. 183–218. Academic Press, London.
8. Fischer, V., Harman, L. S., West, P. R., and Mason, R. P. (1986). *Chem.–Biol. Interact.*, **60**, 115.
9. Mason, R. P. (1982). In *Free radicals in biology* (ed. W. A. Pryor), Vol. V, pp. 161–222. Academic Press, London.
10. Mason, R. P. and Chignell, C. F. (1981). *Pharmacol. Rev.*, **33**, 189.
11. Mason, R. P. and Holtzman, J. L. (1975). *Biochemistry*, **14**, 1626.
12. Docampo, R., Mason, R. P., Mottley, C., and Muniz, R. P. A. (1981). *J. Biol. Chem.*, **256**, 10930.
13. Perez-Reyes, E., Kalyanaraman, B., and Mason, R. P. (1980). *Mol. Pharmacol.*, **17**, 239.
14. Mottley, C. and Mason, R. P. (1989). In *Biological magnetic resonance* (ed. L. J. Berliner and J. Reuben), Vol. 8, pp. 489–546. Plenum, New York.
15. Buettner, G. R. and Mason, R. P. (1990). In *Methods in enzymology* (ed. L. Packer and A. N. Glazer), Vol. 186, pp. 127–133. Academic Press, London.
16. Knecht, K. T. and Mason, R. P. (1993). *Arch. Biochem. Biophys.*, **303**, 185.
17. DeGray, J. A. and Mason, R. P. (1994). In *Specialist periodical report—Electron spin resonance* (ed. N. M. Atherton, M. J. Davies, and B. C. Gilbert), Vol. 14, pp. 246–301. Royal Society of Chemistry, Cambridge.
18. Halpern, H. J., Yu, C., Barth, E., Peric, M., and Rosen, G. M. (1995). *Proc. Natl Acad. Sci. USA*, **92**, 796.
19. Jiang, J., Liu, K. J., Shi, X., and Swartz, H. M., (1995). *Arch. Biochem. Biophys.*, **319**, 570.

20. Jiang, J.-J., Liu, K. J., Jordan, S. J., Swartz, H. M., and Mason R. P. (1996). *Arch. Biochem. Biophys.* (in press).
21. Lai, E. K., McCay, P. B., Noguchi, T., and Fong, K.-L. (1979). *Biochem. Pharmacol.*, **28**, 2231.
22. Folch, J., Lees, M., and Stanley, G. H. S. (1957). *J. Biol. Chem.*, **226**, 497.
23. Albano, E., Lott, K. A. K., Slater, T. F., Stier, A., Symons, M. C. R., and Tomasi, A. (1982). *Biochem. J.*, **204**, 593.
24. Kalyanaraman, B., Parthasarathy, S., Joseph, J., and Froncisz, W. (1991). *J. Magn. Reson.*, **92**, 342.
25. Janzen, E. G., Towner, R. A., and Haire, D. L. (1987). *Free Radic. Res. Commun.*, **3**, 357.
26. Zini, I., Tomasi, A., Grimaldi, R., Vannini, V., and Agnati, L. F. (1992). *Neurosci. Lett.*, **138**, 279.
27. Arroyo, C. M., Kramer, J. H., Dickens, B. F., and Weglicki, W. B. (1987). *FEBS Lett.*, **221**, 101.
28. Knecht, K. T. and Mason, R. P. (1988). *Drug Metab. Dispos.*, **16**, 813.
29. Knecht, K. T. and Mason, R. P. (1991). *Drug Metab. Dispos.*, **19**, 325.
30. Sentjurc, M, and Mason, R. P. (1992). *Free Radic. Biol. Med.*, **13**, 151.
31. Hanna, P. M., Kadiiska, M. B., Jordan, S. J., and Mason, R. P. (1993). *Chem. Res. Toxicol.*, **6**, 711.
32. Knecht, K. T., Bradford, B. U., Mason, R. P., and Thurman, R. G. (1990). *Mol. Pharmacol.*, **38**, 26.
33. Burkitt, M. J. and Mason, R. P. (1991). *Proc. Natl Acad. Sci. USA*, **88**, 8440.
34. Kadiiska, M. B., Hanna, P. M., Hernandez, L., and Mason, R. P. (1992). *Mol. Pharmacol.*, **42**, 723.
35. Kennedy, C. H., Hatch, G. E., Slade, R., and Mason, R. P. (1992). *Toxicol. Appl. Pharmacol.*, **114**, 41.
36. Coghlan, J. G., Flitter, W. D., Holley, A. E., Norell, M., Mitchell, A. G., Ilsley, C. D., and Slater, T. F. (1991). *Free Radic. Res. Commun.*, **14**, 409.
37. Tortolani, A. J., Powell, S. R., Mišík, V., Weglicki, W. B., Pogo, G. J., and Kramer, J. H. (1993). *Free Radic. Biol. Med.*, **14**, 421.

3

Detection of free radical reaction products by high-field nuclear magnetic resonance spectroscopy

D. P. NAUGHTON, E. LYNCH, G. E. HAWKES, J. HAWKES,
D. R. BLAKE, and M. GROOTVELD

1. Introduction

Nuclear magnetic resonance (NMR) spectroscopy uses the absorption of energy from the radiofrequency region of the electromagnetic spectrum to detect changes in the alignment of nuclear magnets during exposure to a powerful external magnetic field. The absorption frequency is dependent on the magnetic (and, therefore, chemical) environment of nuclei. Furthermore, the appearance (multiplicity) of a resonance in the ^1H (proton) NMR spectrum of a particular chemical component is influenced by neighbouring hydrogen nuclei in a well-characterized way. Hence, much useful information about the molecules present in a biological sample can be obtained from NMR spectroscopic techniques.

The recent development of high field NMR spectrometers with increased resolution, dynamic range, and sensitivity has permitted the rapid and simultaneous determination of a wide range of components present in biological samples or alternative complex multicomponent systems. The technique is generally non-invasive since it involves only minimal sample preparation (addition of an internal standard and deuterated solvent, the latter to provide a field frequency lock) prior to analysis, and previous high resolution proton (^1H) NMR investigations of human and animal biofluids (e.g. blood plasma, urine, and knee-joint synovial fluid), cell culture media and perchloric acid ($HClO_4$) extracts of cultured cells have provided much useful biochemical and clinical information (1–5). Moreover, the multicomponent analytical ability of high field NMR analysis offers many advantages over alternative methods since chemical shifts, coupling patterns, and coupling constants of resonances detectable in spectra of such samples offer much valuable information regarding the structures of biomolecules present. The broad overlapping resonances which arise from any macromolecules present in untreated samples are routinely suppressed by the application of spin-echo pulse sequences, resulting in spectra which contain many well-resolved, sharp signals attributable to a variety of low-molecular-mass metabolites together with the mobile portions of macromolecules (typically >40 metabolites per spectrum of human plasma at an operating frequency of 600 MHz).

Although NMR spectrometers of operating frequencies >400 MHz are costly and require specialist technical support staff, the technique provides a broad 'picture' of the chemical modifications arising from the reactions of free radicals or related oxidants in complex, multicomponent samples such as intact biofluids, tissue sample extracts, pharmaceuticals, and or foodstuffs, and hence enables the analyst to decide which products (i.e. 'marker' molecules) to determine by cheaper, alternative techniques where required. The sensitivity of this spectroscopic technique can be improved by a factor of 100-fold or greater by the application of pre-concentration procedures (e.g. solid-phase or organic solvent extraction, where appropriate) prior to analysis.

Although there is currently a wide diversity of literature available dealing with the nature, mechanisms, and extent of reactive oxygen species-mediated oxidative damage to various biomolecules (e.g. DNA, polynucleotides, proteins, carbohydrates, and lipids), such studies are only of limited value since they involve assessments of the reactions of selected reactive oxygen species with isolated, single component chemical model systems. However, high-resolution NMR spectroscopy has the capacity to evaluate simultaneously oxidative damage to a wide range of endogenous (or, where appropriate, exogenous) components present in intact biofluids or tissue sample extracts, providing much useful molecular information regarding the relative radical- or H_2O_2-scavenging activities of antioxidants therein.

2. NMR analysis of chemical model systems

2.1 α-Keto acid anions as antioxidants

The reactivity of endogenous and exogenous α-keto acid anions (β-hydroxy-pyruvate, β-phenylpyruvate, 2-ketobutyrate, and 2-ketoglutarate) with hydrogen peroxide (H_2O_2) has been assessed under physiologically relevant conditions (6). Using ^1H NMR spectroscopy, the rate and extent of these reactions were evaluated for the above α-keto acid anions present at a concentration of 1.00 mM. At all H_2O_2 concentrations used, the order of reactivity of the α-keto acid anions was β-hydroxypyruvate > β-phenylpyruvate > 2-ketobutyrate > 2-ketoglutarate. These results are in agreement with a proposed mechanism for these reactions involving nucleophilic attack of the mono-deprotonated peroxide species (HO_2^-) at the C-2 carbonyl group carbon centre. Thus, α-keto acids have a potential use as therapeutic agents in clinical conditions where H_2O_2 has been shown to play a critical role in the disease process, that is, those involving 'oxidative stress'.

2.2 Lipid peroxidation products

NMR spectroscopy has been used to detect a wide range of polyunsaturated fatty acid (PUFA)-derived peroxidation products (e.g. conjugated dienes, epoxides, and oxysterols) in samples of oxidized low density lipoprotein (7). More recently, this technique has been applied to the detection of Cu(II)-induced cholesterol oxidation products (7-ketocholesterol and the 5α,6α- and 5β,6β-epoxides) in isolated samples of plasma low density lipoprotein (8).

Thermal stressing of model PUFA compounds (e.g. alkyl esters of linoleic and linolenic acids) generates a variety of aldehydic components (*n*-alkanals, *trans*-2-alkenals, and *trans,trans*- and *cis,trans*-alka-2,4-dienals) which arise from the degradation of their conjugated hydroperoxydiene precursors, and these lipid peroxidation products are readily detectable by ^{1}H NMR spectroscopic analysis (9). For example, methyl linoleate (2.00 g) was heated in a 10 mm diameter sample tube (sample surface area 0.79 cm^2) and samples, removed at various time-points, were stored in the dark at ambient temperature for 96 h prior to analysis. The unheated (control) sample was stored in the same manner for an equivalent length of time. Samples were analysed using *Protocol 1*. Typical 600 MHz spectra are shown in *Figure 1*. The arrows in the spectra shown in (b) and (c) denote a sharp singlet resonance (δ = 7.97 p.p.m.) generated during episodes of thermal stressing at 180°C.

NMR-detectable conjugated hydroperoxydiene species were also detectable in corresponding ^{1}H NMR spectra of control (unheated) samples of PUFA esters which were allowed to autoxidize in the dark at ambient temperature in the presence of atmospheric O$_2$. As expected, the thermally induced production of aldehydes in such samples was synchronous with a reduction in the intensities of resonances attributable to the conjugated diene systems of *cis,trans*- and *trans,trans*-conjugated hydroperoxydienes.

Protocol 1. Single-pulse and Hahn spin-echo NMR spectroscopic analysis of biofluids

Reagents and equipment

- Sodium 3-(trimethylsilyl)-[2,2,3,3-^{2}H$_4$]propionate (TSP, δ = 0.00 p.p.m.) (Sigma)
- ^{2}H$_2$O (Sigma)
- Millipore 'Ultrafree' (CUFC4 LCC 25) ultrafiltration devices (Millipore UK Ltd)

A. Method

(For Hahn spin-echo NMR spectroscopy omit step 2)

1. Centrifuge biofluid samples to remove all cells and debris (5000 *g* for 15 min).

2. Ultrafilter samples using pre-washed Millipore 'Ultrafree' ultrafiltration devices (5000 *g* for 1 h).

Protocol 1. *Continued*

3. To 0.60 ml of biofluid ultrafiltrate add 0.07 ml of ^2H_2O (to provide a field frequency lock) containing TSP (1.00 mM) as a chemical shift reference. Transfer samples to 5 mm diameter NMR tubes.

4. Determine spectra at 25°C and reference to internal TSP.

5. Typical pulse sequence for the acquisition of spectra is:

   (a) single-pulse: 45° pulse angle, 2.730 sec acquisition time, 2.270 sec pulse delay;

   (b) Hahn spin-echo: (D[90° *x-t-*180°*y-t-*collect]), with *t*=68 msec.

6. Suppress the intense H_2O signal by the application of secondary irradiation at the water frequency.

2.3 Heparin degradation

The c. 50% reduction in anti-factor Xa activity exhibited by heparin resulting from the free radical reaction involving Cu^{2+}, H_2O_2, and ascorbate has been investigated by NMR spectroscopy (10). The site of free radical attack appears to be located adjacent to α-L-iduronate 2-sulphate residues, generating a product with subtle differences from the control (untreated) heparin preparation.

2.4 DNA–protein crosslinks

Margolis *et al.* (11) have used ^1H NMR spectroscopy to investigate spurious radiation-induced DNA–protein crosslink formation. High-performance liquid chromatography (HPLC) analysis of aqueous mixtures of tyrosine and thymine subjected to hydroxyl radical (˙OH) attack resulted in the isolation of novel thymine–tyrosine dimers. ^1H and ^{13}C NMR spectroscopy, and high-resolution mass spectrometric analysis demonstrated that the product was covalently bonded via the thymine methyl group and the C-3 carbon of the tyrosine aromatic ring. Further experiments are required to extend this first characterization of ˙OH-induced DNA–protein crosslinks.

3. NMR analysis of biofluids and tissues

3.1 H_2O_2 scavengers in healthy and diseased human biofluids

Herz *et al.* (12) utilized high field ^1H NMR spectroscopy to assess the hydrogen peroxide (H_2O_2)-scavenging antioxidant capacities of endogenous metabolites present in healthy human blood serum and inflammatory knee-joint synovial fluid. ^1H NMR analysis demonstrated that prolonged equilibration (24 h) of normal serum ultrafiltrates with H_2O_2 gave rise to the oxidative

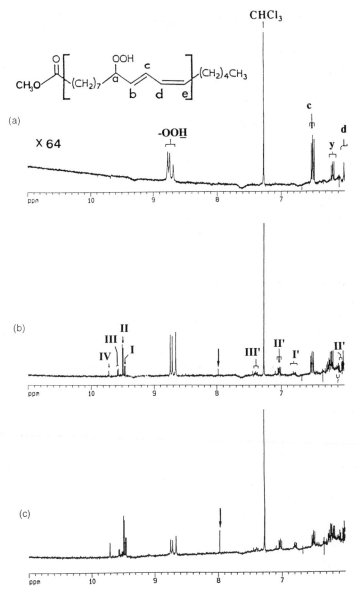

Figure 1. Expanded 6.00–11.00 p.p.m. regions of 600 MHz ^{1}H NMR spectra of a commercially available sample of methyl linoleate acquired (a) before and at (b) 30 and (c) 60 min after heating at a temperature of 180°C in the presence of atmospheric O_2. CHCl$_3$, residual chloroform; c and d, vinylic proton resonances of the conjugated diene systems of 13- and/or 9-hydroperoxy-substituted octadecadienoylglycerol species (the *cis*-9,*trans*-11 and *trans*-10,*cis*-12 isomers, respectively) as denoted in (a); y, a resonance attributable to the vinylic protons of *trans,trans*-conjugated hydroperoxydienes; I, II, III, and IV, aldehydic group (=CHO) protons of *trans*-2-alkenals, *trans,trans*-alka-2,4-dienals, *cis,trans*-alka-2,4-dienals, and *n*-alkanals, respectively; I′, II′, and III′, vinylic proton resonances of *trans*-2-alkenals, *trans,trans*-alka-2,4-dienals, and *cis,trans*-alka-2,4-dienals, respectively; =OOH, hydroperoxy group proton resonances of conjugated hydroperoxydienes.

29

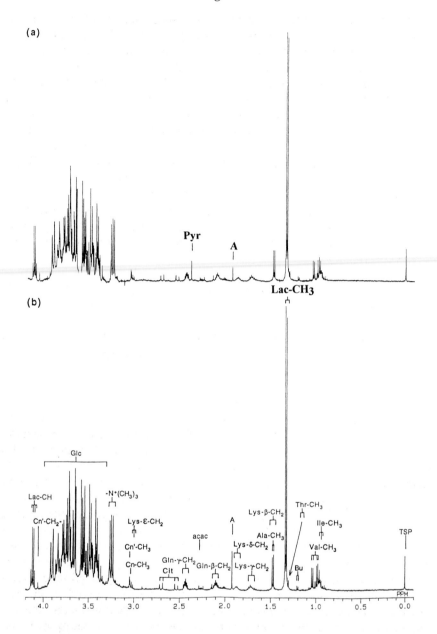

Figure 2. High field (aliphatic) region of 500 MHz single-pulse ^{1}H NMR spectra of a normal serum ultrafiltrate sample obtained (a) before and (b) after equilibration with 2.00 × 10^{-4} mol/dm^3 mM H_2O_2 at ambient temperature for a period of 24 h. Typical spectra are shown. A, acetate; Pyr, pyruvate; Ala, alanine; Lac, lactate; Glc, glucose; acac, acetoacetate; Lys, lysine; Cn, creatine; Cn', creatinine; Gln, glutamine; Cit, citrate; Thr, threonine; Ile, isoleucine; Val, valine; TSP, sodium 3-(trimethylsilyl)-[2,2,3,3-^{2}H$_4$]propionate.

decarboxylation of pyruvate to acetate (*Figure 2*), the rise in acetate concentration being quantitatively accounted for by the reduction in that of pyruvate for all samples investigated. Approximately half of the synovial fluid ultrafiltrate samples examined also showed an increase in acetate concentration that was equivalent to the decrease in that of pyruvate following treatment with H_2O_2 in the above manner. However, for *c.* 35% of the synovial fluid ultrafiltrates the increase in acetate concentration was greater than the reduction in that of pyruvate observed, results indicating that certain of these samples contain a 'catalytic' source with the ability to promote the production of $^{\bullet}OH$ from added H_2O_2, consistent with previous reports (13, 14) that approximately 40% of inflammatory synovial fluids contain trace levels of bleomycin-detectable iron complexes with the capacity to generate $^{\bullet}OH$ from phagocytically generated H_2O_2. Indeed, the elevation in acetate concentration which is not accounted for by the direct reaction of added H_2O_2 with pyruvate appears to arise from the consecutive two-step reaction sequence detailed in eqns 1 and 2.

$$CH_3CHOHCO_2^- \quad -^{\bullet}OH \longrightarrow \quad CH_3COCO_2^- \qquad (1)$$
$$\text{lactate} \qquad\qquad\qquad\qquad\qquad \text{pyruvate}$$

$$CH_3COCO_2^- \quad -^{\bullet}OH\backslash H_2O_2 \longrightarrow \quad CH_3CO_2^- + CO_2 \quad (2)$$
$$\text{pyruvate} \qquad\qquad\qquad\qquad\qquad \text{acetate}$$

3.2 γ-Radiolysis of human biofluids

Grootveld *et al.* (15) employed this technique to investigate radiolytic damage to biomolecules present in human body fluids. Gamma-radiolysis of healthy or rheumatoid human serum (5.00 kGy) in the presence of atmospheric O_2 gave rise to reproducible elevations in the concentration of NMR-detectable acetate which are predominantly ascribable to the prior oxidation of lactate to pyruvate by $^{\bullet}OH$ radical followed by oxidative decarboxylation of pyruvate by radiolytically generated H_2O_2 and/or further $^{\bullet}OH$ (Equations 1 and 2, see *Protocol 1*). Moreover, substantial radiolytical-mediated elevations in the concentration of serum formate, arising from the oxidation of carbohydrates present, by $^{\bullet}OH$ radical, were also detectable. In addition to the above modifications, gamma-radiolysis of inflammatory knee-joint synovial fluid generated a low-molecular-mass oligosaccharide species derived from the radiolytic fragmentation of hyaluronate. The radiolytical-mediated production of acetate in synovial fluid samples was markedly greater than that observed in serum samples, a consequence of the much higher levels of $^{\bullet}OH$ radical-scavenging lactate present. Indeed, increases in synovial fluid acetate concentration were detectable at doses as low as 48 Gy. In most of the biofluids examined, a resonance located at 5.40 p.p.m. attributable to the single NMR-detectable proton of allantoin was produced following gamma-radiolysis, consistent with the ability of endogenous urate to scavenge $^{\bullet}$OH radical as previously suggested by Grootveld and Halliwell (16).

As expected, treatment of lactate with H_2O_2 in aqueous solution did not give rise to any NMR-detectable products (12). However, exposure of lactate to sources of $^{\bullet}$OH radical generated pyruvate, and subsequently acetate and CO_2, providing evidence for its powerful $^{\bullet}$OH radical scavenging ability. In view of the high levels of lactate present in inflammatory synovial fluid and the extremely high second-order rate constant for its reaction with $^{\bullet}$OH radical (4.8×10^9/M sec) (17), this scavenger may play an important role in neutralizing the toxic effects of $^{\bullet}$OH radical arising from any residual H_2O_2 (i.e. that which escapes consumption by pyruvate) in this matrix.

Hence, high-field ^{1}H NMR analysis provides much useful information regarding the relative radioprotectant abilities of endogenous components and the nature, status, and levels of radiolytic products generated in intact biofluids. Interestingly, NMR-detectable radiolytic products with associated toxicological properties (e.g. formate) may play an important role in contributing to the deleterious effects observed following exposure of living organisms to sources of ionizing radiation.

NMR-detectable hyaluronate-derived saccharide fragments are present in untreated aqueous extracts of human synovial membrane obtained from a patient with rheumatoid arthritis (RA) (*Figure 3; Protocol 1*), an observation providing evidence for oxygen radical activity (specifically, that of $^{\bullet}$OH radical) in the inflamed rheumatoid joint.

3.3 Redox-active transition metal ion speciation

Redox-active transition metal ions (specifically those of iron and copper) have been credited with a role in mediating $^{\bullet}$OH radical production from H_2O_2 in biological systems. ^{1}H NMR spectroscopy is an ideal technique for 'speciation' of these ions in biofluids since their paramagnetic nature gives rise to selective broadenings of resonances of endogenous ligands or chelators to which they bind. Using this approach, the authors have demonstrated that addition of the strong iron(III) chelator desferrioxamine to isolated rheumatoid synovial fluid samples enhances the intensity of the citrate resonance in some of the samples investigated in this manner, suggesting that the citrate signal had been pre-broadened by complexation of low-molecular-mass 'catalytic' iron ions therein (18). Iron-citrate chelates readily stimulate the adverse production of oxygen-derived free radical species from H_2O_2 (19).

Low-molecular-mass copper(II) species have been detected and quantified in ultrafiltrates of rheumatoid synovial fluid by a highly sensitive HPLC-based assay system (20). High-field ^{1}H NMR spectroscopy demonstrated that addition of aqueous Cu(II) to isolated samples of rheumatoid synovial fluid resulted in complexation primarily by histidine, formate, alanine, and threonine, and, at higher concentrations, by lactate, tyrosine, and phenylalanine (*Figure 4; Protocol 1*). Circular dichroism (CD) spectra of similar solutions exhibited

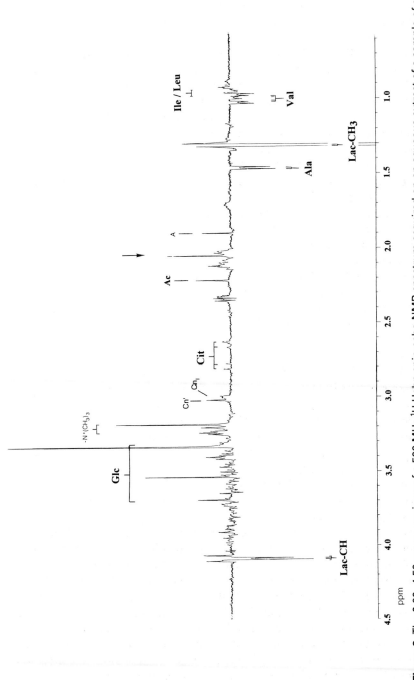

Figure 3. The 0.00–4.50 p.p.m. regions of a 500 MHz ^{1}H Hahn spin-echo NMR spectrum acquired on an aqueous extract of a sample of synovial membrane obtained from a patient with rheumatoid arthritis. A typical spectrum is shown. The arrow denotes the acetamidomethyl group (=NHCOCH_3) protons of a hyaluronate-derived oligosaccharide fragment. Ac, acetone; other abbreviations as in *Figure 2*.

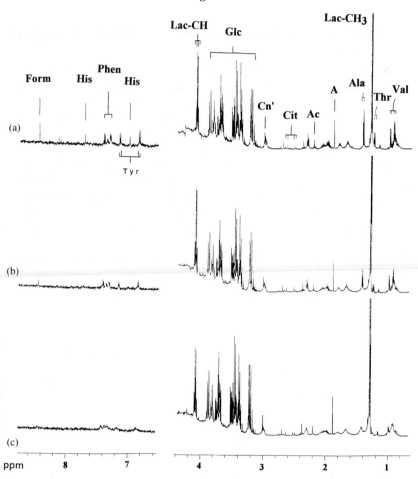

Figure 4. 500 MHz ^{1}H NMR spectra of (a) ultrafiltrate of synovial fluid from a rheumatoid arthritis patient and the same sample following equilibration with Cu(II) at concentrations of (b) 1.40×10^{-5} and (c) 7.20×10^{-3} mol/dm^3. Form, formate; other abbreviations as in *Figure 2*.

absorption bands typical of copper(II)–albumin complexes, in addition to a band attributable to a low-molecular-mass histidinate complex. Since both albumin and histidine are potent ˙OH radical scavengers, these results indicate that any ˙OH radical generated from bound copper ions will be 'site-specifically' scavenged. Hence low-molecular-mass copper complexes with the ability to promote the generation of ˙OH radical which can then escape from the co-ordination sphere of the metal ion (and in turn cause damage to critical biomolecules) appear to be absent from inflammatory synovial fluid.

3.4 Detection of lipid peroxidation products in biological samples

Recent high resolution ^{1}H NMR investigations conducted by the authors have revealed the presence of conjugated hydroperoxydiene species in samples of human atheroma (necrotic gruel isolated from the interior of advanced atherosclerotic plaques in the aorta), confirming results previously acquired by Carpenter *et al.* (21) who utilized a tandem gas chromato-graphic–mass spectrometric technique to determine lipid peroxidation products in such materials. Resonances attributable to *cis,trans*-9- and 13-hydroperoxy-octadecadienoylglycerols were detectable in the 400 MHz ^{1}H NMR spectra of deuterated chloroform (C^2HCl_3) extracts of eight out of nine atheroma samples examined. Corresponding C^2HCl_3 extracts of matched normal artery samples did not contain any NMR-detectable, PUFA-derived conjugated diene adducts.

We have also employed ^{1}H NMR analysis to detect conjugated hydroxydiene species in samples of synovial membrane obtained from patients with inflammatory joint diseases (*Figure 5*).

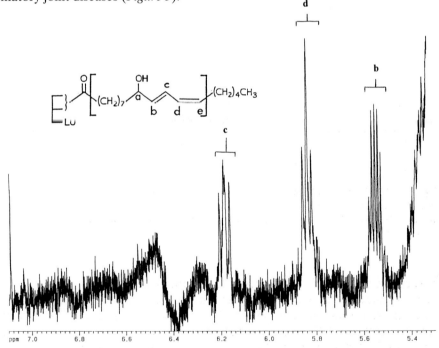

Figure 5. Expanded 5.20–7.10 p.p.m. regions of the 600 MHz ^{1}H NMR spectrum of a C^2HCl_3 extract of synovial membrane obtained from a rheumatoid arthritis patient. Abbreviations are as in *Figure 1*, with b, c, and d representing the vinylic protons of the conjugated diene systems of 13- and/or 9-hydroxy-substituted octadecadienoylglycerol species (the *cis*-9,*trans*-11 and *trans*-10,*cis*-12 isomers, respectively) as denoted for the latter.

3.5 Therapeutic applications of free radical species

Intriguingly, the therapeutic applications of the oxidizing free radical chlorine dioxide ($ClO_2^\bullet$, stabilized via an effective delocalization of its unpaired electron) to periodontal diseases such as marginal gingivitis and halitosis has recently attracted much interest from free radical chemists and biochemists engaged in clinical research. Indeed, a commercially available oral rinse preparation, containing chlorite anion (ClO_2^-) with the capacity to generate significant levels of this stable free radical species at pH values which reflect those of the acidotic oral environment, has been formulated. Chlorine dioxide is liberated via a reaction sequence involving the disproportionation of chlorous acid (eqns 3 and 4),

$$ClO_2^- + H^+ \rightleftharpoons HClO_2 \tag{3}$$

$$4HClO_2 \longrightarrow 2ClO_2^\bullet + ClO_3^- + Cl^- + 2H^+ + H_2O \tag{4}$$

and we have recently applied high resolution ^1H NMR spectroscopy to demonstrate the oxidative consumption of salivary pyruvate by this product, a reaction consistent with *Equation 5* (Lynch *et al.* (unpublished data)).

$$4CH_3COCO_2^- + 2ClO_2^\bullet \longrightarrow 4CH_3CO_2^- + 4CO_2 + Cl_2 \tag{5}$$

However, at pH values closer to neutrality (i.e. those of saliva with a mean pH of 5.96), chlorite anion itself can effect the oxidative decarboxylation of pyruvate to acetate and CO_2 (eqn 6).

$$CH_3COCO_2^- + ClO_2^- \longrightarrow CH_3CO_2^- + CO_2 + OCl^- \tag{6}$$

4. Applications to cell culture systems

4.1 Application of methionine as a reactive oxygen species 'target' molecule

Exposure of the amino acid methionine to H_2O_2 or an $\cdot OH$ radical flux generates methionine sulphoxide as a major product, and Stevens *et al.* (4) have demonstrated the applications of high-field ^1H NMR analysis to the detection of methionine sulphoxide in neutrophil- or endothelial cell-conditioned culture media subsequent to stimulation with phorbol esters (*Figure 6*). These data have supplied valuable information regarding the nature and levels of reactive oxygen species generation by these cells, and the relative susceptibility of culture medium components to oxidative damage.

4.2 Assessment of free radical-based drug resistance in leukaemic cell lines

High-field (^1H) NMR spectroscopy has been employed to investigate and compare the metabolic profiles of vinblastine-sensitive and vinblastine-resistant

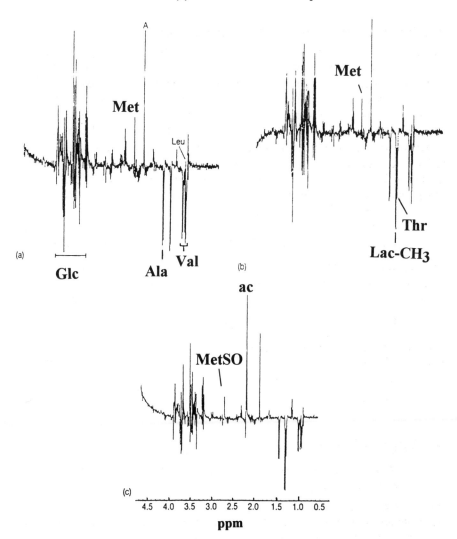

Figure 6. High field (aliphatic) regions of 400 MHz ^{1}H Hahn spin-echo NMR spectra of (a) E-199 culture medium; (b) as (a), but after a 2 h incubation with neutrophils at 37 °C; (c) as (b), but incubated in the presence of phorbol 12-myristate 13-acetate (PMA). The 2.245 p.p.m. singlet detectable in spectrum (c) is attributable to the $=CH_3$ group protons of acetone, the solvent in which PMA was solubilized. Met, methionine; MetSO, methionine sulphoxide; other abbreviations as in *Figure 2*.

T-lymphoid leukaemic cell lines (CCRF-CEM and CEM/VLB$_{100}$ respectively) (5). A significantly lower taurine content in the resistant CEM/VLB$_{100}$ subline (expressed relative to that of its drug-sensitive parental counterpart, *Protocol 2*) was detected. These data suggest differences in the nature and relative involvements of taurine biosynthetic pathways between the two cell lines, a phenomenon that may be related to their differing sensitivities towards chemotherapeutic agents, such as adriamycin, which promote the generation of cytotoxic reactive oxygen species (ROS) *in vivo*. However, the ^{1}H NMR data obtained provided no evidence for an increased metabolic consumption of hypotaurine (a metabolic precursor of taurine with powerful ˙OH radical scavenging properties) in CCRF-CEM cells since differences observed in the hypotaurine:taurine concentration ratio between the drug-sensitive and -resistant cell lines were not statistically significant. Furthermore, hypotaurine is unlikely to compete with alternative endogenous ˙OH radical scavengers present such as lactate since its level in either of the two cell lines investigated (*c.* 6.0×10^{-8} mol/10^8 cells) is insufficient for it to act as an antioxidant in this context. No significant differences in the levels of alternative NMR-detectable endogenous antioxidants were detected.

Interestingly, Kirk and Kirk (22) have recently suggested that the human *mdr*1 gene product P-glycoprotein (PGP), a 170 kDa transmembrane protein which is overexpressed in multidrug-resistant tumour cells, may control the volume regulatory release of taurine from a human lung cancer cell line.

Protocol 2. Single-pulse ^{1}H NMR spectroscopic analysis of perchloric acid extracts of cell lines

Reagents and equipment

- Distilled and deionized water
- NaCl, HClO$_4$, NaOH (Sigma)

- All other reagents and equipment as in *Protocol 1*.

Method

1. Decant cell culture medium, wash cells twice with 0.9% (w/v) NaCl and freeze at $-70\,°$C.

2. Scrape cells into an ice-cold 0.90 M aqueous solution of HClO$_4$ (3.0 ml).

3. Vortex the mixture (1 min) and, after a further 15 min, centrifuge (5000 *g*) at 4°C for 5 min.

4. Remove clear supernatant and neutralize with 9.0 mol/dm^3 NaOH (record final volume).

5. Centrifuge (8000 *g*) at 4°C for 15 min.

6. Lyophilize extracts and store at $-70\,°$C prior to NMR analysis.

7. Reconstitute in 0.60 ml of ^{2}H$_2$O, and add 0.07 ml of ^{2}H$_2$O containing TSP (1.00×10^{-3} mol/dm^3) which serves as a chemical shift reference. The ^{2}H$_2$O provides a field frequency lock.

8. Record spectra in 5 mm NMR tubes at 25°C and refer to internal TSP.

4.3 Protection against paraquat toxicity via thiourea antioxidant activity

The protective effects of thiourea and superoxide dismutase (SOD) on paraquat toxicity in an HL60 cell culture system has been investigated (23). The sulphydryl compound resulting from the reaction of thiourea with superoxide was identified by spectroscopic techniques including ^{13}C NMR analysis.

4.4 Investigation of the mechanism of reperfusion-induced myocardial injury

Myocardial injury, resulting from post-ischaemic reperfusion injury, was investigated in isolated rat ventricular myocytes (24). Exposure to H_2O_2 and Fe(III)-nitrilotriacetate resulted in the production of the highly reactive ˙OH radical (detected by electron spin resonance (ESR) spectroscopy). ^{31}P NMR spectroscopy was used to demonstrate that free radical-induced cellular calcium loading preceded the depletion of ATP. This decrease in intracellular ATP concentration was coupled to the oxidant-induced inhibition of glycolysis, a process giving rise to cellular energy depletion and cell death.

5. NMR analysis of food

5.1 ^{1}H NMR analysis of gamma irradiated food

The development of effective test systems for detecting the irradiation status of foodstuffs is an essential requirement for the establishment of legislative control and consumer choice. Treatment of food with ionizing radiation initially generates extremely reactive free radical species (hydroxyl radicals, hydrated electrons, and hydrogen atoms for food consisting mainly of water) which may generate unique radiolytic products derived from aromatic compounds, DNA, amino acids, carbohydrates, fatty acids, etc. ^{1}H NMR spectroscopy has been used to detect raised formate levels in powdered milk samples which have been subjected to gamma irradiation (25). Formate is a well-known 'end-product' of the radiolytically mediated oxidative deterioration of some common carbohydrates.

The authors have obtained 500 MHz ^{1}H Hahn spin-echo NMR spectra of coded control and gamma-irradiated (5.00 kGy) portions of the aqueous supernatant ($n = 5$ in each group) obtained from a batch of commercially available fresh prawns in order to determine the nature and extent of radiolytic damage to the chemical constitution of these shellfish. The major difference observed between the spectra of control and gamma-irradiated prawn sample supernatants consisted of substantial radiolytical-dependent decreases in the methionine methyl (-S-CH$_3$) group singlet and γ-CH$_2$- group resonances at 2.13 and 2.635 p.p.m. respectively (*Protocol 2*), consistent with the radiolytic depletion of this amino acid. Moreover, the corresponding radiolytic production of a previously undetectable methionine sulphoxide

(-SO-CH_3) group singlet resonance at 2.752 p.p.m., slightly downfield of the intense dimethylamine (DMA) resonance in spectra of gamma-irradiated supernatants, demonstrates the oxidation of NMR-detectable (non-protein-bound) methionine to its corresponding sulphoxide by radiolytically generated ˙OH radical. These data indicate that methionine is a very useful 'target' molecule for radiolytical-generated ˙OH radical and that NMR-detectable methionine sulphoxide is a radiolytic product which might be employed for distinguishing between irradiated and non-irradiated shellfish. Indeed, it was possible to distinguish between control and gamma-irradiated prawn sample supernatants by the absence or presence, respectively, of the methionine sulphoxide -SO-CH_3 group resonance in their ^{1}H Hahn spin-echo spectra. This was achieved without any prior knowledge of the irradiation status of the samples investigated, nor the requirement of an appropriate non-irradiated control sample by the investigators.

The radiolytic depletion of prawn sample methionine and the corresponding production of methionine sulphoxide was also observed in gamma-irradiated (5.0 kGy) supernatants obtained from commercially available frozen prawns (*Figure 7*). For a typical batch of these samples, normalization of the methionine -S-CH_3 group resonance to that of the succinate -CH_2- group gave mean relative signal intensities (methionine -S-CH_3 group signal area/succinate -CH_2- group signal area) of 2.07 $\pm$ 0.06 (mean $\pm$ standard error) for control supernatants ($n = 5$) and 1.20 $\pm$ 0.03 for gamma-irradiated supernatants ($n = 5$).

5.2 ^{1}H NMR analysis of the thermally-induced oxidative deterioration of glycerol-bound polyunsaturates in culinary oils and fat

The most important reaction involved in the thermally induced oxidative deterioration of lipids is the radical-mediated autoxidation of PUFAs, primarily generating conjugated hydroperoxydiene (CHPD) species. Degradation of CHPDs at high temperatures gives rise to numerous aldehydic species which have the capacity to exert toxic effects via their high reactivity with critical biomolecules. We have recently used ^{1}H NMR spectroscopy to monitor the nature and levels of aldehydes generated in culinary oils and fats subjected to episodes of thermal stressing according to standard frying practices (9). Thermal stressing of culinary oils generated high levels of NMR-detectable *n*-alkanals, *trans*-2-alkenals; *trans,trans*- and *cis,trans*-alka-2,4-dienals, and 4-hydroxy-*trans*-2-alkenals via decomposition of their conjugated hydroperoxydiene precursors. For example, a 25 g quantity of groundnut oil was heated in a 100 ml volume conical flask (oil surface area 24.64 cm^2) and a sample removed at 90 min was stored in the dark at ambient temperature for 912 h prior to ^{1}H NMR analysis (the unheated control sample was stored in the same manner for an equivalent length of time). Results are shown in *Figure 8a* and *b*. In *Figure 8c* a 400 MHz two-dimensional COSY ^{1}H NMR

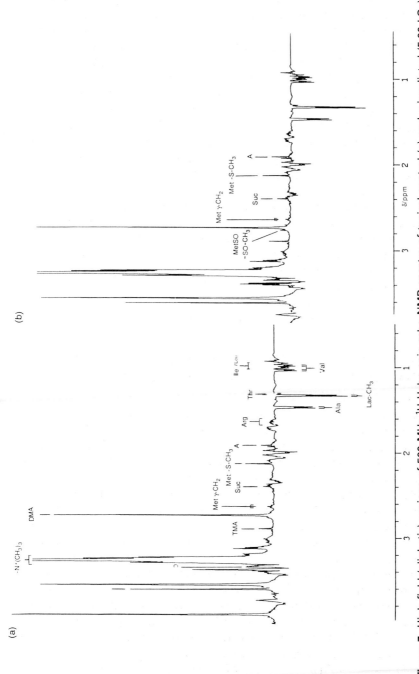

Figure 7. High field (aliphatic) regions of 500 MHz ^{1}H Hahn spin-echo NMR spectra of typical control (a) and γ-irradiated (5.00 kGy) (b), aqueous supernatants derived from a commercially available sample of fresh prawns. TMA, trimethylamine; Suc, succinate; Arg, arginine; other abbreviations as in *Figure 2*.

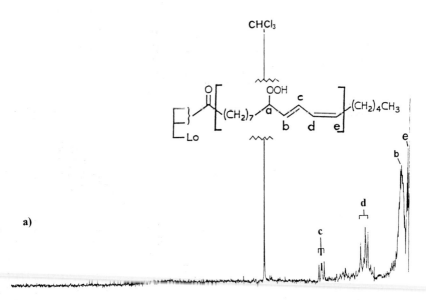

a)

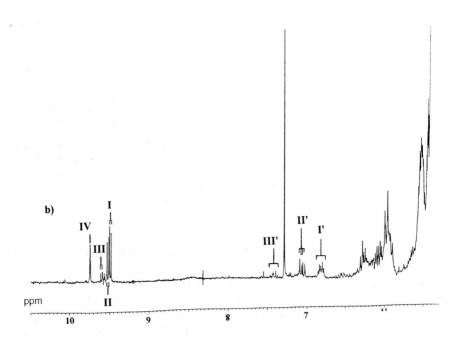

b)

ppm

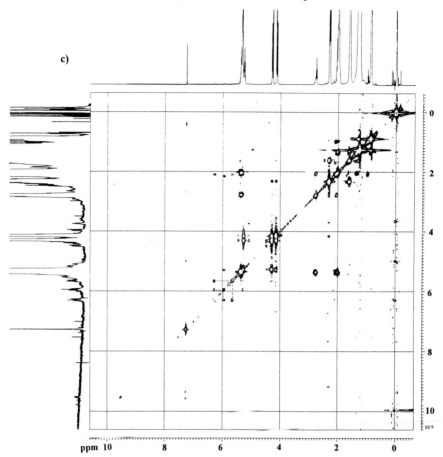

Figure 8. Expanded low field (5.25–10.50 p.p.m.) regions of the 400 MHz ^{1}H NMR spectra of a commercially available sample of groundnut (peanut) oil obtained (a) before and (b) after subjection to an episode of thermal stressing at 180°C for a period of 90 min in the presence of atmospheric O_2. (c) 400 MHz two-dimensional COSY ^{1}H NMR spectrum of a sample of pork fat heated in a domestic fan-assisted oven at 180°C for a period of 90 min. Refer to text for details.

spectrum of a sample of pork fat heated in a domestic fan-assisted oven at 180°C for a period of 90 min is shown. This spectrum reveals clear connectivities between the conjugated diene system proton multiplets centred at 5.93 (*dd*, *j* = 11.0, 10.1 Hz), 6.28 (*dd*, *j* = 13.9, 10.1 Hz) and 5.64 p.p.m. corresponding to the 10-, 11- and 12-position vinylic protons, respectively, of 13-hydroxy-*cis*-9,*trans*-11-octadeca-dienoylglycerol. The solvent employed for the above materials was C^2HCl_3.

Hence, multicomponent analysis of control (unheated) and heated culinary oils and fats by high resolution ^{1}H NMR spectroscopy provides much valuable information regarding the molecular nature and concentrations of PUFA-derived peroxidation products present. Such information is a critical primary requirement for future explorations of the toxicological hazards putatively associated with the frequent consumption of these materials.

With the exception of damage to the gastrointestinal epithelium, the toxicological effects putatively exerted by aldehydes generated in culinary oils and fats during episodes of thermal stressing are, of course, dependent on the rate and level of their *in vivo* absorption from the gut into the systemic circulation. To investigate this phenomenon, we conducted high-field ^{1}H NMR analysis of biofluids, faeces, and tissues obtained from experimental animals (male Wistar albino rats) orally dosed with a typical autoxidized PUFA-derived *trans*-2-alkenal (*trans*-2-nonenal) and found that this α,β-unsaturated aldehyde was indeed absorbed, metabolized, and excreted in the urine as water-soluble 3-position sulphur-substituted mercapturate conjugates (unpublished data). Information previously available in this research area comprises the *in vivo* absorption, metabolism, and urinary excretion of malondialdehyde (MDA) and acrolein, $CH_2{=}CH.CHO$ (the latter being an industrially derived air pollutant which is also present in cigarette smoke); major urinary metabolites arising from the oral administration of these agents to experimental animals are an *N*-acetyl-lysine Schiff base adduct (26), and 3-position-substituted mercapturate conjugates (27), respectively.

The above observations indicate that the *in vivo* absorption of autoxidized PUFA-derived aldehydes is likely to exert a major influence on results obtained from bioanalytical investigations which utilize such molecules as 'markers' of *in vivo* lipid peroxidation (for example, the detection and quantification of *n*-alkanals, MDA, etc. in human biofluids and tissues). A major effort is required to evaluate this phenomenon further.

Acknowledgements

We are very grateful to the Commission of the European Communities (Agriculture and Agroindustry, including Fisheries) and the Arthritis and Rheumatism Council (UK) for financial support and to the University of London Intercollegiate Research Services (Queen Mary and Westfield College and King's College) for the provision of NMR facilities.

References

1. Naughton, D. P., Whelan, M., Smith, E. C., Williams, R., Blake, D. R., and Grootveld, M. (1993). *FEBS Lett.*, **317**, 135.
2. Naughton, D. P., Haywood, R., Blake, D. R., Edmonds, S. E., Hawkes, G. E., and Grootveld, M. (1993). *FEBS Lett.*, **332**, 221.
3. Grootveld, M., Henderson, E. B., Farrell, A., Blake, D. R., Parkes, H. G., and Haycock, P. (1991). *Biochem. J.*, **273**, 459.

4. Stevens, C. R., Bucurenci, N., Abbot, S. E., Sahinoglu, T., Blake, D. R., Naughton, D. P., and Grootveld, M. (1992). *Free Radic. Res. Commun.*, **17**, 143.
5. Jiang, X. R., Yang, M., Morris, C. J., Newland, A. C., Naughton, D. P., Blake, D. R., Zhang, Z., and Grootveld, M. (1994). *Free Radic. Res.,* **19**, 335.
6. Pererra, A., Parkes, H., Herz, H., Haycock, P., Blake, D. R., and Grootveld, M. (1994). *Free Radic. Res.* (in press).
7. Barenghi, L., Bradamante, S., Giudici, G. A., and Vergani, C. (1990). *Free Radic. Res. Commun.*, **8**, 175.
8. Bradamante, S., Barenghi, L., Giudici, G. A., and Vergani, C. (1992). *Free Radic. Biol. Med.*, **12**, 193.
9. Claxson, A. W. D., Hawkes, G. E., Richardson, D. P., Naughton, D. P., Haywood, R., Chander, C. L., Atherton, M., Lynch, E. J., and Grootveld, M. (1994). *FEBS Lett.*, **355**, 81.
10. Liu, Z. and Perlin, A. S. (1994). *Carbohydr. Res.*, **255**, 183.
11. Margolis, S. A., Coxon, B., Gajewski, E., and Dizdaroglu, M. (1988). *Biochemistry*, **27**, 6353.
12. Hertz, H., Blake, D. R., and Grootveld, M. (1995). *Free Radic. Res.*, (in press).
13. Rowley, D., Gutteridge, J. M., Blake, D. R., Farr, M., and Halliwell, B. (1984). *Clin. Sci.,* **66**, 691.
14. Gutteridge, J. M. (1987). *Biochem. J.*, **245**, 415.
15. Grootveld, M., Herz, H., Haywood, R., Hawkes, G. E., Naughton, D. P., Perera, A., Knappit, J., Blake, D. R., and Claxson, A. W. D. (1994). *Radiat. Phys. Chem.*, **43**, 445.
16. Grootveld, M. and Halliwell, B. (1987). *Biochem. J.,* **243**, 803.
17. Anbar, M. and Neta, P. (1967). *Int. J. Appl. Radict. Isot.,* **18**, 493-7.
18. Parkes, H. G., Allen, R. E., Furst, A., Blake, D. R., and Grootveld, M. (1991). *J. Pharm. Biomed. Anal.*, **9**, 29.
19. Gutteridge, J. M. C. and Hou., Y. Y. (1986). *Free Radic. Res. Commun.*, **2**, 143.
20. Naughton, D. P., Knappit, J., Fairburn, K., Blake, D. R., and Grootveld, M. (1995) *FEBS Lett.*, **361**, 167.
21. Carpenter, K. L. M., Taylor, S. E., Ballantine, J. A., Fussell, B., Halliwell, B., and Mithinson, M. J. (1993). *Biochim. Biophys. Acta*, **1167**, 121.
22. Kirk, K. and Kirk, J. (1993). *FEBS Lett.*, **336**, 153.
23. Kelner, M. J., Bagnell, R., and Welch, K. J. (1990). *J. Biol. Chem.*, **265**, 1306.
24. Josephson, R. A., Silverman, H. S., Lakatta, E. G., Stern, M. D., and Zewier, J. L. (1991). *J. Biol. Chem.,* **266**, 2354.
25. Grootveld, M., Jain, R., Claxson, A., Naughton, D. P., and Blake, D. R. (1990). *Trends Food Sci. Technol.,* **1**, 7.
26. McGirr, L. G., Hadley, M., and Draper, H. H. (1985). *J. Biol. Chem.*, **260**, 15427.
27. Draminski, W., Eder, E., and Henschler, D. (1983). *Arch. Toxicol.* **52**, 243.

4

Pulse radiolysis

J. BUTLER and E. J. LAND

1. Introduction

The technique of pulse radiolysis was introduced in 1960 simultaneously by Keene (1) (in these laboratories), Boag and Steel (2), Matheson and Dorfman (3), and McCarthy and McLaughlan (4), ten years after the invention of its light equivalent, flash photolysis, by Porter (5).

The principle of pulse radiolysis is illustrated in *Figure 1*. The absorbance change induced in the sample solution by the pulse of ionizing radiation is monitored by an analysing light beam passing through the sample and reaching a detector (photomultiplier or photodiode) via a monochromator. The detector converts changes in the analysing light intensity into electrical signals. These are digitized and displayed, then stored and treated by microcomputer.

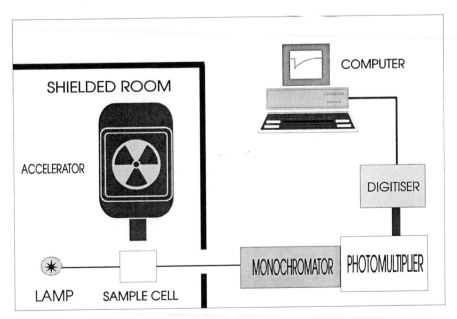

Figure 1. A schematic diagram of pulse radiolysis apparatus.

The transient absorbances thus obtained are calculated as a function of time and wavelength. Conventional absorption spectroscopy remains the most widely used technique for monitoring spectra and kinetics of formation and decay of transient intermediates. However, there are other monitoring techniques such as luminescence, photoacoustic spectroscopy, diffuse reflectance in solids, Rayleigh and Raman scattering, conductivity, polarography, and magnetic resonance. The sources of ionizing radiation include linear accelerators, Febetrons and van de Graaffs. The monitoring light sources are usually xenon arcs, which can be pulsed for studies of very short-lived (<1 μsec) species, or tungsten halide lamps. The sample cells are normally made of quartz with Spectrosil end windows which do not become coloured on irradiation.

Although the technique of pulse radiolysis is of immense value to workers in the field of biologically relevant free radicals, the equipment is very expensive and can be highly specialized. There are thus only a few centres in the world which do this sort of work, and the normal procedure for an interested party is to either have the work done via a collaboration, or go to a centre and do the work under supervision.

2. Design of experiments (see also ref. 6)

The transient species observed on pulse radiolysis are usually due either to excited states or free radicals, the latter tending to be the longer-lived. Sometimes two or more transient species are formed and it is important to be able to resolve the combined absorption spectrum into its individual components. Supplementary analysis by methods such as electron spin resonance (ESR), can sometimes provide hyperfine structure, leading to positive identification of radicals. However, the most important identification method relies upon previous knowledge of the behaviour of the various transient types under the experimental conditions chosen. Such knowledge has been inferred partly from classical fundamental radiation–chemical studies undertaken during the last 40 years, for example, by analysis of stable products. Photolytic studies on similar systems by flash photolysis can also sometimes point the way.

In any dilute solution, it is the solvent which absorbs most of the high energy radiation, and the polarity of the solvent can often aid the assignment of transient species. Polar solvents, e.g. water and methanol, support practically no solute excited states and relatively high yields of radicals, whereas non-polar solvents, e.g. benzene and cyclohexane, support high yields of solute excited states and relatively low yields of radicals. Intermediate solvents such as acetone support moderate yields of both excited states and radicals.

Since the oxygen in aerated solutions can be very reactive towards certain primary and secondary radicals, and indeed excited states, solutions are routinely deoxygenated by flushing with pure nitrogen or argon (unless oxygen is directly involved in the reaction under study). For similar reasons the solvents and solutes which make up the solution under study should be of a high standard of purity.

2.1 Polar solvents

Within 1 nanosecond (10^{-9} sec) of ionizing radiation deposition in water, the following primary radicals are present:

$$H_2O \rightsquigarrow e_{aq}^- + H^{\bullet} + {}^{\bullet}OH \tag{1}$$

Of the three types of radical formed only e_{aq}^- can be observed readily by its broad visible absorption (λ_{max} at 720 nm). The radicals e_{aq}^- and $H^{\bullet}$ are reducing and ${}^{\bullet}OH$ is oxidizing (for yields, see Section 3.2). Thus, for a simple solution of a solute in water, radical products resulting from both one-electron oxidation and reduction of the solute may be expected. There are, however, ways of separating the reductions from the oxidations.

One method of generating exclusively *reducing* radical species is to add a high concentration of sodium formate to water. Formate reacts with oxidizing ${}^{\bullet}OH$ to form $CO_2^{\bullet-}$, which is a reducing species, although less powerful than e_{aq}^-:

$$ {}^{\bullet}OH + HCOO^- \longrightarrow H_2O + CO_2^{\bullet-} \tag{2}$$

Alcohols, e.g. methanol or isopropanol, also convert ${}^{\bullet}OH$ into reducing ${}^{\bullet}CH_2OH$ or $(CH_3)_2CO^{\bullet}H$, respectively. *tert*-Butanol is sometimes used to remove ${}^{\bullet}OH$ since this radical reacts to form ${}^{\bullet}CH_2(CH_3)_2COH$ radicals which are normally very unreactive.

Oxygen is an effective scavenger of both the primary reducing radicals:

$$e_{aq}^- + O_2 \longrightarrow O_2^{\bullet-} \tag{3}$$

$$H^{\bullet} + O_2 \longrightarrow O_2^{\bullet-} + H^+ \tag{4}$$

and a useful way of converting all the primary radicals of water radiolysis into $O_2^{\bullet-}$ is to employ oxygenated formate solutions, $CO_2^{\bullet-}$ rapidly transferring its electron to O_2:

$$CO_2^{\bullet-} + O_2 \longrightarrow CO_2 + O_2^{\bullet-} \tag{5}$$

One method of generating almost exclusively *oxidizing* radical species is to irradiate the aqueous solution saturated with nitrous oxide which converts e_{aq}^- into extra ${}^{\bullet}OH$:

$$e_{aq}^- + N_2O + H_2O \longrightarrow {}^{\bullet}OH + N_2 + OH^- \tag{6}$$

In the presence of N_2O, over 90% of the water radicals formed under such conditions are oxidizing, the remainder being reducing $H^{\bullet}$ atoms.

The radical ${}^{\bullet}OH$ is a vigorous oxidizing agent which tends to add to solutes as well as abstracting electrons or hydrogen atoms. Milder oxidizing agents are the radicals $Br_2^{\cdot-}$ and $N_3^{\cdot}$ which can be produced from $^{\cdot}OH$ by adding a high concentration of the corresponding halide or pseudo-halide:

$$^{\cdot}OH + Br^- \longrightarrow OH^- + Br^{\cdot} \tag{7}$$

$$Br^{\cdot} + Br^- \longrightarrow Br_2^{\cdot-} \tag{8}$$

$$^{\cdot}OH + N_3^- \longrightarrow OH^- + N_3^{\cdot} \tag{9}$$

Thiocyanate can undergo similar reactions (see Section 3.2).

Oxidation, or reduction, of a solute that is insoluble in water alone can sometimes be carried out by adding a small amount (2%) of detergent, e.g. Triton X-100. Radiation-induced one-electron oxidation or reduction of the solute may then occur via electron transfer through the water–detergent interface.

Methanol can be a very useful polar solvent for solutes insoluble in water. A complex mixture of radicals is formed initially on ionizing irradiation of methanol:

$$CH_3OH \rightsquigarrow e_{MeOH}^- + {}^{\cdot}CH_2OH + CH_3O^{\cdot} + H^{\cdot} + {}^{\cdot}OH + CH_3^{\cdot} \tag{10}$$

Many of these radicals react very rapidly with methanol to give more $^{\cdot}CH_2OH$:

$$CH_3O^{\cdot} + CH_3OH \longrightarrow CH_3OH + {}^{\cdot}CH_2OH \tag{11}$$

$$H^{\cdot} + CH_3OH \longrightarrow H_2 + {}^{\cdot}CH_2OH \tag{12}$$

$$^{\cdot}OH + CH_3OH \longrightarrow H_2O + {}^{\cdot}CH_2OH \tag{13}$$

$$CH_3^{\cdot} + CH_3OH \longrightarrow CH_4 + {}^{\cdot}CH_2OH \tag{14}$$

Therefore for studies of dilute solutions in methanol, the 'primary act' reduces to:

$$CH_3OH \longrightarrow e_{MeOH}^- + {}^{\cdot}CH_2OH \tag{15}$$

Both e_{MeOH}^- and $^{\cdot}CH_2OH$ are reducing radicals.

2.2 Non-polar solvents

Some solutes will only dissolve in non-polar solvents, for example, cyclohexane and benzene. Although such solvents mimic the lipid components of cells, radical yields are very low in such media. (Initial radical yields are about $G = 0.01-0.02$ μM/Gy compared with initial excited state yields of about $G = 0.3$ μM/Gy. See section 3.2 for definition of G value.) It is therefore normally only possible to study radicals in such non-polar liquids when the solute radicals formed have very high molar absorption coefficients. Carotenoids are the most important compounds which fit into this category (see Section 4.3).

However, when the triplet excited state of a solute formed pulse-radiolytically can undergo a one-electron reaction with a non-polar solvent (such as cyclohexane), which readily donates H˙ atoms, this can be a source of high yields of, for example, benzophenone ketyl radicals:

$$\phi_2CO^T + C_6H_{12} \longrightarrow \phi_2COH^{\bullet} + C_6H_{11}^{\bullet} \tag{16}$$

This is a key reaction in the energy transfer comparative method for measuring triplet state molar absorption coefficients. Furthermore, if a small amount of duroquinone (10^{-4} M) is added to benzophenone (10^{-1} M) in cyclohexane, the reaction:

$$\tag{17}$$

can be followed. On the other hand, in benzene, which much less readily donates H atoms, benzophenone triplet can react with, for instance, phenol to form the benzophenone ketyl and the phenoxyl radical:

$$\phi_2CO^T + \phi OH \longrightarrow \phi_2COH^{\bullet} + \phi O^{\bullet} \tag{18}$$

3. Physico-chemical parameters

3.1 Handling of raw data from oscilloscopes/digitizers

As mentioned above, the pulse radiolysis detector converts changes in the analysing light intensity into electrical signals which are then digitized, displayed, and stored. The conversion of these signals into absorbance values is achieved by use of the Beer–Lambert–Bouger law in which the initial light intensity before pulsing (I_0) is represented as a voltage on the oscilloscope or digitizer, V_0, and this then changes to another light intensity (I) or voltage V due to a species being formed after pulsing. A typical example* is shown in *Figure 2a*. In this figure, the light level before the pulse is related to a voltage V_0; after the pulse it decreases by 4.3 divisions (i.e. $4.3 \times 50 = 215$ mV). Hence the light level after the pulse is equivalent to $V_0 - 215$ mV. The absorbance after the pulse is given by absorbance $= \log [V_0/(V_0 - 215)]$. In this case, V_0 has been set to 5368 mV, so the absorbance after the pulse is 1.775×10^{-2}.

---

*In the literature, raw oscillograms are sometimes presented, as in *Figure 2*, with decreases in transmission shown downwards, and sometimes with increases in absorption shown as deflections upwards. The philosophy behind the former convention regards a pulse radiolysis apparatus as a time-resolved photometer. The raw data generated thus show changes with time of the light transmitted through an experimental solution, positive changes being depicted upwards and negative changes downwards, as is normal. The latter convention is consistent with the way most absorption spectra are ultimately represented.

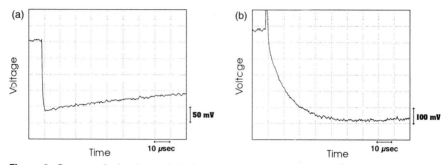

Figure 2. Some typical pulse radiolysis traces. (a) Air-saturated 10^{-2} M KCNS at 500 nm. V_0 = 5368 mV. (b) Oxygen-saturated 50 μM benzoquinone in 0.1 M formate at 400 nm. V_0 = 4860 mV; dose = 6.2 Gy.

It should be noted that, as the majority of pulse data are generated from pulsing a solution in a single optical cell, the reference or blank value is taken from the light level from the solution in the cell before pulsing. Essentially, pulse absorbance measurements always generate difference spectra. In the case of dilute aqueous solutions wherein the solute does not absorb, the spectra would be as seen with a conventional spectrophotometer which has been blanked off against the buffer. However, in many cases, the solute can absorb more than the radicals or excited states and hence 'negative absorptions' are produced. The true absorptions can be obtained from:

$$\Delta Absorbance = optical\ path\ length \times concentration \times (\varepsilon_p - \varepsilon_r)$$

where concentration refers to the concentration of radicals or excited states introduced into the solution; ε_p and ε_r are the molar absorption coefficients of the transient products and the reactants, respectively.

3.2 Dosimetry

Dosimetry is the measurement of the amount of radiation energy absorbed by the medium and is expressed in terms of the Gray (Gy); 1 Gy = J/kg. The concentrations of radicals or excited states produced in the medium as a consequence of this energy absorption are referred to as G values (in units of M/J; can also be simplified to M/Gy). Pulse radiolysis uses highly energetic electrons (typically 1–10 MeV) to irradiate and the G values for the initial radicals and excited states produced in aqueous solutions and in many non-polar solvents have been accurately determined. The G values of the primary radicals formed in a dilute (typically 10^{-2} M) air-saturated aqueous solution at pH 7 are 0.3 μM/Gy for both hydroxyl ($^{\bullet}OH$) radicals and e_{aq}^- and 0.05 μM/Gy for hydrogen atoms ($H^{\bullet}$). These values can vary by as much as 10–20% depending on the solute concentration. The most convenient method in pulse radiolysis of measuring radiation dose and hence radical yields, is to monitor the radiation dose at every pulse using a physical device (e.g. a gas ionization or charge collection chamber). The readings from this device are calibrated at the start of every session with a chemical dosimeter.

Several different chemical dosimeters can be used. However, in our laboratory, the thiocyanate dosimeter is used routinely. This consists of an air-saturated solution of 10^{-2} M potassium thiocyanate (KCNS) in double-distilled water.

In the air-saturated solution, the primary radicals from water (eqn 1) react in this solution:

$$e_{aq}^- + O_2 \longrightarrow O_2^{\bullet-} \tag{3}$$

$$H^{\bullet} + O_2 \longrightarrow O_2^{\bullet-} + H^+ \tag{4}$$

$$^{\bullet}OH + CNS^- \longrightarrow CNS^{\bullet} + OH^- \tag{19}$$

$$CNS^{\bullet} + CNS^- \longrightarrow (CNS)_2^{\bullet-} \tag{20}$$

The absorbance of the $(CNS)_2^{\bullet-}$ is measured at 500 nm (molar absorption coefficient (ϵ) of 7.1×10^3/M/cm). The $O_2^{\bullet-}$ radicals do not absorb in this region (see *Figure 3*). The yield of $(CNS)_2^{\bullet-}$ can be increased by using N_2O saturated solutions as the hydrated electrons are converted into $^{\bullet}OH$ radicals (eqn 6).

Figure 2a is an actual trace produced from pulsing such a solution in an optical cell of 2.5 cm path length. It was determined above that the maximum absorbance is 1.775×10^{-2}, and hence the concentration of $(CNS)_2^{\bullet-}$ radicals can be calculated (using absorbance = concentration $\times$ path length $\times$ ϵ) to be 1.0 μM. As the $(CNS)_2^{\bullet-}$ radicals have been formed from $^{\bullet}OH$ radicals, the G value is the same as $^{\bullet}OH$ in a dilute aqueous solution (ie. 0.3 μM/Gy) and hence the dose given to the solution is 3.3 Gy.

3.3 Spectra

It is obvious from the above examples that the optical spectra of different radicals can readily be obtained from measuring the absorbance of the radicals as a function of wavelength and knowing their concentrations. Many hundreds of such spectra have now been determined. A few relevant examples taken from the literature (7, 8) are given in *Figure 3*.

3.4 Kinetics

One of the main advantages of pulse radiolysis over other techniques is that as the radicals can be directly observed in real time, then the reactivities or kinetics of radical reactions can be directly measured. However, some care must be taken over the design of the experiments.

Consider the reaction of superoxide radicals ($O_2^{\bullet-}$) with a substance, X:

$$O_2^{\bullet-} + X \longrightarrow O_2 + X^{\bullet-}. \tag{21}$$

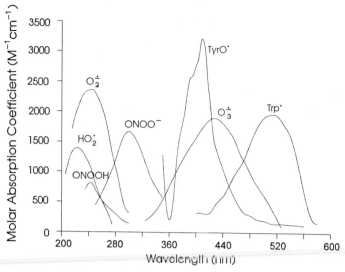

Figure 3. Spectra of various radicals and unstable species.

The $O_2^{\div}$ radicals can be generated from the primary radicals from oxygen-saturated solutions of sodium formate as in eqns 2–5. It is important to ensure that the concentrations are chosen such that neither e_{aq}^-, $^{\bullet}OH$, $H^{\bullet}$, nor the $CO_2^{\div}$ radicals react directly with X. This is achieved by having a relatively high concentration of sodium formate (typically 0.1 M) and an oxygen (1.2 mM) saturated solution. The concentration of X is kept to less than about 1/20 of the oxygen concentration. It is then important to consider the radiation dose given to the solution. If a high dose is given, it is possible that the radicals could react with each other rather than with X. This is not so important with superoxide radicals at physiological pH as the rates are relatively slow. However, it is very important for $^{\bullet}OH$ radical reactions as the rate constant for the $^{\bullet}OH + ^{\bullet}OH$ reaction is 5.5×10^9 M^{-1} sec^{-1}, whereas the rate constants with some compounds can be as low as 10^8 M^{-1} sec^{-1}. In these cases the concentrations of radicals should be kept to at least 50 times less than that of X.

Another problem concerns the wavelengths at which the reactions are to be investigated. *Figure 2b* shows the reaction of $O_2^{\div}$ radicals with benzoquinone (e.g. X in eqn 21) at 400 nm. As neither superoxide radicals nor the parent benzoquinone absorbs significantly at 400 nm, this trace essentially shows the growth of the benzosemiquinone radical. It might be expected that as $O_2^{\div}$ radicals absorb at 240 nm, a trace at this wavelength would show the decay of these radicals over a similar time scale. Unfortunately, benzoquinone and the benzosemiquinone also absorb in this region and the $O_2^{\div}$ radicals absorb relatively weakly (see *Figure 3*). Hence the best wavelength to observe the reaction is at 400 nm.

The rate expressions for reaction 21 are:

$$\frac{-d[O_2^{\bullet}]}{dt} = \frac{-d[X]}{dt} = \frac{d[X^{\bullet}]}{dt} = k\,[O_2^{\bullet}][X] \tag{22}$$

From dosimetry measurements, the concentration of $O_2^{\bullet}$ radicals generated by the pulse was 3.8 μM and these radicals were reacted with 50 μM benzoquinone. It can, therefore, be assumed that the concentration of benzoquinone does not change significantly during reaction and hence pseudo-first order conditions can be applied (X is a constant):

$$\frac{d[X^{\bullet}]}{dt} = k^*\,[O_2^{\bullet}] \tag{23}$$

where, $k^* = k[X]$. As the concentration of $O_2^{\bullet}$ at any time, t, is equivalent to $[(O_2^{\bullet})_0-X^{\bullet}]$ and $(O_2^{\bullet})_0$ is equivalent to the concentration of $X^{\bullet}$ at the end of the reaction $(X^{\bullet})_{\infty}$

$$\frac{d[X^{\bullet}]}{dt} = k^*\,[(X^{\bullet})_{\infty} - (X^{\bullet})_t] \tag{24}$$

The above equation can be integrated to give:

$$\frac{[(X^{\bullet})_{\infty} - (X^{\bullet})_t]}{(X^{\bullet})_{\infty}} = \exp(-k^*t) \tag{25}$$

As absorbance = concentration $\times$ path length $\times$ ϵ, the left-hand expression can be represented as absorbances and hence a plot of ln [absorbance at the end of the reaction (A_{∞}) − Absorbance at any time, t (A_t)] versus time, t, gives a straight line of slope, $-k^*$ and an intercept of $\ln(A_{\infty})$.

From such a plot, a rate of 8.0×10^4/s and hence a rate constant of $1.6 \times 10^9\ M^{-1}\ sec^{-1}$ (the concentration was 50 μM) was derived for the trace in *Figure 2b*. However, to be more certain of the number, the rates at several concentrations of benzoquinone should be determined and a plot of k (sec^{-1}) versus benzoquinone concentration will give a more reliable rate constant.

Some typical rate constants derived by such methods are given in *Table 1*.

3.5 Reduction potentials

Radicals can undergo electron transfer reactions either by donating (reducing) or accepting (oxidizing) electrons. The positions of the equilibria:

$$A^{\bullet} + B \leftrightarrows A + B^{\bullet} \tag{26}$$

or

$$A^{\ddagger} + B \leftrightarrows A + B^{\ddagger} \tag{27}$$

depend simply on the relative ability of the individual couples involved in the reactions to donate or accept electrons. Thus, for example, the equilibrium 26

Table 1. Some relevant rate constants

Reaction	k ($M^{-1}\,sec^{-1}$)
$e_{aq}^- + O_2 \rightarrow O_2^{\cdot -}$	2×10^{10}
$e_{aq}^- + N_2O \rightarrow N_2 + {}^{\cdot}OH + OH^-$	9.1×10^9
${}^{\cdot}OH + {}^{\cdot}OH \rightarrow H_2O_2$	5.5×10^9
${}^{\cdot}OH + HCO_2^- \rightarrow H_2O + CO_2^{\cdot -}$	3.2×10^9
$O^{\cdot -} + O_2 \rightarrow O_3^{\cdot -}$	3.6×10^9
$CO_2^{\cdot -} + O_2 \rightarrow CO_2 + O_2^{\cdot -}$	4.2×10^9
$O_2^{\cdot -} + HO_2^{\cdot} + H^+ \rightarrow H_2O_2 + O_2$ (pH 7)	4.9×10^5
$O_2^{\cdot -} + HO_2^{\cdot} + H^+ \rightarrow H_2O_2 + O_2$ (by SOD)	2.5×10^9
$O_2^{\cdot -} + NO + H^+ \rightarrow ONOOH$	6.7×10^9
${}^{\cdot}OH + NO_2 \rightarrow ONOOH$	4.5×10^9
$ONOOH \rightarrow HNO_3$ (pH 7)	0.5 (sec^{-1})

Values taken from refs 8–10.

is composed of two couples A/A$^{\cdot}$ (meaning A + e$^-$ → A$^{\cdot}$) and B/B$^{\cdot}$ (meaning, B + e$^-$ → B$^{\cdot}$). If B can accept the electron more readily than A, then the equilibrium is to the right-hand side. The relative abilities of the radicals to undergo electron transfer reactions are referred to as reduction potentials. For a simple electron transfer reaction as in eqn 26, the reduction potentials can be defined in terms of the equilibrium constant, K:

$$E(A/A^{\cdot}) = E(B/B^{\cdot}) - RT/F \ln K \qquad (28)$$

where R is the gas constant (8.314 J/K/mol), F is the Faraday constant (96 485 C/mol) and T is the temperature in degrees kelvin. If the units for potential are quoted in millivolts and the temperature is at 25°C, then RT/F is +59 mV.

Pulse radiolysis has been used to determine hundreds of reduction potentials involving many radical couples.* This has been achieved as it is relatively simple to set up equilibria, such as eqn 26 or 27, involving a couple of known reduction potential. As the radicals can be directly observed, the position of the equilibria, and hence the equilibrium constant, can be measured. The reduction potential of the unknown couple can then be calculated by substitution into the above equations.

There are several conditions that have to be considered when setting up an equilibrium. The spectra of the two radicals involved in the equilibrium must first be measured individually and then wavelengths have to be chosen which can be best used to measure the equilibrium. For an equilibrium such as in

---

*Although it is often done, it is incorrect and confusing to talk about a reduction potential of a radical in isolation. Rather, the reduction potential of the radical couple should be quoted, i.e. (A/A$^{\cdot}$), where A is reduced to A$^{\cdot}$ or (A‡/A) where A is oxidized to A‡. The actual numbers for the reduction potentials of the two couples are normally very different.

Table 2. Some one-electron reduction potentials at pH 7.0

Couple	Potential (mV)
$O_2 + e^- \rightleftarrows O_2^{\cdot-}$	-155
$O_2^{\cdot-} + 2H^+ + e^- \rightleftarrows H_2O_2$	$+940$
$^{\cdot}OH + H^+ + e^- \rightleftarrows H_2O$	$+2200$
$(CNS)_2^{\cdot-} + e^- \rightleftarrows 2CNS^-$	$+1330$
$Br_2^{\cdot-} + e^- \rightleftarrows 2Br^-$	$+1660$
$N_3^{\cdot} + e^- \rightleftarrows N_3^-$	$+1330$
$TrpH^{\cdot+} + e^- \rightleftarrows TrpH$	$+1015$
$TyrO^{\cdot} + e^- + H^+ \rightleftarrows TyrOH$	$+930$
Benzoquinone $+ e^- \rightleftarrows$ benzosemiquinone	$+104$

Values taken from refs 11 and 12.

eqn 26, an ideal measurement would include traces which show the loss of $A^{\cdot}$ at one wavelength and the corresponding growth in $B^{\cdot}$ at another wavelength. The rates of attainment of equilibrium at all wavelengths should be the same. As with measuring rate constants (see above), care must also be taken with the radiation dose to ensure that the radical–radical reactions (i.e. $A^{\cdot} + A^{\cdot}$, $B^{\cdot} + B^{\cdot}$, or $A^{\cdot} + B^{\cdot}$ reactions) do not interfere with the attainment of equilibrium. The reduction potentials of some radical couples (11, 12) are given in *Table 2*.

4. Examples

Many of the radicals related to biological and medical systems that have been studied by pulse radiolysis are described in reference 6. Because of space limitations, only three systems are very briefly referred to here, two aqueous and one organic, to illustrate the type of data that can be obtained.

4.1 Oxygen radicals

Some of the earlier uses of pulse radiolysis in the biologically relevant free radical field involved studies on the superoxide radical. Pulse radiolysis was able to show conclusively that superoxide dismutase (SOD) did in fact efficiently dismutate the radicals (13). The method relied on the ability of pulse radiolysis to rapidly produce a known concentration of superoxide radicals in a very short period of time and can be used for many different studies on these radicals.

A solution of 0.1 M sodium formate and 10^{-5} M EDTA in 10^{-2} M potassium phosphate buffer is made up in doubly distilled water containing 10^{-6}– 10^{-4} M reactant, i.e. SOD. The reactions that occur on pulsing this solution are described in Section 2.1. EDTA is included in the solutions to complex possible metal ion impurities, in particular copper, which could also react with the radicals. The superoxide radicals can be observed at 260 nm.

In the case of the SOD study, it was shown that by choosing a low O_2^-:protein ratio (<1:10), it was possible to measure a limit for the initial rate of reaction of O_2^- with the Cu^{2+} of SOD ($k > 10^9$ M^{-1} s^{-1}). It was also shown that the rate of decay of the radicals did not change even when the O_2^-:protein ratio was increased to > 10:1. This proved that SOD is indeed a very good catalyst for the dismutation of superoxide radicals.

The same solutions as above have also been used to show that O_2^- radicals are relatively unreactive towards most biological materials although the more reactive $^{\bullet}OH$ radicals can be generated from them when the O_2^- radicals react with metal ions.

4.2 Radical transfer in proteins

It is acknowledged that free radicals can damage proteins and many studies have been carried out using different free radical generating systems to show that enzymes can be inactivated by radicals and that there are efficient proteolytic systems for removing damaged proteins. However, pulse radiolysis and other radiation techniques can be used to quantify these damaging processes. A simple experiment would involve dissolving a small concentration of an enzyme (a few micromolar) in a non-reactive aqueous buffer (e.g. phosphate) and subjecting the solution to a known dose of radiation (and hence known concentration of radicals; see Section 3.2). If the solution is air saturated, then, due to eqns 1, 3, and 4, the only species initially produced in the solution will be $^{\bullet}OH$ and O_2^- radicals. The loss of enzyme activity as a function of the concentration of these radicals can then be plotted and an estimate of the number of radicals necessary to inactivate the enzyme can be quantified. The individual contributions from $^{\bullet}OH$ and O_2^- can be determined by adding a few micromoles of SOD to the solution or by purging the solution with N_2O (see eqn 6).

For all of the enzyme systems investigated so far, it has been found that several $^{\bullet}OH$ radicals (typically 10–50) are necessary to inactivate one molecule of enzyme. This is as a consequence of the high reactivity of $^{\bullet}OH$ radicals, as they tend to react essentially at the site of collision, and due to the fact that not all reactions are necessarily damaging as far as the enzyme reaction is concerned. The question arises as to which amino acids are more readily damaged than the others. This has been partially answered from pulse radiolysis studies on oxidizing radicals with proteins at a concentration of 10^{-5}–10^{-3} M in N_2O-saturated 10^{-2} M potassium phosphate buffer containing 0.1 M sodium azide (or bromide.)

A similar recipe can be used to selectively generate oxidizing radicals to study other systems.

Essentially, from irradiating N_2O-saturated solutions of proteins containing bromide or azide the corresponding oxidizing radical is generated almost exclusively (see eqns 6–9). In the case of proteins, the oxidizing radicals can selectively oxidize particular amino acid residues. Interestingly, it was found (14 and references therein) that within individual proteins, these oxidized amino acid radicals can transfer their 'damage' to other sites such that, for example, tryptophan radicals (λ_{max} = 510 nm, see *Figure 3*), can oxidize nearby tyrosines to produce tyrosine radicals (λ_{max} = 405 nm) and these can, in some proteins, transfer to cysteines. The order of radical transfer processes within proteins is methionine $\rightarrow$ tryptophan $\rightarrow$ tyrosine $\rightarrow$ cysteine.

The order is consistent with the one electron reduction potentials of the amino acid$^{\bullet+}$/amino acid couples and is consistent with the well-known fact that cysteines are the most readily oxidized of the amino acids.

4.3 Carotenoid radical ions

As alluded to in Section 2.2, the irradiation of solutes in non-polar hydrocarbon solvents can lead to high yields of solute triplet-excited states, formed via fast (<1 nsec) geminate ion recombination, and low yields of solute radical ions, originating from ions which escape such recombination (15). A typical solution contains 10^{-4} reactant, e.g. β-carotene, in argon or N_2O saturated purified hexane.

The dominating feature of the transient spectrum observed after pulse radiolysis of 10^{-4} M β-carotene in argon-flushed hexane is the triplet excited state absorption, λ_{max} = 515 nm, ϵ = 2.4 $\times$ 10^5 M^{-1} cm^{-1}. In the range 600–1280 nm, however, two further peaks are discernible at 880 and 1040 nm, both of which have time characteristics that are quite different from the triplet (16). The rates of formation of these two infrared peaks are in turn themselves quite different from each other. The 880 nm peak was fully developed immediately after a 50 nsec pulse, whereas the 1040 nm peak grew in over 1 μsec after the pulse. The second-order decay rates of the two, however, were very similar. Further evidence that these species were due to radical ions was obtained by showing that the absorption decays matched the rate of decay of the induced conductivity. The 880 nm absorption peak was assigned to the β-carotene radical anion since, (a) it was formed at the very rapid rate ($k \sim 10^{12}$ M^{-1} cm^{-1}) expected for a reaction with mobile electrons in hexane, and (b) it was removed on saturation of the solution with N_2O. The 1040 nm peak, which was unaffected by N_2O, was assigned to the β-carotene radical cation formed by the positive charge transfer from hexane radical cations. Taking into account the known low solvent free ion yields in hexane (see Section 2.2), the molar absorption coefficients of the β-carotene radical anion and cation were estimated to be 4.4 $\times$ 10^5 (880 nm) and 2.2 $\times$ 10^5 (1040 nm) M^{-1} cm^{-1}, respectively.

Due to their extended conjugation, carotenoids may be expected to undergo charge-transfer reactions readily. Indeed, by using pulse radiolysis, a number of carotenoid radical ions have been shown to transfer charge to porphyrins at near diffusion-controlled rates. Carotenoid radical cations can also be detected in 'blocked' photosynthetic reaction centres even though no role has been established in normal photosynthesis. Carotenoids do, however, exhibit anticancer properties, and this role may be related to their ability to trap a variety of oxygen radicals which would otherwise cause DNA damage within cells, leading to carcinogenesis.

Acknowledgement

We are grateful to the Cancer Research Campaign for the support of our studies referred to in this chapter.

References

1. Keene, J. P. (1960). *Nature*, **188**, 843.
2. Doag, J. W. and Steel, R. E. (1960). *British Empire Cancer Campaign Report*, **38**, Part II, 251.
3. Matheson, M. S. and Dorfman, L. M. (1960). *J. Chem. Phys.*, **32**, 1870.
4. McCarthy, R. L. and McLaughlan, A. (1960). *Trans. Faraday Soc.*, **56**, 1187.
5. Porter, G. (1950). *Proc. Roy. Soc., A*, **200**, 284.
6. Bensasson, R. V., Land, E. J., and Truscott, T. G. (1993). *Excited states and free radicals in biology and medicine: contributions from flash photolysis and pulse radiolysis*. Oxford University Press, Oxford.
7. Hug, G. L. (1981). *National Stand. Ref. Data Ser.-NBS*, **69**.
8. Løgager, T. and Sehested, K. (1993). *J. Phys. Chem.*, **97**, 6664.
9. Buxton, G. V., Greenstock, C. L., Helman, W. P., and Ross, A. B. (1988). *J. Phys. Chem. Ref. Data*, **17**, 513.
10. Swallow, A. J. (1978). *Prog. React. Kinet.*, **9**, 197.
11. Wardman, P. (1989). *J. Phys. Chem. Ref. Data*, **18**, 1637.
12. Koppenol, W. H. and Butler, J. (1985). *Free Radic. Biol. Med.*, **1**, 91.
13. Fielden, E. M., Roberts, P. B., Bray, R. C., Lowe, D. J., Mautner, G. N., Rotilio, G., and Calabrese, L. (1974). *Biochem. J.*, **139**, 49.
14. Prutz, W. A., Butler, J., Land, E. J., and Swallow, A. J. (1989). *Int. J. Radiat. Biol.*, **55**, 539.
15. Bensasson, R. V., Land, E. J., and Truscott, T. G. (1983). *Flash photolysis and pulse radiolysis: contributions to the chemistry of biology and medicine*. Pergamon Press, Oxford.
16. Dawe, E. A. and Land, E. J. (1975). *J. Chem Soc., Faraday Trans. 1*, **71**, 2162.

Further Reading

1. Asmus, K.-D. (1984). Pulse radiolysis methodology. In *Methods in enzymology* (ed. L. Packer), Vol. 105, p. 167. Academic Press, London.
2. Bensasson, R. V., Land, E. J., and Truscott, T. G. (1983). *Flash photolysis and pulse radiolysis: contributions to the chemistry of biology and medicine*. Pergamon Press, Oxford.

3. Bensasson, R. V., Land, E. J., and Truscott, T. G. (1993). *Excited states and free radicals in biology and medicine: contributions from flash photolysis and pulse radiolysis*. Oxford University Press, Oxford.
4. Swallow, A. J. (1978). *Reactions of free radicals produced from organic compounds in aqueous solution by means of radiation*. In *Progress in reaction kinetics* (ed. K. R. Jennings and R. B. Cundall), Vol. 9, p. 195. Pergamon Press, Oxford.
5. Tabata, Y., Ito, Y., and Tagawa, S. (1991). *Handbook of radiation chemistry*. CRC Press, Boca Raton, FL.
6. Von Sonntag, C. (1987). *The chemical basis of radiation biology*. Taylor and Francis, London.

Part II
Free radical detection: biochemical methods

5

Intrinsic (low-level) chemiluminescence

YURY A. VLADIMIROV

1. Introduction

The first attempts to measure the intrinsic low-level ultraviolet emission ('Mitogenetic rays') of living cells and biological tissues were made by the Russian scientist Alexander Gurvich who used different dividing cells as biological detectors of ultraviolet (UV) radiation. More recently, gas-discharging quartz tubes with inert metal cathodes as UV-photon counters have been used (see ref. 1 for review).

Today the intrinsic low-level light emission (chemiluminescence, CL) by living cells and tissues is measured by very sensitive photomultipliers (PMs). Previously these were cooled either by liquid nitrogen (2) or by solid carbon dioxide (3) to reduce PM dark current noise and, in this way, increase the instrument sensitivity. Today PMs are sensitive enough to detect cell and tissue CL without cooling and are thus easier to use. During the last 40 years, different terms have been used (ultra-weak luminescence (2), dark luminescence, low-level CL, ultraweak CL, biophoton emission, ultraweak photon emission, intrinsic CL) to denote what we now call intrinsic cell and tissue CL (4).

2. Instruments

In most measurements of low-level CL performed up to now, various 'homemade' instruments have been used. In our laboratory we used the device shown in *Figure 1*. The first version of this instrument was described in 1972 (5). It can be used in all experiments described in this chapter except those in Section 3.1.

The following elements are essential for accurate measurement of intrinsic low-level CL of tissue homogenates and suspensions of cells, cell organelles, biomembranes, blood lipoproteins and other biochemical objects:

1. *a cuvette* of glass or transparent colourless plastic, with low after-glow level after exposure to daylight;

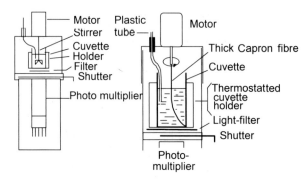

Figure 1. The main parts of the chemiluminometer used to measure the kinetics of low-level chemiluminescence in different biochemical systems.

2. *a thermostatted cuvette holder* with a built-in electronic device to allow the pre-set temperature to be maintained with an accuracy of $\pm 0.2°C$;

3. *a mechanical stirrer*. Unfortunately, it is impossible to use a magnetic stirrer, because the alternative magnetic field, rotating the stirring element, influences the electron focusing in the PM situated close to the cuvette. In case of small volumes a thick plastic fibre clamped on to the axle of a small electric motor (*Figure 1*, right) produces adequate mixing;

4. *a small black plastic tube* to make additions or withdraw samples for biochemical analyses during the experiment;

5. *light filters* to isolate the desired spectral region of CL (see Section 6 for details);

6. *a shutter* bound to an electromechanical system: the cover of the cuvette department cannot be open when the shutter is open and the PM switched on;

7. *a photomultiplier* connected to a photon counter system.

A photon-counting regime is favoured over a constant-current regime in respect of the signal-to-noise ratio. In addition, if a computer is used for data storage and processing, photon counting allows all information contained in the temporary photon distribution in the measured luminescence to be extracted. In some commercial luminometers a system for automatic sequential measurement of a large amount of samples is used; this is important for routine CL assays. For scientific application the sensitivity, good aeration, and sampling during the experiment are more important.

3. Experimental material

3.1 Whole organs

Since 1961, when the light emission of mouse liver *in situ* was first measured (6), many scientists have used a sensitive light detector to study the exposed

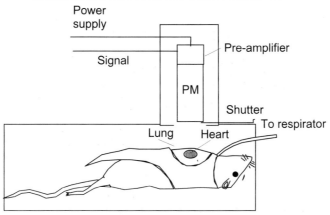

Figure 2. A system for the measurement of whole organ chemiluminescence *in situ* (11–13) PM, photomultiplier.

organs of laboratory animals. For example, Boveris *et al.* measured rat liver CL both *in situ* and in perfused organs (7–10). The light emission by lung both perfused and *in situ* has also been studied (11–13). The system for such measurements is shown in *Figure 2*.

Due to the low intensity of the light emitted by living organs it is measured by very sensitive PMs, responsive in the range 400–900 nm. In a typical experiment, the animal is fixed in a light-tight chamber supplied with necessary physiological equipment. Lung or liver is exposed to the PM placed as close as possible to the organ surface (no more than 10 cm). In order to avoid CL from other organs (including skin), the animal is covered with aluminium foil, and only the organ of interest is exposed to the PM. Lipid peroxidation is obviously involved in the photon emission, since CL increases in vitamin E- and selenium-deficient rats (14), and in animals treated with CCl_4 (10) and paraquat (12). On the other hand, it has been found that a competitive inhibitor of NO synthetase, nitro-L-arginine, decreases the spontaneous light emission by perfused lung in a dose-dependent manner (15). In experimental models, the reaction of peroxynitrite ($ONOO^-$) with some proteins, including bovine serum albumin, produces CL which is more than 20 times more intensive than that observed after addition of organic hydroperoxides to the same substrates (15). Peroxynitrite is formed in the reaction of the superoxide (dioxide) radical $O_2^{\bullet}$ with nitric oxide (NO). Thus peroxynitrite reactions may be responsible for the main body of intrinsic CL in normal tissue. However, when inflammation in the tissue is provoked by adding the leucocyte stimulant, phorbol myristate acetate (PMA), to the perfusate fluid, the tissue CL becomes insensitive to NLA, but is totally abolished by superoxide dismutase (13, 15). Thus reactive oxygen species and lipid radicals became primary contributors to tissue CL under conditions of inflammation.

3.2 Tissue homogenates, mitochondria, and microsomes

Tissue homogenates incubated in aerated medium emit increasing amounts of light with time (10, 16).

By using *Protocol 1* the CL intensity depends on the degree of tissue fragmentation. *Figure 3* shows that the CL of small pieces of liver is greater than that of homogenates prepared from these pieces. The main source of the CL in crude homogenates is the supernatant (16).

Protocol 1. CL of rat liver homogenates according to ref. 16 with minor modifications

Equipment and reagents

- Phosphate-buffered saline (PBS): 5 mM Na_2HPO_4, pH 7.4; 0.9% (w/v) NaCl
- Laboratory homogenizer
- Thermostat set at 39°C

A. *Preparation of liver homogenates*

1. Chill an open Petri dish on ice; isolate the liver from the rat and transfer it to the dish.

2. Put 1 g of the liver tissue in the homogenizer and add 9 ml of PBS cooled to 0°C to give a 1:10 (w/v) suspension (final).

3. Homogenize the liver tissue for 10 sec ('crude homogenate') or 3 min ('fine homogenate').

B. *Measurement of the homogenates chemiluminescence*

1. Place 10 ml of crude or fine liver homogenate in the cuvette of the chemiluminometer (*Figure 1*).

2. Switch on the chemiluminometer, the stirrer, and the thermostat. Set the temperature to 39°C. Wait 2–3 min for temperature equilibration.

3. Select the maximal instrument sensitivity and adjust the signal zero level.

4. Record the dark current for 1 min with the shutter closed.

5. Open the shutter and record the chemiluminescence for 1 h. Record the chemiluminescence as counts per minute (c.p.m.) or mV, depending on the registration device. Continue recording the chemiluminescence, if the intensity of chemiluminescence does not decrease.

6. Close the shutter and record the dark current once more. (Typical results are shown in *Figure 3*.)

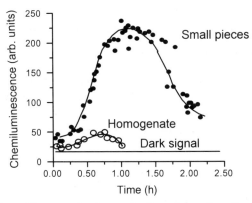

Figure 3. The effect of homogenization on the liver tissue chemiluminescence (16).

Though tissue homogenates can easily be prepared from small amounts of isolated tissue and even biopsy material, it is not so easy to interpret the data obtained, since the presence of different pro- and antioxidant compounds may influence free radical reactions and associated light emission. Mitochondrial and microsomal preparations are much simpler, and thus more appropriate. Early experiments based, in particular, on addition of ATP and substrates to the suspension, suggested that spontaneous mitochondrial CL depended on metabolism.

More thorough experiments show, however, that CL in mitochondrial and microsomal suspensions in fact depends on the current concentration of divalent iron ions in the system (see refs 4 and 17 for review). Even if iron salts are not deliberately added, there are always some Fe^{2+} ions present in the suspension. This is because:

1. Many reagents, in particular ATP and ADP, contain iron ions as a contaminant.
2. Reducing agents, e.g. ascorbate, cysteine, and glutathione, can reduce iron and extract Fe^{2+} ions from electron-transporting chains in mitochondria and microsomes (17).
3. Mitochondria are able to reduce Fe^{3+} in Fe-ATP (Fe-ADP)-containing suspensions, the process being dependent on the state of the respiratory chain (17).

The addition of Fe^{2+} chelators, such as EDTA, to the suspension immediately inhibits CL. Though lipid peroxidation (LPO) is an obvious source of CL, in aerated suspensions of mitochondria (4, 17) and microsomes (8, 18) the relationship between the reaction rate and CL intensity is not so simple. During peroxidation, the CL quantum yield grows and the emission spectrum is shifted to the red spectral regions due to accumulation of some product(s) (19). It has recently been proposed that it is the interaction of such products with Fe^{3+} which gives rise to emission of rather intense luminescence at wavelengths exceeding 600 nm (20).

3.3 Blood plasma

Blood plasma, as well as whole blood, emits spontaneous CL (21), which is altered in some human diseases (22). The intensity of this emission is very low and the system is too complex for proper analysis of the results. Addition of high concentrations of hydrogen peroxide causes a burst of CL, the amplitude of which can be used for diagnostic purposes (23). In our laboratory systematic studies have been carried out on blood plasma and blood lipoprotein suspensions to which Fe^{2+} was added to initiate LPO using *Protocol 2* (24). A typical CL curve is shown in *Figure 4*.

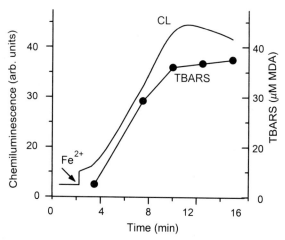

Figure 4. The chemiluminescence (CL) and accumulation of LPO products (TBARS) in dog blood plasma (24). Fe^{2+} was added at the time marked by an arrow.

Protocol 2. The CL of blood plasma and apo-B lipoproteins
(according to ref. 25, with minor modifications)

Equipment and reagents

- Human blood (from donors) stabilized with 1% crystalline heparin (for plasma), or blood without heparin (for serum); in this case let blood clot before separation.
- PBS: 140 mM NaCl, 40 mM K_2HPO_4, pH 7.5
- $FeSO_4 \cdot 7H_2O$, 25 mM acidified with 0.01 M HCl to pH 2–3 to prevent Fe^{2+} auto-oxidation (prepare the solution on the day of experiment)

- Heparin, 0.25% aqueous solution
- $CaCl_2$, 0.5 M
- Trichloroacetic acid (TCA), 30%
- Butylated hydroxytoluene (BHT)
- 2-Thiobarbituric acid (TBA), 120 mM
- Laboratory centrifuge
- Boiling-water bath

5: Intrinsic (low-level) chemiluminescence

A. *Preparation of blood plasma*

1. Centrifuge 5 ml of the blood at 900 g for 5 min.
2. Carefully withdraw the upper yellow supernatant (blood plasma or serum) and store on ice.

B. *Preparation of apo-B lipoproteins*

1. Put 0.1–0.5 ml of blood plasma or serum, 0.1 ml of 0.25% heparin solution, and 0.5 ml of 0.5 M $CaCl_2$ in the centrifuge tube and add H_2O to a final volume of 4 ml.
2. Centrifuge the mixture at 900 g for 20 min
3. Gently remove the supernatant and dissolve the sediment in 1.0 ml of PBS.

C. *Measurement of chemiluminescence*

1. Transfer 1 ml of plasma or apo-B lipoprotein solution and 19 ml of PBS to the chemiluminometer cuvette (*Figure 1*).
2. Perform steps 2–4 in *Protocol 1B*.
3. Open the shutter and record the background CL of the sample for 1–2 min.
4. Add 1 ml of $FeSO_4$ to the cuvette and record the CL for 20 min. A typical result is shown in *Figure 4*.

D. *Measurement of LPO products*

1. Take 2.0 ml of the solution from the cuvette using the plastic tube (*Figure 1*) immediately after addition of $FeSO_4$ and then every 4 min.
2. Add 1.0 ml of TCA and 0.1 ml of BHT to each sample. Shake the mixture.
3. Centrifuge the samples at 1800 g for 20 min.
4. Take 2.0 ml of the supernatant, add 2.0 ml of TBA, and place the mixture in the boiling-water bath for 15 min.
5. Cool the solution to room temperature, add 2.0 ml of chloroform, and mix the suspension.
6. Take the upper solution (aqueous) and measure its absorbance (A) at 532 nm in a 1 cm cuvette.
7. Calculate the concentration (c) of MDA in the cuvette from the equation: c (μM) = 0.156 $\times$ A.

In addition to the development of CL there was oxidation of Fe^{2+} and accumulation of LPO products, including thiobarbituric acid reactive substances (TBARS), suggesting that LPO caused the CL. It was shown that blood lipoproteins, mainly low density lipoproteins (LDL), were responsible for serum CL. Other components, including caeruloplasmin, inhibit Fe^{2+}-initiated CL (25). The CL in plasma thus depends on both materials, the concentration of which differ in plasma (or serum) and in a lipoprotein suspension, thus explaining the different results obtained in both. In support of this is that in the presence of 7.7 M NaN_3, a caeruloplasmin inhibitor, there is no difference between an apo-B lipoprotein suspension and serum (25).

3.4 Phospholipid vesicles

A suspension of phospholipid vesicles (liposomes) is the simplest system for studying CL associated with LPO. The main features of CL in mitochondrial suspensions (4, 17) and liposomes (26) are similar; both types show a strong dependence on the concentrations of Fe^{2+}, phosphate, and oxygen, and they have similar reaction spectra and kinetics. The results with liposomes are considered in more detail in the next section.

4. Reaction kinetics

The kinetics of the CL after LPO initiation with Fe^{2+} were first studied in rat liver mitochondria (4). Very similar results have also been obtained with liposomal suspensions (26, 27). A typical experiment performed using *Protocol 3* is shown in *Figure 5*. Homogenates and blood plasma show simpler CL kinetics (*Figures 3* and *4*).

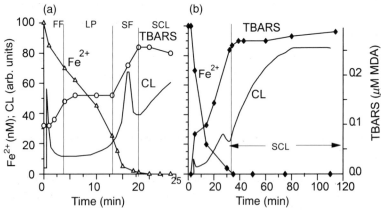

Figure 5. Chemiluminescence kinetics of a suspension of phospholipids (liposomes) (26). (a) and (b) different duration of the CL measurements. FF, fast flash; LP, latent phase; SF, slow flash; SCL, stationary CL.

Protocol 3. The measurement of CL kinetics in liposome suspension (26)

Equipment and reagents

- Tris–HCl buffered saline (TBS): 20 mM Tris–HCl buffer, 0.1 M KCl, pH = 7.4 at 20°C
- 2-Thiobarbituric acid (TBA), 0.5% (w/v)
- *o*-Phosphoric acid
- *n*-Butanol
- *o*-Phenanthroline, 10 mM
- Butylated hydroxytoluene (BHT), 10 mM in 96% ethanol
- $FeSO_4 \cdot 7H_2O$, 1 mM
- Egg yolk lecithin (in chloroform), 100 mg/ml
- Chemiluminometer (*Figure 1*)
- Rotary vacuum vaporizer (dryer)
- Ultracentrifuge (with bucket rotor)
- Pipettes, pipette tips, and glass tubes (>10 ml)
- Thermostat for the glass tubes (100°C)

A. *Preparation of liposome suspension*

1. Transfer 2–5 ml of the lecithin solution into a round-bottomed rotary dryer flask and evaporate off the solvent to obtain a dry lipid film on the wall of the flask.

2. Add TBS and shake the vessel energetically until all the lecithin is suspended. The final concentration of the phospholipid should be 2 mg/ml.

3. Disperse the liposome suspension by freezing in liquid nitrogen and then thawing to room temperature. Repeat five times.

B. *Measurement of chemiluminescence kinetics*

1. Put 4.5 ml of the liposome suspension in the cuvette of chemiluminometer.

2. Switch on the chemiluminometer (*Figure 1*); turn the stirrer on. Close the light-tight chamber of the device and open the shutter. Record the background CL over 1 min.

3. Add 0.5 ml of $FeSO_4$ using a slender black plastic tube (see *Figure 1*).

4. Record the time-course of CL development for 80–120 min to follow the kinetic stages, i.e. fast flash, lag period, slow CL flash, and stationary CL (*Figure 5*).

C. *Lipid peroxidation product kinetic measurement (simultaneously with procedure B)*

1. Take 0.1 ml aliquots of the liposome suspension with a micropipette from the cuvette using the plastic tube before and every 2 min after addition of $FeSO_4$.

2. Transfer the probe into a glass tube and add 1 ml of *o*-phosphoric acid and 0.025 ml of BHT.

Protocol 3. *Continued*

3. After completion of the lipid peroxidation kinetic measurement add 1.5 ml of TBA to each tube.

4. Heat the probes at 100°C for 45 min and then cool them to room temperature.

5. Add 2.0 ml of *n*-butanol to each tube and shake to mix the butanol and water phases.

6. Separate phases by centrifugation (1000 *g*, 5 min).

7. Withdraw 1 ml of the upper butanol solution and measure the absorbency at 532 and 600 nm. The difference between these values ($A_{532} - A_{600}$) is a measure of the concentration of the final products of LPO (TBARS) (*Figure 5*).

4.1. Analysis of the kinetic curves

The kinetics of CL associated with Fe^{2+}-induced LPO in liposomes using *Protocol 3* are shown in *Figure 5*.

The kinetic curves are rather complex and consist of several phases: the fast flash of CL, the latent phase, the slow flash of CL, and stationary CL (4, 26).

Most of the results of the kinetic study of CL can be explained by assuming that lipid dioxide radicals are responsible for CL due to the disproportionation reaction:

$$LO_2^{\cdot} + LO_2^{\cdot} \rightarrow L{=}O^* + LOH + {}^1O_2 \ldots \rightarrow \text{chemiluminescence.}$$

In its turn, lipid dioxide radicals are formed as a result of the reaction sequence shown in *Figure 6* (4, 17, 28).

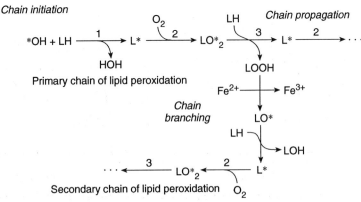

Figure 6. LPO reactions in the presence of Fe^{2+} ions.

Different reactions dominate each of the phases of the kinetics shown in *Figure 5*, and by careful measurement of the kinetics at different stages it is possible to determine the reaction rates of each of these reactions (28).

5. Main factors affecting low-level chemiluminescence

5.1 Iron ions

Virtually any incubation system contains sufficient traces of iron ions necessary for CL appearance. These ions are participants of a number of reactions involved in LPO and light-producing events. In our experiments with liposomes, lipoproteins, mitochondria, and microsomes a certain amount of Fe^{2+} is required to initiate LPO and CL (4, 17, 28). The results of a typical experiment with liposomes with various amounts of Fe^{2+} added to the suspension are shown in *Figure 7*.

The analysis of the kinetic curves presented in *Figure 7* has proved very useful in helping us to understand the complicated dependence of CL intensity (and LPO product level) on ferrous ion concentration. We know that:

1. At low Fe^{2+} concentrations, these ions have a pro-oxidant effect, and the CL intensity increases with Fe^{2+} concentration.

2. At higher concentrations of Fe^{2+}, the antioxidant action prevails, manifesting itself in growing latent period of CL development (and time-to-peak τ in *Figure 7*).

The time-dependence of the accumulation of lipid hydroperoxides (LOOH) and dioxide radicals (LOO·) can be described by eqn 1.

$$[LOO^\bullet] = [LOO^\bullet]_o \, e^{\gamma t} \text{ and } I = I_o \, e^{2\gamma t}; \gamma = k_p \, ([Fe^{2+}]^* - [Fe^{2+}]), \qquad (1)$$

where k_p is the reaction rate constant of chain branching and $[Fe^{2+}]^*$, called the critical iron concentration, is also a constant depending on the rate constants of chain propagation and termination (4, 28).

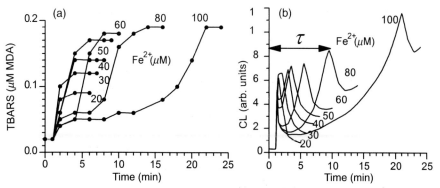

Figure 7. The effect of Fe^{2+} concentration (μM added) on the accumulation of lipid peroxidation products (TBARS) and chemiluminescence in a suspension of liposomes (29).

The decay of FF in *Figures 5* and *7b* corresponds to the situation when $[Fe^{2+}] > [Fe^{2+}]^*$. The increasing CL in the first phase of the slow flash represents the phase at which $0 < [Fe^{2+}] < [Fe^{2+}]^*$. After all the Fe^{2+} added to the suspension has been oxidized, the SL decays as a result of a decrease of lipid radical from the different reactions of chain termination. The results of computer simulation of the reaction kinetics fit very well to the experimental curves for liposomes and mitochondria at the fast and slow flash stages of CL, including an intermediate latent phase (28).

It should be borne in mind that the most important reactions of Fe^{2+} occur inside the lipid phase or on the surface of membranes, and for this reason it is the membrane-bound iron, rather than the total iron concentration in the suspension, that determines the kinetics of LPO and concomitant CL. The decrease in liposome concentration brings about alterations of CL kinetics, resembling those at higher Fe^{2+} concentrations (26). On the other hand, the addition of competing ions, such as Fe^{3+}, Ca^{2+}, or Eu^{3+}, affects the CL in the same way as the decrease in the concentration of Fe^{2+} (26).

Let us return to the CL kinetics in liposomes. Using *Protocol 3*, *Figure 5b* shows that after virtually all the Fe^{2+} has been oxidized and LPO products, namely TBARS, have ceased to accumulate, a new, slowly growing luminescence starts to develop, called stationary CL (4). The increase in the initial amount of Fe^{2+} (added at the beginning of the experiment) results in a proportional augmentation of the stationary CL amplitude. A similar increase in the stationary CL is observed when Fe^{3+} is added at the beginning in addition to the 'background' Fe^{2+} (29). It is postulated that a reaction of Fe^{3+} with LPO products (possibly hydroperoxides) is responsible for the stationary CL. This reaction is accompanied by photon emission at longer wavelengths and has essentially higher quantum yield than the reaction of $LOO^•$ radicals.

In the majority of early experiments scientists did not differentiate between the two CL reactions accompanying LPO, and hence, the effects of Fe^{2+} and Fe^{3+} were sometimes not interpreted correctly.

5.2 Aeration

Both LPO and low-level CL of tissue homogenates (15), mitochondria (16, 17), and microsomes (18) are strongly affected by the concentration of dissolved oxygen. In experiments performed with the instrument shown in *Figure 1*, a mechanical stirrer is used to provide agitation and sufficient aeration of the reaction mixture. In our early experiments on tissue homogenates, aeration was performed by bubbling air through the homogenate (1, 15). This method was rejected because of surface denaturation of proteins indicated by the appearance of insoluble flakes.

There are several reasons why oxygen is essential for CL:

1. LPO chain propagation (in which new LOOH is formed (*Figure 6*), but not new lipid radicals) can develop only in an oxygen-containing atmosphere.

2. In the reaction of Fe^{2+} with lipid hydroperoxides, where new lipid radicals (including $L^\bullet$ and $LO^\bullet$ in *Figure 6*) are formed, $LOO^\bullet$ (apparently responsible for the light-producing reaction) can be formed only in the presence of O_2.

3. Emission in the red region of the spectrum may be a result of formation of excited oxygen dimers:

$$Fe^{3+} + LOOH \Rightarrow {}^3P; {}^3P + {}^3O_2 \Rightarrow P + {}^1O_2,$$
$${}^1O_2 + {}^3O_2 \Rightarrow (O_2)_2{}^* \Rightarrow (O_2)_2 + h\nu.$$

Here ${}^3P$ is a reaction product in excited (triplet) state, ${}^1O_2$ is singlet oxygen, $(O_2)_2{}^*$ are oxygen excimers, and $h\nu$ is a photon with wavelength 631 nm emitted by dioxygen excimers. The excited oxygen species can also be formed only in a dioxygen-containing atmosphere.

5.3 Temperature

The development of LPO and CL depends very strongly on temperature, as can be shown for rat liver homogenates using *Protocol 1*. By taking the maximal slope at the beginning of the curves (see *Figure 3*) as the reaction rate, the activation energy at temperatures of 37–41 °C has been calculated as 55–57 kcal/mol, a rather high value (see ref. 16 for more details). This is in agreement with the findings of other authors (12, 30).

6. Chemiluminescence spectra

The measurement of CL spectra with a monochromator and sensitive light detector can be performed only in the case of relatively intense luminescence. A crude spectrum can be obtained by using a series of 'cut-off' filters, i.e. filters with a sharp, short-wavelength boundary of light-transparency (27, 31). The method of measurement is shown schematically in *Figure 8*. The

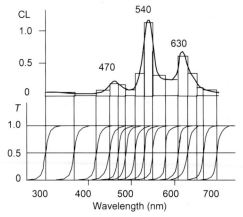

Figure 8. Scheme for the measurement of chemiluminescence spectra. Below, cut-off light filter transparency (*T*). Above, the spectrum of liposomes. See text for more details.

wavelength at which the transparency (*T* in *Figure 8*) is 50%, is taken as the short-wavelength boundary λ_n of the filter *n*. Then the CL intensity is measured through the series of filters in the same conditions. The intensity measured through the filter *n* is denoted as I_n. The difference between the luminescence intensities for two filters in series $\Delta I = I_{n+1} - I_n$ is proportional to:

- the real CL intensity I_λ at the wavelength $\lambda = (\lambda_{n+1} + \lambda_n)/2$
- the photomultiplier sensitivity at this wavelength K_λ
- the wavelength difference between transparency boundaries $\Delta\lambda = \lambda_{n+1} - \lambda_n$.

In other words, $\Delta I = K_\lambda \times \Delta_\lambda \times I_\lambda$. Hence,

$$I_\lambda = \frac{\Delta I}{K_\lambda \times \Delta_\lambda} \tag{2}$$

A column diagram in which I_λ is plotted as a function of λ, is called a CL spectrum; it looks like a histogram (*Figure 8*) but can also be drawn as a curve.

Protocol 4. Measurements of CL spectra according to the methods described in refs 31 and 27 with minor modifications

Equipment and reagents

- All reagents and equipment as in *Protocol 3*
- Spectrophotometer
- Cut-off light filters (yellow, orange, and red filters with sharp, short-wavelength transmission boundary)

A. *Measurement of transparencies of light-filters*

1. Place each filter in the spectrophotometer in turn and measure the transparency (*T*) of each of them as a function of wavelength λ.

2. On the short-wavelength side of the curve, find the wavelength (λ_n for the filter number *n*), at which the transparency *T* is 50% (*Figure 8*). Assign numbers *n* to the filters ($f_1, f_2, f_3, ... f_n$) in such a way that λ_n should increase gradually with increasing *n*. Calculate $\Delta\lambda = \lambda_{n+1} - \lambda_n$ (the difference in wavelength) between each, numerically ordered, pairs of filters in series, i.e. $f_2 - f_1, f_3 - f_2$, etc.

3. Place the first filter f_1 between the shutter of the chemiluminometer and the cuvette (*Figure 1*).

4. Measure the CL kinetics according to *Protocol 3* (or *Protocol 1 or 2* for different objects). Determine the amplitude of the CL I_n for the measurement with the filter f_1.

5. Repeat step 4 with all filters.

6. Calculate the difference between I_{n+1} and I_n for all filter pairs. Denote these values as $\Delta I(\lambda)$.

7. Calculate the mean CL intensity I_λ for each interval of wavelengths $\Delta\lambda$ using eqn 2. If K_λ (the relative PM sensitivity at wavelength λ) is not stated in the PM documents, assume K to be constant (e.g. $K = 1$).

8. Plot the graph I_λ as a function of λ. *Figure 8* illustrates how to do this.

In oxidized unsaturated fatty acids, mitochondria, and microsomes the emission spectra contain maxima attributable to ketone emission (400–480 nm) and singlet oxygen emission (520, 575–580, and 630–640 nm; see refs 28, 31–33). The spectra measured in different laboratories are, however, rather different. It has been shown in our experiments on mitochondria (1, 19) and liposomes (4, 29) that the spectra changed during the LPO reaction (see *Figure 9*).

The accumulation of LPO products that may participate in the second CL reaction or serve as sensitizers of the CL may be responsible for this effect.

7. The sensitization phenomenon

Some luminescent compounds, when added to an oxidized organic substrate, enhance the CL intensity and change the emission spectra. This CL enhancement was first used in studies of liquid-phase hydrocarbon oxidation. In biological systems CL sensitizers were also applied to increase the sensitivity of the method and for the investigating the nature of primary excited products (34, 35).

Several classes of luminescent compounds (brominated polycyclic hydrocarbons, rare-earth ion chelates, rhodamine, porphyrin, and coumarin derivatives) have been used to enhance the CL accompanying LPO in

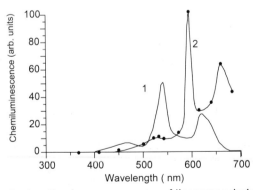

Figure 9. Changes in chemiluminescence spectra of liposomes during lipid peroxidation (29). 1, spectrum at the slow flash stage of chemiluminescence (cf. *Figure 5a*); 2, spectrum at the stationary state stage of chemiluminescence, in 40 min after Fe^{2+} addition (cf. *Figure 5b*).

Table 1. Maximal enhancement of chemiluminescence in liposomes peroxidized in the presence of Fe^{2+}

Compound	EF[a]	c(μM)	F_{max}(nm)	Ref.
9,10-Dibromoanthracene	5	6	440	(36)
Eu^{3+}–tetracycline	1120	1	620	(27,36)
$TbCl_3$	10	100	545	(36)
Protoporphyrin IX	16	100	635	b
Bacteriopheophorbide	360	2	680	b
Bacteriopheophytin	80	3.3	680	b
meso-Tetraphenylporphyrin	7	100	650	b
Rhodamine G	37	60		(38)
Quinolysine-coumarins:				
C-525	1620	20	512	(37)
C-510	992	52	510	(37)
C-338	733	0.53	510	(37)
C-153	350	2.6	508	(37)
C-504	585	92	506	(37)

[a]EF, Enhancement factor: the ratio of the amplitude of chemiluminescence at the 'slow flash' maximum in the presence of the enhancing compound to that without the sensitizer. c, the concentration showing maximal effect. F_{max}, the position of the maximum in the fluorescence spectrum of the sensitizer.
[b]Data obtained in our laboratory by V. S. Sharov and E. S. Driomina.

biomembranes and blood plasma lipoproteins (27, 36, 37). The most efficient enhancers are a Eu^{3+}–tetracycline complex (EuT) and coumarin laser dyes (see *Table 1*).

Rhodamines, although useful in some applications, are less potent (37). CL spectra in the presence of EuT and coumarins are close, if not identical, to the photoluminescence spectra of the sensitizers (36, 37). The amount of TBARS accumulated by the end of the Fe^{2+}-induced chain reactions (slow flash) were not changed by any of these sensitizers. All of this may be taken as evidence that the mechanism of CL enhancement is the excited energy transfer to the sensitizer, according to the following equations.

$$\text{Primary CL reaction: } LOO^{\bullet} + LOO^{\bullet} \rightarrow P^{*}$$

$$\text{Non-enhanced CL: } P^{*} \rightarrow P + \varphi_P h\nu_P$$

$$\text{Energy transfer: } P^{*} + A \rightarrow P + A^{*}$$

$$\text{Enhanced CL: } A^{*} \rightarrow A + \varphi_A h\nu_A; (\varphi_A \gg \varphi_P)$$

where φ_P and φ_A are the emission quantum yields of primary excited product and sensitizer, respectively.

8. Conclusion

Low-level CL is an indirect, but rather sensitive method of measuring the stationary state concentration of reactive species formed in cells and tissues.

Table 2. The reactions responsible for tissue chemiluminescence

No.	Reaction	Excited species	Reference
1	$LOO^\bullet + LOO^\bullet$	L = O (ketones) and $^1O_2$ (singlet oxygen)	(31, 32)
2	$Fe^{3+} + H_2O_2$	$^1O_2$	(39)
	$Haem(Fe^{2+}] + H_2O_2$	$^1O_2$	(40, 41)
	$Fe^{3+} + LOOH$	$^1O_2$	(20)
3	$ONOO^- + proteins$	unidentified	(15)

However, the interpretation of data depends on our understanding of the nature of the reactions responsible for the luminescence in each particular case. *Table 2* summarizes our current knowledge of the reactions responsible for intrinsic (low-level) tissue CL. Additional biochemical methods can be used to determine which of these reactions dominates in each situation.

References

1. Vladimirov, Yu. A. (1966). In *Sverkhslabyje svechenija pri biokhimicheskikh reakzijach* (*Ultraweak luminescence following biochemical reactions*, in Russian) (ed. G. M. Frank). Nauka, Moscow.
2. Vladimirov, Yu. A. and Litvin, F. F. (1959). *Biofizika*, **4**, 601.
3. Colli, L. and Faccini, U. (1954). *Nuovo Cimento*, **12**, 150.
4. Vladimirov, Yu. A. (1994). In *Free radicals in the environment, medicine and toxicology* (ed. H. Nohl *et al.*), pp. 345–73. Richelieu Press, London.
5. Vladimirov, Yu. A. and Archakov, A. I. (1972). *Perekisnoe okislenie lipidov v biologicheskikh membranakh* (*Lipid peroxidation in biological membranes*, in Russian) (ed. G. M. Frank). Nauka, Moscow.
6. Tarusov, B. N., Polivoda, A. I., and Zhuravlev, A. I. (1961). *Radiobiologia*, **1**, 150.
7. Boveris, A., Cadenas, E., Reiter, R., Filipkowski, M., Nakase, Y., and Chance, B. (1980). *Proc. Natl Acad. Sci. USA*, **77**, 347.
8. Boveris, A., Llesuy, S. F., and Fraga, C. G. (1985). *Free Rad. Biol. Med.*, **1**, 131.
9. Fraga, C. G., Arias, R. F., Llesuy, S. F., Koch, O. R., and Boveris, A. (1987). *Biochem. J.*, **242**, 383.
10. Fraga, C. G., Martino, V. S., Ferraro, G. E., Coussio, J. D., and Boveris, A. (1987). *Biochem. Pharmacol.*, **36**, 717.
11. Cadenas, E., Arad, I. D., Fisher, A. B., Boveris, A., and Chance, B. (1980). *Biochem. J.*, **192**, 303.
12. Turrens, J. F., Giulivi, C., Pinus, C. R., Lavagno, C., and Boveris, A. (1988). *Free Radic. Biol. Med.*, **5**, 319.
13. Barnard, M. L., Gurdian, S., and Turrens, J. F. (1993). *J. Appl. Physiol.*, **75**, 933.
14. Samuel, D., McLauchlin, J., and Taylor, A. G. (1990). *J. Biolumin. Chemilumin.*, **5**, 179.
15. Turrens, J. F. (1994). In *International conference on clinical chemiluminescence*, Berlin, 25–28 April 1994 (ed. I. Popov), p. O2. Humboldt University, Berlin.
16. Vladimirov, Yu. A. and L'vova, O. F. (1965). In *Biofizika kletki* (*Cell Biophysics*, in Russian) (ed. G. M. Frank), pp. 74–83. Nauka, Moscow.

17. Vladimirov, Yu. A., Olenev, V. I., Suslova, T. B., and Cheremisina, Z. P. (1980). *Adv. Lipid Res.*, **17**, 173.
18. Aleksandrova, T. A., Archakov, A. I., Vladimirov, Yu. A., Olenev, V. I., and Panchenko, L. F. (1971). *Biofizika*, **16**, 946.
19. Olenev, V. I. and Vladimirov, Yu. A. (1973). *Studia Biophys.*, **38**, 131.
20. Vladimirov, Yu. A. (1994). In *International conference on clinical chemilumin-escence*. Berlin, 25–28 April 1994 (ed. I. Popov), p. A1. Humboldt University, Berlin.
21. Bondarev, I. M., Zhuravlev, A. I., and Shpolianskaya, A. M. (1971). *Probl. Tuberk.*, **49**, 71.
22. Tarusov, B. N., Ivanov, I. I., and Petrusevich, Y. M. (1967). *Sverchlaboje svechenije biologicheskikh sistem* (*Ultraweak luminescence of biological systems*, in Russian) (ed. B. N. Tarusov). MGU, Moscow.
23. Vladimirov, Yu. A. and Sherstnev, M. P. (1992). *Soviet Med. Rev. B*, **2**, 1.
24. Vladimirov, Yu. A., Sharov, A. P., and Maliugin, E. F. (1973). *Biofizika*, **18**, 148.
25. Klebanov, G. I., Teselkin, Yu. O., and Vladimirov, Yu. A. (1988). *Biofizika*, **33**, 512.
26. Driomina, E. S., Sharov, V. S., and Vladimirov, Y. A. (1993). *Free Radic. Biol. Med.*, **15**, 239.
27. Vladimirov, Yu. A., Sharov, V. S., and Suslova, T. B. (1981). *Photobiochem. Photobiophys.*, **2**, 272.
28. Vladimirov, Yu. A. (1986). In *Free Radicals, Aging, and Degenerative Diseases* (ed. J. E. Johnson, Jr, *et al.*), pp. 141–95. Alan R. Liss, New York.
29. Sharov, V. S., Driomina, E. S., and Vladimirov, Yu. A. (1996). *J. Biolumin. Chemilumin.* (in press).
30. Sies, H. and Cadenas, E. (1985). *Phil. Trans. R. Soc. Lond. Biol.*, **311**, 617.
31. Inaba, H. (1988). *Experientia*, **44**, 550.
32. Cadenas, E. (1984). *Photochem. Photobiol.*, **40**, 823.
33. Murphy, M. E. and Sies, H. (1990). *Anal. Proc.*, **27**, 217.
34. Ivanov, I. I., Buzas, S. K., Gol'dshtein, N. I., and Tarusov, B. N. (1971). *Biofizika*, **16**, 735.
35. Sharov, V. S., Suslova, T. B., Deev, A. I., and Vladimirov, Yu. A. (1980). *Biofizika*, **25**, 923.
36. Sharov, V. S., Kazamanov, V. A., and Vladimirov, Yu. A. (1989). *Free Radic. Biol. Med.*, **7**, 237.
37. Vladimirov, Yu. A., Sharov, V. S., Driomina, E. S., Gashev, S. B., and Reznitchenko, A. V. (1995). *Free Radic. Biol. Med.*, **18**, 739.
38. Vladimirov, Yu. A., Atanayev, T. B., and Sherstnev, M. P. (1992). *Free Radic. Biol. Med.*, **12**, 43.
39. Shen, X., Xu, W. and Tian, J. (1994). In *International conference on clinical chemiluminescence*. Berlin, 25–28 April 1994 (ed. I. Popov), p. T6. Humboldt University, Berlin.
40. Nohl, H. and Stolze, K. (1993). *Free Radic. Biol Med.*, **15**, 257.
41. Nohl, H. and Stolze, K. (1994). In *International conference on clinical chemilumi-nescence*. Berlin, 25–28 April 1994 (ed. I. Popov), p. O1. Humboldt University, Berlin.

6

Visual assessment of oxidative stress by multifunctional digital microfluorography

MAKOTO SUEMATSU, GEERT W. SCHMID-SCHÖNBEIN,
YUZURU ISHIMURA, and MASAHARU TSUCHIYA

1. Introduction

There has been an increase in the understanding of the potential role of oxidants, such as the superoxide anion ($O_2^{\cdot-}$), hydrogen peroxide (H_2O_2), hydroxyl radicals ($^{\cdot}OH$), as mediators of cell injury in a wide variety of physiological and pathophysiological conditions. These active oxygen metabolites can interact with cellular membrane structures thereby inducing lipid peroxidation, which has been implicated as one of the major mechanisms of 'oxidative stress' in tissue injury. By contrast, the nitric oxide radical ($NO^{\cdot}$), which has been identified as an endothelium-derived relaxing factor, plays an indispensable role in the maintenance of blood perfusion and antithrombotic properties of vascular endothelium, and contributes as a natural antioxidant to scavenging superoxide anion. Under some pathological conditions, the interaction between $NO^{\cdot}$ and $O_2^{\cdot-}$ may yield peroxynitrite ($ONOO^-$) and thereby promote cell injury.

Although this scenario has attracted great interest from oxygen radical researchers investigating the mechanisms of oxidative tissue injury, only limited evidence has been provided to answer important questions in this field. How large is the magnitude of oxidative stress? Which cells produce oxygen radicals? Which cells are susceptible to oxidative stress? When these cells are exposed to oxidants do they lose their viability? What is the topographic correlation between the sites of oxygen radical production and distribution of cell injury? What is the temporal relationship between the two phenomena? The reason why these questions remain unsolved is because of the technical difficulties of investigating the spatial distribution and/or time history of oxidative stress *in vivo* or even *in vitro*. Although there is no perfect experimental system, multifunctional digital microfluorography gives a clue to resolving these questions. This chapter summarizes current methodology in visually monitoring spatial and temporal alterations of oxidative changes in addition to simultaneous monitoring of cell function and viability *in vivo* or *ex vivo*.

2. Intravital fluorescence video microscopy

The intravital video microscope is composed of a microscope (orthostatic/ inverted), an imaging device, and a video recording system including a digital image processor. This system allows the haemodynamics or behaviour of circulating cells such as platelets or leucocytes in various types of organ microcirculation to be observed intravitally and described quantitatively. The range of organs suitable for intravital observation of microvascular changes includes not only mesentery or skin, which have been examined for a number of years (1), but also gastrointestinal tract (2), brain (3, 4), lung (5), heart, liver (6), and kidney. Intravital observation of microvascular changes has many advantages in the understanding of the pathophysiology of oxidant-induced organ injury, since the microcirculation serves as the primary site of rate-limiting phenomena which dictate the severity of oxidative stress: oxygen delivery, regulation of blood supply to the organ parenchyma, and leucocyte–endothelial cell interactions. The transillumination technique allows the observation of the detailed behaviour of leucocytes in microvessels of the mesentery or other parenchymal organs, as long as these organs are thin enough to permit light transmission (e.g. small intestine, spinotrapezius muscle, the wedge of the liver, etc.). Furthermore, in combination with a fluorescence epi-illumination system the technique can be used with thicker portions of organs. By using different filter combinations, several different fluorescent images in the same microscopic field can be recorded by a digital image processor or video recorder. Accordingly, when the tissues or organs are loaded with functional fluorescent probes (see Chapter 7), several different types of information, such as the severity of oxidative impact, mitochondrial energization, and cell viability, can be obtained in the same portion of organ microcirculation, and the time history and topographic distribution of each parameter can be traced. In order to digitize the images and to evaluate the fluorescence signals quantitatively, an image analyser (e.g. Image 1.58) and video board should be installed on a computer.

In most cases, a high-sensitivity camera is required to observe vital organs, since it is important to minimize the intensity and duration of the light to which samples are exposed, in order to avoid photobleaching of fluorescence, which is a particular problem in the presence of oxygen in vital tissues. A silicon intensified target (SIT) camera or a cooled intensified CCD camera should be used. In order to limit the exposure time, an electric shutter-controlling system should be installed between the fluorescent light source and the microscope. This device is needed particularly to visualize weak and rapidly photobleached autofluorescence (e.g. NADH, vitamin A, etc.), because of the need to limit light exposure during long-term observation (7, 8).

Table 1 Functional fluorescent probes useful for evaluation of oxidative tissue injury and their excitation/emission wavelengths.

Type of probe	E_m/E_x (nm)
Autofluorescence	
NAD(P)H, vitamin A (retinoids)	360/450
Dichlorofluorescein and its derivatives (peroxide-sensitive)	
Dichlorofluorescin (dihydrodichlorofluorescein: DCFH) diacetate	480/530
Carboxydichlorofluorescin (CDCFH) diacetate	
Carboxydichlorofluorescin (CDCFH) diacetate *bis* acetoxymethyl ester	
Dichlorofluorescin diacetate succinimidyl ester	
Other oxidant-sensitive fluorescence probes (peroxide-sensitive)	
Dihydroethidium bromide (hydroethidine)	530/590
Dihydrorhodamine 123	480/530
Dihydrorhodamine G	530/590
Dihydrotetramethylrosamine	530/590
Probes for mitochondrial membrane potential (potential-driven fluorochromes)	
Rhodamine 123	480/530
JC-1	
Probes for cell viability	
(a) *Stain viable cells*	
Carboxyfluorescein diacetate	480/530
bis-Carboxyethyl carboxyfluorescein acetoxymethyl ester (BCECF)	450/510
(b) *Stain damaged cells*	
Propidium iodide	530/590
Ethidium bromide	530/590

3. Fluorescent probes for assessment of oxidative stress

The choice of fluorescent probe is important for the effective assessment of oxidative tissue injury using multifunctional digital microfluorography. *Table 1* lists such probes, their fluorogenic properties, and the related functional parameters which each fluorescence response indicates. Occasionally, measurement of tissue autofluorescence is a useful tool for assessing oxidative metabolism.

3.1 Autofluorescence

Tissue autofluorescence derived from reduced pyridine nucleotides (e.g. NADH) has been examined as an index of redox metabolism, since an increase in NADH autofluorescence occurs as a consequence of alterations in oxygen metabolism during hypoxia or increased NAD^+ reduction during xenobiotic metabolism (e.g. ethanol metabolism by alcohol dehydrogenase).

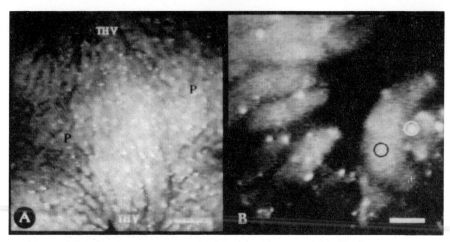

Figure 1. Vitamin A autofluorescence in the rat hepatic microcirculation *in vivo*. Bars in A and B represent 200 μm and 50 μm, respectively. Black and white circles in B indicate parenchymal NADH and Ito cell-derived patchy vitamin A autofluorescence, respectively. (From ref. 7, with permission.)

The method of determining NADH autofluorescence (λ_{ex} = 360 nm, λ_{em} = 450 nm) is based on reflectance spectrophotometry (9), in which a microoptic fibre is placed on the surface of a perfused organ to check the ultraviolet-excited autofluorescence under epi-illumination. NADH autofluorescence has also been studied in perfused rat liver by visualizing directly using intravital fluorescence microscopy. This methodology has several advantages over reflectance spectrophotometry. Particularly, topographic evaluation of NADH by reflectance spectrophotometry is not reliable when the overlying effects of other sources of autofluorescence with the same wavelength are not taken into account. This serious problem occurs in the liver, where there is an abundant source of vitamin A autofluorescence (λ_{ex} = 360 nm, λ_{em} = 450 nm) in perisinusoidal Ito cells (7, 8) (*Figure 1*). With our more recent method, exposure of the surface of the perfused liver to a short period of ultraviolet epi-illumination preferentially abolishes vitamin A autofluorescence which allows intralobular NADH distribution to be evaluated without interference from the Ito cell-associated autofluorescence (8) (*Figure 2*).

3.2 Dichlorofluorescein and its derivatives

Several lines of fluorescent probes are sensitive to oxidative changes in cells. Dichlorofluorescin (DCFH) diacetate and its derivatives are such fluorochrome precursors. Since these fluorescent probes are hydrophobic, because of their ester structure, the probes can enter an intracellular compartment as non-fluorescent hydrophilic precursors. In the presence of oxidants, these precursors can be oxidized to yield fluorochromes. These hydrophilic fluoro-

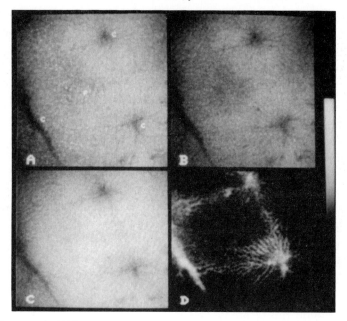

Figure 2. Increase in NADH autofluorescence in response to low-flow hypoxia in the perfused rat liver. A, control image, showing both vitamin A and NADH autofluorescence. B, the image after eliminating vitamin A autofluorescence. After a short exposure to UV light, a 10–20 min stabilizing period is required to obtain steady-state NADH autofluorescence. Note that NADH autofluorescence in the pericentral region (C) is greater than that in the periportal region (P). C, 10 min after the start of 25% low-flow hypoxia. NADH shows panlobular saturation, showing homogeneous patterns of NADH elevation. D, Microangiography in the same region by injection of FITC-labelled BSA to the perfusion circuit which is useful to confirm the state of perfusion and lobular landmarks. (From ref. 8, with permission.)

chromes can be accumulated in cells, and the fluorescence signal is elevated as long as cell viability is maintained. Once the cytoplasmic membrane is damaged, the fluorochrome rapidly leaks out of the cell.

The DCFH method is based upon the original work of Keston and Brandt (10), in which non-fluorescent DCFH is oxidized to the fluorescent dichlorofluorescein (DCF) by H_2O_2 in the presence of peroxidase or haem. Bass *et al.* (11) quantified intracellular oxidizing species in individual neutrophils using the ester-type precursor, DCFH diacetate. This reagent enters living cells and is converted into DCFH by esterase. Although the DCF technique is a powerful tool that is widely used for assessing oxidative burst of neutrophils using fluorocytometry, it is possible that DCFH may be oxidized not only by H_2O_2 but also by organic hydroperoxides such as lipid hydroperoxides (12, 13). Since it has been established (14) that stimulated neutrophils primarily release O_2^- to extracellular space and that 100% of H_2O_2, a membrane-permeable oxidant, is its secondary product, a part of the DCF

increase in stimulated neutrophils may involve intracellular lipid peroxidation which is induced in response to stimuli. The majority of recent reports
have thus applied this technique to assess intracellular hydroperoxide formation in isolated cells (15–17) rather than in neutrophils.

We have recently applied this technique to investigate the topographic distribution and temporal alteration of oxidative stress in isolated perfused liver
microcirculatory units. This method has a number of advantages over other
methods of evaluating oxidative stress.

1. Non-fluorescent DCFH is thought to be oxidized stoichiometrically with
   equimolar hydroperoxides, but not with endoperoxides.

2. By combining with another functional fluoroprobe that has a different
   emission wavelength, the mechanism of tissue injury can be investigated
   by tracing multifunctional parameters simultaneously.

3. An increase of DCF fluorescence in response to intracellular oxidative
   events can be monitored only when cells are viable. Once cells become non-
   viable, the fluorochrome leaks out. In other words, the causal relationship
   between cell injury and oxidative changes can be assessed by this technique.

In order to apply this technique to various experimental models, several technical points should be checked:

• whether cell or tissue autofluorescence is low enough relative to the loading level of DCF fluorescence

• whether the observed tissue contains colour-quenching materials (e.g.
  haemoglobin, erythrocytes, etc.) which may disturb digital fluorometry

• clearance and extracellular leakage of the fluorescence probe from the tissue

Investigators should also pay attention to the design of control experiments
to confirm whether DCF elevation *in vivo* or *in vitro* may result from a specific event related to oxidative stress. The authors usually combine a control
study using carboxyfluorescein diacetate with the original DCF study as a
separate set of experiments.

3.2.1 Assessment of oxidative stress in organs

DCFH diacetate has been used to monitor oxidative stress visually in the rat
perfused liver microcirculation. DCFH diacetate is perfused at 5 μM for
20 min and rinsed with the precursor-free perfusate for 10 min. The DCF
fluorescence (exitation, 480 nm; emission, 530 nm) labels the entire lobules,
but periportal regions are labelled more strongly than pericentral regions,
showing an intralobular gradient of the fluorochrome. After the dye loading,
parenchymal hepatocytes seem to be a major source of DCF fluorescence
signals, inasmuch as these cells occupy the majority of the total cell population
and can be preferentially stained by ester-type fluorochromes as compared
with other non-parenchymal cells such as endothelial cells (unpublished

observation). Since the intracellular fluorochrome undergoes intrahepatic metabolism involving biliary excretion, the hepatic DCF fluorescence decreases time-dependently without altering the intralobular gradient of the fluorescence. By adding the second fluorochrome, propidium iodide (PI) in the perfusate (1 μM), it is possible simultaneously to monitor intracellular oxidative changes and cell injury. PI is a DNA-intercalating reagent which stains the nuclei of inviable cells with a concomitant red fluorescence (590 nm). PI microfluorographs are recorded under epi-illumination at 535 nm. When PI is used to assess cell viability, it is necessary to check the state of perfusion in the microvascular beds observed, since whether PI molecules can reach the regions *in situ* is a limiting step of this method. Accordingly, FITC-labelled albumin solution is usually injected in order to confirm the regional perfusion and lobular landmarks such as terminal portal and hepatic venules.

We have studied the temporal alteration and topographic distribution of oxidative stress and their correlation with cell death in two different models of hepatic injury: low-flow hypoxia-induced and carbon tetrachloride (CCl_4)-induced liver damage. In liver exposed to low-flow hypoxia, despite a relatively reductive state, oxidative stress can be paradoxically observed in marginally oxygenated midzonal regions, but not in anoxic pericentral regions (*Figure 3*). Furthermore, hypoxia-induced midzonal cell death (18) is a pathological consequence of oxidative stress in the same region, inasmuch as the elevation of midzonal DCF fluorescence occurs prior to the onset of PI staining in the same regions. However, the late pericentral necrosis proceeds independent of oxidative events *in situ*, indicating that topographic distribution of oxidative changes does not necessarily correspond to that of cell injury.

Oxidative stress induced by CCl_4 infusion shows a distinct picture from that in hypoxia, although the final cell death pattern (centrilobular necrosis) is quite similar between the two experimental models (19). In liver perfused with 1 mM CCl_4, DCF enhancement occurs first in pericentral regions and then extends to the proximal regions, forming a centrilobular pattern (*Figure 4*). Cell injury (PI fluorescence) follows DCF elevation in the same pericentral regions. Furthermore, the oxidative changes were attenuated by pretreatment with SKF-525A, an inhibitor of cytochrome P450, an oxygenase predominantly distributed in pericentral regions. The mechanisms of CCl_4-induced pericentral oxidative stress seem to result from the localization of cytochrome P450, an enzyme responsible for CCl_4-elicited lipid peroxidation. Such a direct microtopographic correlation between *in situ* oxidant production and cell injury has not been provided by any previous methodology. The detailed mechanisms of early midzonal oxidative stress and subsequent centrilobular necrosis are discussed further in refs 8 and 20.

3.2.2 Assessment of microvascular oxidative stress

The use of DCF microfluorography in the rat mesenteric microcirculation allows the evaluation of the severity of oxidative changes in microvascular

Makoto Suematsu et al.

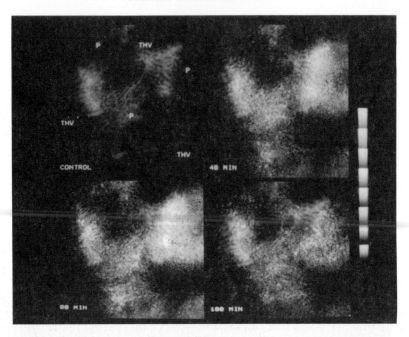

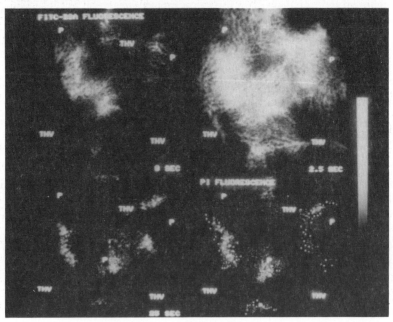

90

Figure 3. (a) Representative images of DCF microfluorographs in hypoperfused rat liver. The microscopic fluorescent images were recorded before, and 40, 60, and 80 min after the 25% low-flow perfusion was started. Note that multiple patchy fluorogenic activities were observed predominantly in the midzonal region at 60 min. (b) Serial pictures of FITC-labelled BSA fluorographs and the intrahepatic PI fluorograph (the right lower part) in the liver treated with the 80 min hypoxia. The FITC-labelled BSA (2%) solution was injected at the end of experiments. Note the massive leakage and pooling of FITC-labelled BSA in the midzonal parenchyma, suggesting the damaged integrity of sinusoidal perfusion *in situ*. PI distribution clearly corresponds to the midzonal FITC-labelled BSA pooling regions. (From ref. 18, with permission.)

endothelium during neutrophil recruitment in post-capillary venules. We have found that it is easy to load DCFH into the tissue, but that the dye can occasionally leak out after completing the loading procedure. Carboxydichlorofluorescein (CDCFH) diacetate *bis* acetoxymethyl ester is one of the best DCF derivatives in that respect, as it can be loaded into the microvascular endothelium and interstitial mast cells, and slightly in parenchymal mesothelial cells. Since the amount of CDCFH in cells after the loading procedure

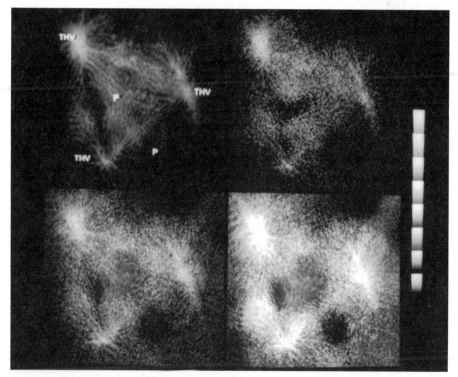

Figure 4. Pericentral oxidative stress induced by CCl_4 in the perfused rat liver. Left upper: microangiography by FITC-labelled BSA; right upper, left lower and right lower parts represent a time-dependent increase in DCF fluorescence (every 10 min after the start of 1 mM CCl_4 administration) which occurs preferentially in pericentral regions. (From ref. 19, with permission.)

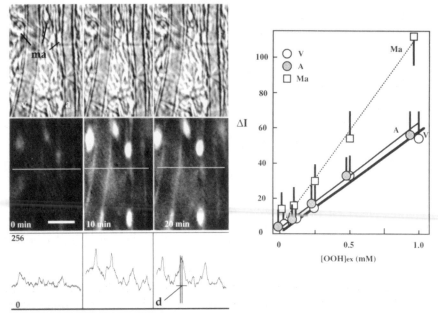

Figure 5. Left: elevation of DCF fluorescence in the mesenteric microcirculation (from ref. 21, with permission) in response to *tert*-butyl hydroperoxide at 1.0 mM. ma: mast cells. Using known concentrations of *tert*-butyl hydroperoxide, calibration curves showing the relationship between the concentration of hydroperoxide ($[OOH]_{ex}$) and the increase in grey levels can be established. Right: representative calibration curves in arteriole (A), venule (V), and mast cells (Ma). Note that the steepness of the curve varies among different cells such as arteriolar and venular endothelium, and mast cells. (From ref. 21, with permission.)

differs between cells, presumably because of variation in intracellular esterase activity, the fluorescence signal should be calibrated by superfusing known concentrations of H_2O_2 or standard lipid hydroperoxide (e.g. *tert*-butylhydroperoxide (*t*-BH)) for a constant time. In fact, the fluorogenic response to *t*-BH is different among different sites of the mesentery. Calibration procedures in DCF-assisted microfluorography for assessing intracellular oxidative changes in the rat mesenteric microcirculation *in vivo* are described below and in *Figure 5*.

(a) *Dye loading*. Dissolve 5–10 μM DCFH precursor (DCFH diacetate, CDCFH diacetate, etc.) in Krebs–Henseleit buffer (saturated with 95% nitrogen and 5% CO_2). Superfuse the mesentery with the buffer for 25–40 min.

(b) *Dye rinsing*. Superfuse the mesentery with a fluorochrome-free buffer for 5–10 min. It is usual to avoid continuous dye loading in order to prevent the dye from diffusing into the blood-stream, which occasionally stain circulating leucocytes. Fluorescently-tagged leucocytes often prevent the accurate evaluation of the fluorescence intensity, especially when they adhere to microvessels in the area being observed.

(c) Although DCFH diacetate can be loaded well into the tissue, it may often leak out quickly during the rinsing period. The loading properties of CDCFH diacetate *bis* AM ester seem to be better.

When the rat mesentery is superfused with bacterial chemotactic peptide (FMLP, 100 nM), there is a marked increase in the number of neutrophils adhering to venular endothelium. In CDCFH-loaded mesentery, there is a significant increase in the CDCF fluorescence in post-capillary venules in parallel with the increasing number of the adherent cells. Arteriolar endothelium or interstitial mast cells which do not interact directly with neutrophils exhibit no increase in fluorescence. In this model, pre-treatment with a monoclonal antibody against neutrophil-associated adhesion molecules such as CD18 attenuates the FMLP-elicited oxidative changes by reducing the number of adherent neutrophils (21).

Nitric oxide (NO) has been considered a modulator of endothelial cell—neutrophil interactions in microvessels (22), although the mechanisms by which NO lowers the adhesivity of endothelial cells *in vivo* remains to be identified. Since NO can scavenge $O_2^{\cdot-}$ or modulate the function of several iron-containing enzymes such as lipoxygenase or cyclooxygenase (23), it can be speculated that constitutive levels of NO generation might contribute to down regulation of oxidative stress or lipid peroxidation as an endogenous antioxidant. On the other hand, NO might potentiate oxidative stress by interacting, with $O_2^{\cdot-}$ and resultant active oxidants such as $ONOO^-$ or $\cdot OH$ (24). By using CDCFH-assisted microfluorography, it has been shown that endogenous NO suppression elicits an increase in leucocyte adhesion by enhancing hydroperoxide formation in endothelial cells (25). *Figure 4* illustrates the effects of N^{ω}-methyl-L-arginine methyl ester (L-NAME), an NO synthase inhibitor, on leucocyte adherence and oxidant formation in the rat mesenteric microcirculation. As shown in the transmission images, L-NAME superfusion (100 μM, 60 min) evokes a significant increase in neutrophil adherence in post-capillary venules. However, the topographic distribution of oxidative stress shows quite a different picture from that in the FMLP-treated microcirculation, in which the increase in CDCF fluorescence can be observed only in venular segments co-localized with adherent neutrophils. In the L-NAME-treated mesentery, CDCF fluorescence increases not only in venules but also in arterioles and interstitial mast cells (*Figure 5*). The oxidative impact in venules induced by L-NAME is equivalent to that elicited by superfusion with approximately 400 μM *t*-BII for 10 min. Furthermore, fluorescence enhancement occurs as early as 15–30 min after the start of L-NAME administration, which precedes the onset of neutrophil adhesion. Even in rats treated with several monoclonal antibodies against ICAM-1, P-selectin, or CD18, L-NAME-elicited CDCF enhancement cannot be abolished, suggesting that the early oxidative stress by endogenous NO suppression does not involve neutrophil-dependent processes (*Figure 6*). Interestingly, an iron chelator (desferrioxamine) significantly attenuates L-NAME-elicited CDCF enhancement and subsequent increase in leucocyte recruitment (25). Since lipid hydroperoxides are known to be a potent inducer of P-selectin expression in endothelial cells (26), it is likely that endogenous NO suppression may activate iron-dependent lipid peroxidation and thereby upregulate P-selectin expression (*Figure 6*) and neutrophil recruitment. These results prompt us to investigate the possibility that NO generated by microvascular endothelial cells might attenuate the oxidative impact occurring in themselves and thereby regulate functions like cell adhesivity. There is evidence to show such a possibility using DCFH diacetate-loaded human umbilical venous endothelial cells (27). The authentic sources of superoxides or lipid peroxides in endothelial cells are quite unknown. Although there was possible involvement of mitochondrion-derived oxidants indicated by the attenuating effects of sodium azide, the data interpretation should be carefully evaluated, as CDCFH oxidation by hydroperoxides *in vitro* does not occur in the absence of a catalyst such as peroxidase or haem, the function of which can be eliminated by the application of azide.

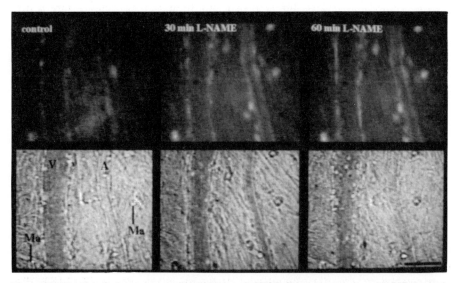

Figure 6. Spatial and temporal distribution of CDCF fluorescence in L-NAME-treated mesenteric microcirculation. The figure shows the data recorded (a) before, (b) 30 min, and (c) 60 min after the start of L-NAME superfusion (100 μM). Top, representative transmission images in the same microcirculatory units; Bottom: representative CDCF microfluorographs before and after the start of 100 μM L-NAME superfusion. Ma, A, and V denote mast cells, arteriole, and venule, respectively. Note that CDCF fluorescence in arterioles, venules, and mast cells was elicited by L-NAME application prior to venular leucocyte adherence. Bar, 50 μm. (From ref. 25, with permission.)

In similar experiments *in vitro*, it has been shown that epithelial cells from fetal rat heart (28) and porcine aorta can be loaded with CDCFH diacetate *bis* acetoxymethyl ester without a serious dye leakage at least 1 h after the loading procedure. In rat heart microvascular endothelial cells, superfusion with L-NAME (100 μM), but not with D-NAME, produces a significant increase in CDCF fluorescence. The response of the fluorogenic reaction varies in each cultured cell. On average, the oxidative impact elicited by 100 μM L-NAME for 30 min is equivalent to that induced by 50 μM *t*-BH superfusion for 10 min. The present system can provide multiple information on cell function by combining with another fluorochrome possessing different emission spectra. For example, 8-diacetyl low-density lipoprotein (a red fluorescence-conjugated LDL) can be applied at the end of experiments in order to confirm the specificity of each cultured cell. This procedure is important particularly when primary cultured cells are used for measurements. As in the study using perfused liver, propidium iodide can be co-applied to the culture dishes in order to check cell viability. It is noteworthy that the *t*-BH concentration required to elicit a given amount of CDCF enhancement is very different for the intravital mesenteric preparation and for cultured epithelial cells. One of the reasons for the variability in oxidative vulnerability between *in vivo* and *in vitro* systems might be that the oxidant-scavenging capacity of the two systems are very different (e.g. due to plasma proteins, erythrocytes, regional oxygen tension, etc.).

4. Other applications of digital microfluorography: leucocyte-dependent tissue injury *in vivo*

Leucocytes are one of the important cellular components which determine the severity of oxidative tissue injury. It is therefore important to investigate the behaviour of circulating leucocytes and its correlation with cell injury in disease conditions. Although the roles of leucocytes in oxidant-dependent injury mechanisms have been examined for many years, it is still unclear whether leucocytes *cause* cell injury at the site of adherence or whether leucocyte adherence is just a *consequence* of the injury. Multifunctional digital microfluorography can assess the temporal and spatial correlation between tissue leucocyte recruitment and cell injury in several disease models. Carboxyfluorescein diacetate succinimidyl ester (CFSE) is useful for staining circulating leucocytes and platelets. By using rat spinotrapezius muscle preparation *in vivo*, the behaviour of circulating leucocytes has been examined during haemorrhagic hypotension followed by reperfusion while monitoring cell viability by superfusing the muscle preparation with propidium iodide (29). In skeletal muscle microcirculation exposed to ischaemia–reperfusion, leucocytes are trapped in two different ways, namely by capillary plugging (30) and by venular adherence. Interestingly, the onset of significant cell injury occurs rapidly after the start of reperfusion, and leucocytes accumulate

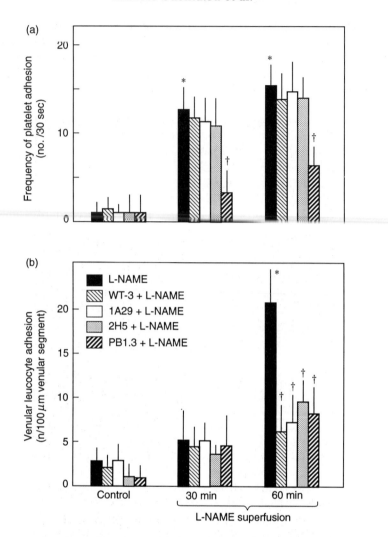

Figure 7. The effects of anti-ICAM-1 (1A29) and anti-CD18 (WT-3) monoclonal antibodies on (b) venular oxidative changes and (a) adherent cell density in the L-NAME-superfused microcirculation. L-NAME and FMLP were superfused at 100 μM and 100 nM, respectively. Note that pretreatment with any of three monoclonal antibodies (1A29, WT-3, or PB1.3) did not attenuate venular hydroperoxide generation evoked by 30 min L-NAME superfusion. Polymorphonuclear cell depletion by methotrexate (MT + L-NAME) attenuated leucocyte adhesion and hydroperoxide generation elicited by 60 min L-NAME superfusion. In these animals, however, L-NAME induced a significant increase in hydroperoxide formation at 30 min even in the absence of adherent cells, suggesting that the early oxidative change is independent of leucocyte adhesion. Statistically significant ($P < 0.05$) compared with †control values before application of L-NAME. *group treated with L-NAME. # group treated with FMLP. (From ref. 25, with permission.)

following the cell injury (*Figure 7*). Furthermore, the distribution of cell injury (as judged by the presence of propidium iodide-positive nuclei) is topographically dissociated from that of adherent or plugging leucocytes, suggesting that early cell injury in post-ischaemic tissue may be independent of leucocyte activation. In this experimental model, early cell injury after

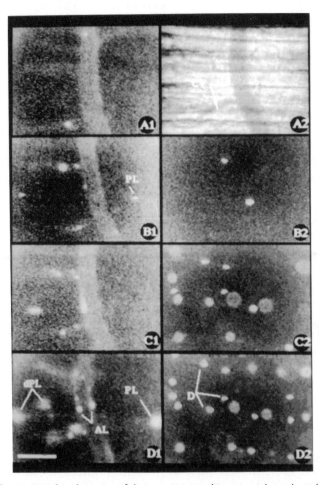

Figure 8. Representative images of leucocyte recruitment and nuclear injury in the spinotrapezius muscle microcirculation exposed to hypotension followed by reperfusion. A1 and A2: CFSE-labelled leucocyte fluorograph and a transillumination image in the control period, respectively. Note that plasma in the collecting venule was faintly visualized without showing any diffusion in the interstitium. B1, 2, Cl, 2, and D1, 2: CFSE and Pl fluorographs at 30 min hypotension, and 30 min and 60 min after reperfusion, respectively. PL, plugging leucocyte in the capillary. The Pl fluorographs (B2, C2, D2) were digitally processed as binary images. D, damaged nuclei stained by Pl; dPL, defocused plugging leucocytes; AL, adherent leucocytes in collecting venules. Bar, 50 μm. (From ref. 29, with permission.)

reperfusion seemed to be leucocyte-independent, and the leucocyte recruit-ment in capillaries and venules may be a consequence of cell injury.

Another experimental model in which tissue leucocyte sequestration may play an important role in the mechanism of tissue injury is post-transplantation graft injury. Menger *et al.* (31, 32) used rhodamine G, a fluorochrome for intravital leucocyte staining, and dual-colour intravital microfluorography to study leucocyte recruitment and microvascular perfusion during acute organ failure after transplantation. Since this fluorochrome has a red fluorescence, the second functional fluoroprobe with a different colour (e.g. FITC–albumin/dextran (green fluorescence) for assessment of microvascular per-meability) may be used to study the role of leucocytes in tissue injury.

5. Conclusion

In vivo (or *ex vivo*) oxidative stress can be visualized by multifunctional digi-tal microfluorography using several oxidant-sensitive fluoroprobes. Although the present technique can provide semiquantitative information on the fluor-escence signals, a major problem which remains to be answered is to establish the method for quantitative measurements of specific oxidant species gener-ated in extra- or intracellular space. Extensive refinement of photonic probes which have an excellent specificity with low background signals would be necessary to establish this methodology.

Acknowledgements

The authors thank Professor Michael Menger for fruitful discussion regard-ing intravital digital microfluorography, and Dr Nobuhito Goda for the study of cultured endothelial cells. Part of this work was supported by the Interna-tional Program of a Grant-in-Aid for Scientific Research from the Ministry of Education, Science, and Culture of Japan, and by a grant from Keio Univer-sity School of Medicine.

References

1. Chambers, R. and Zweifach, B. W. (1994). *Am. J. Anat.*, **75**, 173.
2. Nagata, H. and Guth, P. H. (1983). *Am. J. Physiol.*, **245**, G201.
3. Kontos, H. A., Wei, E. P., Navari, R. M., Levasseur, J. E., Rosenblum, W. I., and Patterson, J. R. Jr (1978). *Am. J. Physiol.*, **234**, H371.
4. Ohshima, N. and Sato, M. (1987). *Microvasc. Res.*, **34**, 250.
5. Sohara, Y. (1987). *Jpn. Coll. Angiology*, **27**, 113.
6. McCuskey, R. S., Reilly, F. D., and McCuskey, P. A. (1979). *Bibl. Anat.*, **18**, 73.
7. Suematsu, M., Oda, M., Suzuki, H., Kaneko, H., Furusho, T., Masushige, S., and Tsuchiya, M. (1993). *Microvasc. Res.*, **46**, 28.
8. Suzuki, H., Suematsu, M., Ishii, H., Kato, S., Miki, H., Mori, M., Ishimura, Y. Nishino, T., and Tsuchiya, M. (1994). *J. Clin. Invest.*, **93**, 155.

9. Quistorff, B., Chance, B., and Takeda, H. (1983). In *Frontiers of biological energetics*. Vol. 2, p. 1478. Academic Press, New York.
10. Keston, A. S. and Brandt, R. B. (1965). *Anal. Biochem.*, **11**, 1.
11. Bass, D. A., Parce, J. W., DeChatelet, L. R., Szejda, P., Seeds, M. C., and Thomas, M. (1983). *J. Immunol.*, **130**, 1910.
12. Cathcart, R., Schwiers, E., and Ames, B. N. (1983). *Anal. Biochem.*, **134**, 111.
13. Gores, G. J., Flarsheim, C. E., Dawson, T. L., Nieminen, A. L., Herman, B., and Lemasters, J. J. (1989). *Am. J. Physiol.*, **257**, C347.
14. Makino, R., Tanaka, T., Iizuka, T., Ishimura, Y., and Kanegasaki, S. (1986). *J. Biol. Chem.*, **261**, 11444.
15. Scott, J. A., Homcy, C. J., Khaw, B. A., and Rabito, C. A. (1988). *Free Radic. Biol. Med.*, **4**, 79.
16. Kurose, I., Fukumura, D., Saito, H., Suematsu, M., Miura, S., Morizane, T., and Tsuchiya, M. (1991). *Cancer Lett.*, **59**, 201.
17. Suzuki, H., Suematsu, M., Miura, S., Asako, H., Kurose, I., Ishii, H., and Tsuchiya, M. (1993). *Pancreas*, **8**, 465.
18. Suematsu, M., Suzuki, H., Ishii, H., Kato, S., Yanagisawa, T., Suzuki, H., and Tsuchiya, M. (1992). *Gastroenterology*, **103**, 994.
19. Suematsu, M., Kato, S., Ishii, H., Yanagisawa, T., Asako, H., Suzuki, H., Oshio, C., and Tsuchiya, M. (1991). *Lab. Invest.*, **64**, 167.
20. Suematsu, M., Suzuki, H., Ishii, H., Kato, S., Hamamatsu, H., Miura S., and Tsuchiya, M. (1992). *Lab. Invest.*, **67**, 434.
21. Suematsu, M., Schmid-Schönbein, G. W., Chavez-Chavez, R. H., Yee, T. T., Tamatani, T., Miyasaka, M., DeLano, F. A., and Zweifach, B. W. (1993). *Am. J. Physiol.*, **264**, H881.
22. Kubes, P., Suzuki, M., and Granger, D. N. (1991). *Proc. Natl Acad. Sci. USA*, **88**, 4651.
23. Kanner, J., Harel, S., and Granit, R. (1992). *Lipids*, **27**, 46.
24. Beckman, J. S., Beckman, T. W., Chen, J., Marshall, P. M., and Freeman, B. A. (1990). *Proc. Natl Acad. Sci. USA*, **87**, 1620.
25. Suematsu, M., Tamatani, T., DeLano, F.A., Miyasaka, M., Forrest, M., Suzuki, H., and Schmid-Schönbein, G. W. (1994). *Am. J. Physiol.*, **266**, H2410.
26. Patel, K. D., Zimmerman, G. A., Prescott, S. M., McEver, R. P., McEver, R. P., and McIntyre, T. M. (1992). *J. Biol. Chem.*, **267**, 15168.
27. Xiao-fei, N., Smith, C. W., and Kubes, P. (1994). *Circ. Res.*, **74**, 1133.
28. Kasten, F. (1972). *In Vitro*, **8**, 128.
29. Suematsu, M., DeLano, F. A., Poole, D., Engler, R. L., Miyasaka, M., Zweifach, B. W., and Schmid-Schönbein, G. W. (1994). *Lab. Invest.*, **70**, 684.
30. Engler, R. L., Schmid-Schönbein, G. W., and Pavelec, R. S., (1983). *Am. J. Pathol.*, **111**, 98.
31. Gonzalez, A. P., Sepulveda, S., Massberg, S., Baumeister, R., and Menger, M. D. (1994). *Transplantation*, **58**, 403.
32. Menger, M. D. and Lehr, H-A. (1993). *Immunol. Today*, **14**, 519.

7

Salicylic acid and phenylalanine as probes to detect hydroxyl radicals

HARPARKASH KAUR and BARRY HALLIWELL

1. Introduction

It is now widely accepted that reactive oxygen species (ROS) and reactive nitrogen species (RNS) (including free radicals) are involved in the pathogenesis of several inflammatory and degenerative disease states (for a recent review see ref. 1). In spite of their existence these species have not as yet been measured in clinical diagnosis. This is partly due to the lack of agreed formats and standardized methodology for measuring oxidative stress and/or levels of free radicals in humans

1.1 Detection of free radicals

The only technique which can detect radicals directly is electron spin resonance (ESR) spectroscopy. Although the technique is highly sensitive (thresholds of 10^{-8} M spins for signals with one Gauss linewidth have been detected), it is not directly applicable to the study of biological oxidations or to the majority of radical chemistry. A more successful technique permitting ESR investigation of short-lived reactive free radicals, by transforming them into more persistent species is the so-called 'spin trapping' method.

The most common spin traps in general use are nitrones and c-nitroso compounds though the latter have found limited application in biological studies. Nitrones such as phenyl-*tert*-butylnitrone (PBN) and 5,5-dimethyl-1-pyrroline N-oxide (DMPO) have been used most frequently since they give the longest-lived spin adducts with oxygen centred radicals. The results obtained with these traps have often suggested the trapping of hydroxyl radical ($^{\bullet}$OH).

1.2 The hydroxyl radical

The hydroxyl radical is the most reactive oxygen radical known. The earliest, well-documented studies on its properties were carried out by radiation chemists (2).

The hydroxyl radical reacts at, or close to, a diffusion-controlled rate with all biological molecules; its half-life in cells has been estimated to be 10^{-9} sec. It can be produced when water is exposed to ionizing radiation, fragmenting the oxygen–hydrogen covalent bond, leaving a single electron on hydrogen and one on oxygen (Reaction 1):

$$H—O—H \xrightarrow{\text{intermediate stages}} H^{\cdot} + {\cdot}OH \qquad (1)$$

$H^{\cdot}$ is the hydrogen radical (hydrogen atom) and ${\cdot}OH$ is the hydroxyl radical. Much of the ${\cdot}OH$ generated *in vivo*, except during excessive exposure to ionizing radiation, probably comes from the metal ion-dependent breakdown of hydrogen peroxide (H_2O_2) according to the following reaction:

$$M^{n+} + H_2O_2 \longrightarrow M^{(n+1)+} + {\cdot}OH + {}^{-}OH \qquad (2)$$

in which M^{n+} is a metal ion. Examples of M^{n+} are titanium(III), copper(I), iron(II) or cobalt(II). The iron(II)-dependent formation of ${\cdot}OH$ is better known as the Fenton reaction (3). Hydrogen peroxide and superoxide radical (with some other reducing agents, such as ascorbate) can assist ${\cdot}OH$ formation by reducing iron in the Haber–Weiss reaction (4, 5), reaction 3:

$$O_2^{\cdot} + H_2O_2 \xrightarrow{\text{Fe/Cu}} {\cdot}OH + {}^{-}OH + O_2 \qquad (3)$$

Interest in nitric oxide ($NO^{\cdot}$), a vasodilator radical produced by several cell types, including phagocytes and vascular endothelial cells, has led Beckman *et al.* (6) to postulate the formation of ${\cdot}OH$ via the interaction between $O_2^{\cdot}$ and $NO^{\cdot}$ (see Reactions 4–6)

$$NO^{\cdot} + O_2^{\cdot} \longrightarrow ONOO^{-} \text{ (peroxynitrite)} \qquad (4)$$

$$ONOO^{-} + H^{+} \longrightarrow ONOOH \qquad (5)$$

$$ONOOH \longrightarrow {\cdot}OH + NO_2^{\cdot} \qquad (6)$$

This reaction has been shown to occur *in vitro* and its physiological significance, especially the detection of nitrated products resulting from the decomposition of peroxynitrite, are under rigorous investigation. It seems unlikely that stoichiometric quantities of ${\cdot}OH$ are produced (discussed later in this chapter).

1.3 Detection of hydroxyl radicals: 'trapping assays'

Trapping assays imply the reaction of ${\cdot}OH$ with a foreign 'detector' molecule to form specific products (spin trapping is, of course, an example). Other methods used have included the ability of ${\cdot}OH$ to oxidize methional and related compounds to ethene (5), degrade tryptophan (7), convert dimethyl-sulphoxide into methanesulphinic acid (8), decarboxylate [*carboxyl*-14C] benzoic acid to $^{14}CO_2$ (9), and oxidize deoxyribose into thiobarbituric acid-reactive material (10). However, the two trapping methods with the greatest potential for specifically identifying hydroxyl radical are ESR spin trapping (11) and aromatic hydroxylation (12).

Although the spin-trapping method can be a definitive way to trap ˙OH, artefacts exist and a great deal of expertise is required to interpret the experimental results. In all biological studies using currently available spin traps, high and potentially toxic concentrations of the spin traps have been necessary to achieve results. This makes them impossible to use in humans, and often difficult to use in animals. An alternative in *in vivo* studies is to measure trapping of ˙OH by its reaction with a relatively non-toxic aromatic compound, either an exogenous compound such as the non-steroidal anti-inflammatory drug aspirin or a naturally occurring compound such as the amino acid phenylalanine. Similar compounds can be used for *in vitro* studies.

2. Aromatic hydroxylation

Aromatic compounds undergo addition reactions with hydroxyl radicals producing characteristic products of hydroxylation. This is not as simple as it sounds due to the possibility of other complex oxidation reactions of aromatic compounds. For example, there is a wealth of literature on the oxidation of benzene and derivatives of benzene by metal ion–H_2O_2 systems (13). Benzene [I] itself reacts with ˙OH, at diffusion-controlled rates, leading to the formation of the hydroxycyclohexadienyl radical [II].

This radical can undergo several reactions, such as dimerization [III] followed by elimination of water to give biphenyl [IV].

The radical [II] can also undergo a disproportionation reaction to give a mixture of phenol [V] and benzene [I].

The overall chemistry is affected by pH and the presence of oxidizing agents such as O_2, Fe^{3+}, or Cu^{2+} yielding increased products of hydroxylation. Studies have demonstrated that, under physiologically relevant conditions (pH 7.4, metal ions, and O_2 present) the predominant products formed are due to hydroxylation. Aromatic hydroxylation as an assay for ˙OH in biological systems, was first demonstrated by measuring ˙OH in a system generating $O_2^{\bullet-}$ and H_2O_2 (from a mixture of xanthine and xanthine oxidase) using salicylic acid as the detector molecule (14). The products of hydroxylation were quantified using a colorimetric method that measured *o*-dihydric phenol. The assay was improved, resulting in greater sensitivity, by Richmond *et al.* (14), and has often been used as a simple test-tube assay of ˙OH formation in activated phagocytic cells, isolated hepatocytes, soluble enzymes, ischaemia/reperfusion studies, and similar biological systems. *Protocol 1* outlines the method for qualitative assessment of hydroxyl radical production (it does not measure hydroxylated products that do not show up in the colorimetric assay).

Protocol 1. Aromatic hydroxylation for detection of hydroxyl radicals in xanthine oxidase system using salicylate as detector molecule (14)

Equipment and reagents

All chemicals and solvents used in this protocol can be purchased from Sigma and BDH unless otherwise stated.

- $FeCl_3$ 5 mM: prepare fresh before use
- EDTA 5 mM
- KH_2PO_4 buffer, 150 mM, pH 7.4
- KH_2PO_4–KOH buffer, 150 mM, pH 7.4
- Hypoxanthine, 2 mM: prepare stock initially in 50 mM NaOH and dilute with KH_2PO_4–KOH buffer, pH 7.4
- Salicylate, 20 mM in KH_2PO_4–KOH buffer, pH 7.4
- Xanthine oxidase: diluted just before use into buffer to give 0.4 enzyme units/ml
- HCl, 11.6 M
- NaCl, solid
- Diethyl ether (ethoxyethane)
- Double-distilled water
- Trichloroacetic acid, 10% (w/v) in 0.5 M HCl
- Sodium tungstate, 10% (w/v) aqueous solution
- Sodium nitrite, 0.5% (w/v) aqueous solution, made up fresh daily
- KOH, 0.5 M
- 2,3-dihydroxybenzoate (2,3-DHB)
- Glass test tubes
- Vortex mixer
- Water bath

Method

1. Prepare the reaction mixture (final volume 2 ml) containing the following:

$FeCl_3$, 5 mM	40 μl
EDTA, 5 mM	40 μl
hypoxanthine, 2 mM	200 μl
salicylate, 20 mM	200 μl
KH_2PO_4 buffer	1.48 ml

2. Start the reaction by adding 40 μl of 0.4 units/ml xanthine oxidase.

3. Incubate the tubes with gentle shaking at 25°C for 90 min.

4. Stop the reaction by adding 80 μl of 11.6 M HCl and 0.5 g of solid NaCl followed by 4 ml of chilled diethyl ether (ethoxyethane). Vortex for 30 sec.

5. Carefully remove 3 ml of the upper (ether) layer to a boiling tube using a syringe and evaporate to dryness at 40°C in a fume cupboard.

6. (a) Dissolve the residue in 0.25 ml of double-distilled water and add the following reagents in the order stated:

    (i)   1.25 μl 10% (w/v) trichloroacetic acid dissolved in 0.5 M HCl
    (ii) 0.25 ml 10% (w/v) aqueous sodium tungstate
    (iii) 0.25 ml 0.5% (w/v) aqueous sodium nitrite.

    (b) Mix the reagents well.

7. Allow the mixture to stand for 5 min.

8. Add 0.5 ml of 0.5 M KOH and read the absorbance of the pink complex at 510 nm after 60 sec.

9. Prepare standard curves using 2,3-DHB carried through the same ether extraction and colorimetric assay.

A complete hypoxanthine–xanthine oxidase system should give a final A_{510} of about 0.65, corresponding to 150–200 nmol of hydroxylated product. Formation of hydroxylated products can be inhibited almost completely by superoxide dismutase, catalase, or the iron chelator desferrioxamine.

This method gives an underestimate (no more than 50–70%) of products of hydroxylation, and thus of hydroxyl radical production, since only a single hydroxylated product is being measured. Hydroxylation products can also be measured from their fluorescence spectra, but this is limiting as only the fluorescent products can be estimated. In the past, gas–liquid chromatography was used with an electron-capture detector or coupled with mass spectrometry to improve the identification of products detected, but this requires the conversion of products into volatile material and may result in loss of product (13). Hence for a more quantitative approach, high-performance liquid chromatography (HPLC) is recommended, and to gain a true measure of ˙OH in simple and complex biological systems the need for a non-toxic aromatic compound is stressed (13). For *in vivo* studies such a compound could already be present in humans and animals or it could be administered to them in established doses, for example non-steroidal anti-inflammatory drugs administered to patients with rheumatoid arthritis (RA). The products measured should arise as a direct result of hydroxylation by ˙OH and be distinct from possible enzyme-produced hydroxylated metabolites.

2.1 Salicylate

Aspirin (*o*-acetylsalicylic acid) is a widely used analgesic for self-medication, and much larger doses are sometimes used in the treatment of RA. Aspirin is also increasingly used as a prophylactic agent against thrombotic vascular disease. Aspirin administered to humans or to animals is rapidly hydrolysed to salicylate, which can be attacked by ˙OH at pH 7.4 to give three products (2,3-DHB, 2,5-dihydroxybenzoate (2,5-DHB), and catechol), as shown in *Figure 1*, which can be identified by separation using HPLC (15-17). Salicylate reaches its peak in plasma about 1.5 h after the oral intake of aspirin and is in part metabolized by conjugation with glycine (liver glycine *N*-acetylase), by conjugation with glucuronic acid to form salicyl-acyl and salicyl phenolic glucuronide, and by hydroxylation (liver microsomal hydroxylases) to form 2,5-DHB. About 60% of the salicylate remains unmodified and can undergo ˙OH attack (*Figure 1*).

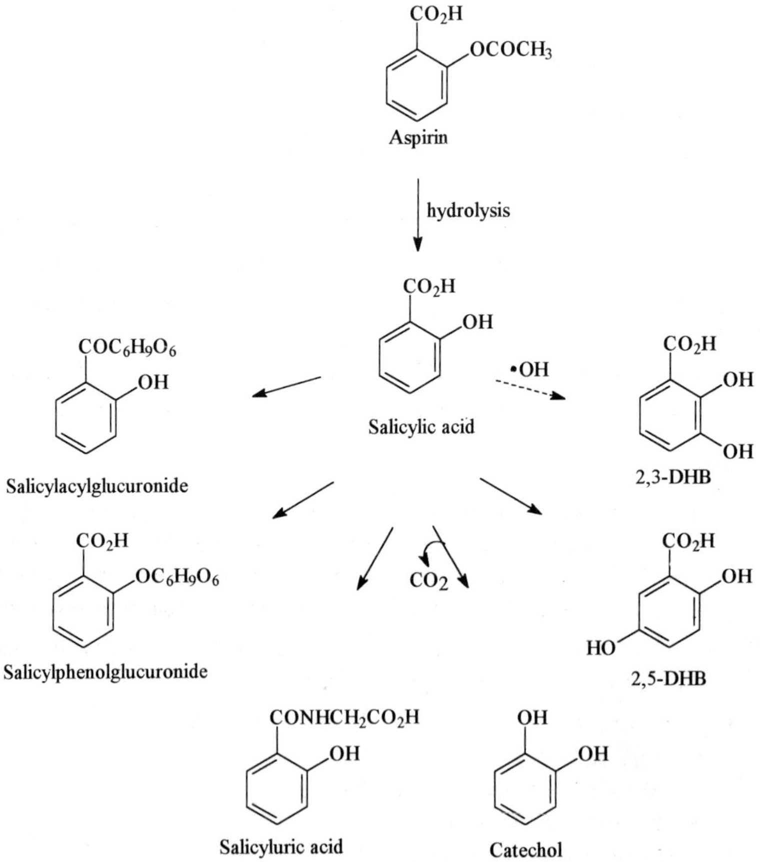

Figure 1. Metabolites and products of ˙OH attack and hydroxylation of salicylic acid.

Protocol 2 describes a suitable method for analysis by HPLC for the products of *in vivo* ˙OH attack of salicylate (aspirin). Typical results of the chro-

---

Protocol 2. HPLC method used to detect the products of salicylic acid hydroxylation

Equipment and reagents

- Salicylic acid
- 2,3-DHB
- 2,5-DHB
- Resorcinol (Aldrich Chemical Co.)
- HCl, 1 M
- Citric acid
- Sodium acetate (anhydrous)
- Sodium hydroxide (BDH Chemicals)
- HPLC grade water
- Ethyl acetate
- Methanol
- Heparinized blood tubes

- Spherisorb 5 ODS column (HPLC Technology): 25 cm × 4.6 mm with Hibar guard column and C-8 cartridge (BDH)
- HPLC system: Rheodyne injector valve with 20 μl loop, isocratic pump (Polymer Laboratories Ltd), and electrochemical detector (EDT LCA 15) equipped with a glassy carbon working electrode and an Ag/AgCl reference electrode
- Glass stoppered tubes (10 ml)
- Vortex mixer
- Centrifuge
- Water bath set at 55°C

A. *Plasma samples*

1. Collect 2.0 ml of blood into heparinized tubes.
2. Centrifuge the blood (1000 *g*, 10 min, 4°C) to obtain plasma.
3. Add resorcinol (1.0 μM) and 1 M HCl (50 μl) to each plasma sample.
4. Extract with ethyl acetate (2 × 8 ml) on a vortex mixer for 2 min.
5. Bulk the ethyl acetate layers and evaporate to dryness in a water bath at 55°C.
6. Reconstitute the residue in HPLC-grade water (200 μl) and 1 M HCl (50 μl).
7. If samples are not to be analysed immediately, store at −20°C.

B. *HPLC*

1. Prepare standard solutions of 2,3-DHB, 2,5-DHB, salicylic acid, and resorcinol in HPLC-grade water containing 50 mM HCl to a maximum concentration of 5 mM. These solutions are stable at 4°C for a period of 1 week; allow to reach room temperature and vortex mix before use.
2. Prepare HPLC eluent (34 mM sodium citrate/27.7 mM acetate buffer (pH 4.75)/methanol 97.2:2.8, v/v) as follows.
   (a) Dissolve the following in 2.5 litres of HPLC grade water,
      - citric acid (15.5 g)
      - anhydrous sodium acetate (6.68 g)
      - sodium hydroxide (5.0 g)
   (b) Adjust to pH 4.75 with HCl.
   (c) Mix with methanol 97.2:2.8 (v/v).

107

3. Inject 20 μl of the sample or standard solutions via the Rheodyne injector valve, elute at a flow rate of 0.9 ml/min, and monitor (with an electrochemical detector) at +0.65 V (see *Figure 2* for typical chromatograms).

matography of standards and extracts of samples are shown in *Figure 2*. Results shown are for the separation of an internal standard mix, namely 1 μM each of 2,3-DHB, 2,5-DHB, and resorcinol (RS) (*Figure 2a*), and of plasma sample extracts from a healthy volunteer either not consuming (*Figure 2b*), or taking (*Figure 2c*), aspirin. The extra peaks marked SA and SU represent salicylate and salicylurate respectively. *Figure 2d* shows the chromatographic separation of an extract of a plasma sample from an RA

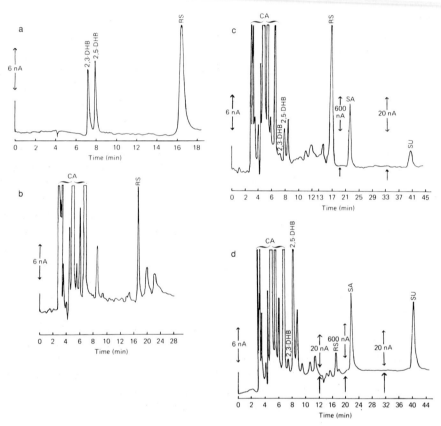

Figure 2. Separation by HPLC of the hydroxylated products of salicylate (adapted from ref. 17). (a) An internal standard mix of 2,3-DHB, 2,5-DHB, and RS; (b)–(d) extracts of plasma samples from a healthy volunteer not consuming (b) and taking (c) aspirin, and a rheumatoid arthritis patient treated with aspirin (d). DHB, dihydroxybenzoate; RS, resorcinol; CA, catecholamine; SA, salicylate; SU, salicylurate.

patient receiving treatment with aspirin. The large peaks marked CA at the start of the chromatograms of plasma extracts are probably catecholamines but, as shown in *Figure 2d*, these do not interfere with the dihydroxyben-zoates (DHBs). The arrows on each chromatogram indicate changes in sensitivity of the electrochemical detector.

The DHBs are stable to air oxidation in body fluids and tissue extracts only if the pH is kept low. The metabolites of aspirin or salicylate (*Figure 1*) administered to humans are reported in the literature to include 2,5-DHB but not 2,3-DHB. This suggests that the hydroxylation of salicylate can be used *in vivo* to assess ˙OH formation, by measuring the amount of 2,3-DHB only. Although 2,5-DHB is also formed (*Figure 1*), this product can also be generated from salicylate by enzymes such as cytochromes P450 (15).

Raised concentrations of 2,3-DHB have been detected in the body fluids of RA patients taking aspirin (16), in subjects consuming excess alcohol (T. J. Peters, personal communication), and in rats treated with the redox-cycling drug adriamcyin (17), subjected to hyperoxia (18), or exposed to ionizing radiation. Furthermore, recent reports show significantly higher (23%) plasma levels of 2,3-DHB in carefully controlled diabetic patients than in control subjects following the administration of 1 g of aspirin (19). The presence of trace amounts of 2,3-DHB in the plasma of healthy subjects after aspirin intake could be due to the base-line rate of intracellular ˙OH formation from ionizing radiation and Fenton reactions.

The most serious disadvantage of using salicylate and catechols is that they have high affinities for iron(III) and copper(II) (20), which might perturb iron- and copper-dependent ˙OH generation. In addition, salicylate can exert other effects; for example, it is a weak inhibitor of cyclooxygenase and thereby interferes with the production of prostaglandins. This is detrimental to the use of salicylate to study the formation of ˙OH during ischaemia and reperfusion, since the cyclooxygenase pathway is a potential source of radicals and prostaglandins appear to be involved in ischaemia/reperfusion injury as well as myocardial stunning. Salicylate also inhibits phospholipase C and suppresses glycosaminoglycan synthesis by cartilage. In animal studies, high salicylate levels can cause cardiovascular toxicity (e.g. in the dog (21)). A recent review (22) has presented the case for salicylate at only 100 μM, as the trapping molecule of choice (offering an alternative to spin trapping) in probing the role of ˙OH in cardiac oxidative injury in isolated perfused heart preparations. However, much higher levels than this will be required for *in vivo* experiments, since blood contains many more competing ˙OH scavengers than the buffers used in isolated heart perfusions.

2.2 Phenylalanine

Another potential trap for ˙OH is the essential amino acid, phenylalanine. *In vivo* the L-isomer (but not the D-isomer) of phenylalanine is hydroxylated by phenylalanine hydroxylase at position 4 on the aromatic ring to give L-*p*-

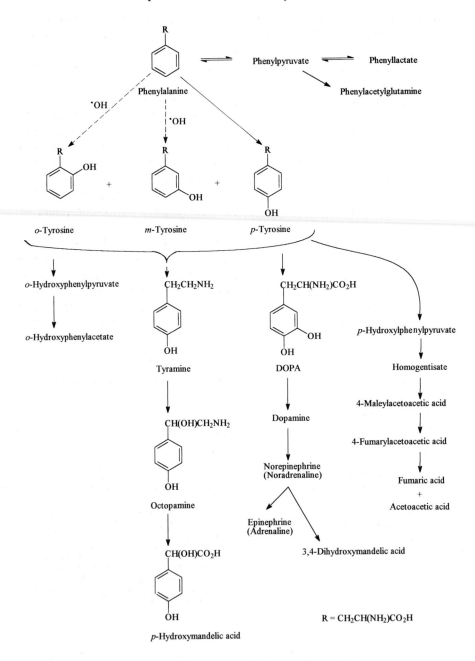

Figure 3. Hydroxylated products and metabolites of phenylalanine (diagram compiled from reports in the literature).

tyrosine. Indeed, the metabolism of phenylalanine is complex (*Figure 3*). By contrast, ˙OH cannot distinguish between the L- and D-isomers and will produce *o*-, *m*- and *p*-tyrosines as products of hydroxylation, as illustrated in *Figure 3* (together with the known metabolites of phenylalanine reported in the literature).

Separation of *o*-tyrosine (2-hydroxyphenylalanine), *m*-tyrosine (3-hydroxy-phenylalanine), and *p*-tyrosine (4-hydroxyphenylalanine) has been achieved by HPLC with electrochemical detection (*Protocol 3*). Using an ultraviolet or a fluorescence detector has the further advantage of detecting the phenylalanine precursor (*Figure 4a*) and most of its metabolites (*Figures 3* and *4b*). *Figure 4* shows chromatograms of HPLC separations of a standard mixture of *o*-, *m*-, and *p*-tyrosines (70 μM) and phenylalanine (3.5 mM) on its own (*Figure 4a*) or with metabolites of phenylalanine (70 μM each) as indicated. As can be seen, the reported metabolites of phenylalanine (*Figure 3*) do not interfere with the detection of tyrosines and phenylalanine (*Figure 4*). Separation by HPLC of synovial fluid from an RA patient dropped into saline to measure any endogenous tyrosines or phenylalanine is shown in *Figure 4c* and separation of synovial fluid from an RA patient dropped into phenylalanine (5 mM) in shown in *Figure 4d* to demonstrate the *in vitro* formation of products of hydroxylation. Peaks in the chromatograms are not identified unless they are pertinent to the discussion.

Protocol 3. Detection of phenylalanine hydroxylation products

Equipment and reagents

All reagents from BDH unless stated otherwise.

- Tyrosines: *o*-tyrosine (2-hydroxyphenylala-nine), *m*-tyrosine (3-hydroxyphenylalanine), and *p*-tyrosine (4-hydroxyphenylalanine), all of the highest purity (Sigma)
- DL-phenylalanine
- Octopamine
- Dihydroxyphenylalanine (DOPA)
- Tyramine
- *m*-Hydroxymandelic and *p*-hydroxymem-delic acids
- Uric acid
- KH_2PO_4
- Orthophosphoric acid: H_3PO_4

- HPLC-grade water (Romel Chemicals)
- Methanol (Romel Chemicals)
- Amicon 'microcentrifree' filtration devices to achieve protein-free samples
- Centrifuge
- HPLC system: 25 cm × 4.6 mm Nucleocil 5 μm C-18 column (HPLC Technology) with Hibar guard column and a C-8 cartridge (BDH) Polymer Laboratories Ltd pump and either an ultraviolet detector set at 274 nm (sensitivity 0.02), a fluorescence detector set at the excitation wavelength of 275 nm and emission of 305 nm, or an electro-chemical detector set at 0.82

A. *Sample preparation*

1. Prepare plasma as in *Protocol 2*. Draw synovial fluid, e.g from patients with rheumatoid arthritis, according to the criteria of the American Rheumatism Association.

2. Filter approximately 500 μl plasma or synovial fluid by centrifuging

Protocol 3. *Continued*

through Amicon 'microcentrifree' filtration units at 2000 *g* for 60 min at 10°C.

3. Analyse the ultrafiltrate achieved immediately or store at −20°C.

B. *HPLC*

1. Prepare standard solutions of tyrosines (*o*-, *m*-, and *p*-) in HPLC-grade water to a final concentration of 1 mM. Prepare phenylalanine at a final concentration of 5 mM. These solutions are stable at 4°C for a period of 4 weeks. Prepare mixtures of 70 μM *o*-, *m*-, and *p*-tyrosines, and 3.5 mM phenylalanine with or without 70 μM of each metabolite of phenylalanine ready for injection. Such mixtures are stable at room temperature for up to 2 weeks.

2. Prepare HPLC eluent of 500 mM buffer/ethanol (90/10, v/v) as follows. Dissolve 136 g of KH_2PO_4 in 2 litres of HPLC-grade water and adjust the pH to 3.01 with H_3PO_4. Mix this buffer with HPLC-grade methanol (90:10, v/v).

3. Inject on to HPLC and elute at 1.0 ml/min with 500 mM buffer/ethanol (90/10, v/v).

4. Verify all peaks by running samples to which an appropriate concentration of standards has been added.

Figure 4 shows the chromatograms of synovial fluid from RA patients indicating the presence of little or no *o*- and *m*-tyrosines. However, when fluid is aspirated from the knee joints of patients with active inflammation and added immediately to a solution of phenylalanine, these products appear, suggesting that the fluid is in the process of generating ˙OH (*Figure 4c*). Similarly, when blood from some premature babies ($n = 18$) was added to phenylalanine, production of *o*- and *m*-tyrosines was observed and this decreased by 30–60% when mannitol (100 mM) was included with the phenylalanine (23).

One disadvantage of using phenylalanine may be its rate constant for reaction with ˙OH, at $1.9 \times 10^9\,M^{-1}\,sec^{-1}$. This is about five times less than that of salicylate. However, unlike salicylate, phenylalanine has no known inhibitory effects on biochemical functions and little toxicity at millimolar concentrations. Furthermore, it has been shown that activated neutrophils produce ˙OH only in the presence of added 'catalytic' metal ions (24) using only 5 mM phenylalanine which is much less than the concentrations of DMPO (often up to 100 mM) used in studies of neutrophils.

Bolli *et al.* (21) used phenylalanine to investigate directly the role of ˙OH in the pathogenesis of myocardial stunning in the intact dog by measuring the levels of tyrosines. The results demonstrate that reperfusion after a brief episode of regional myocardial ischaemia is associated with a burst of release

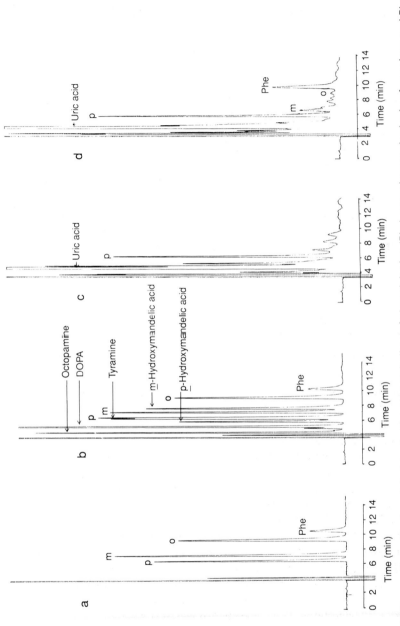

Figure 4. HPLC separation of hydroxylated products and metabolites of phenylalanine (Phe), namely: standard mix of tyrosines and Phe in the absence (a) and presence (b) of Phe metabolites; and synovial fluid extract from a rheumatoid arthritis patient dropped into (c) saline to measure any endogenous tyrosines or Phe, and (d) Phe to demonstrate the *in vitro* formation of products of hydroxylation. P, m, and o indicate *p*-, *m*-, and *o*-tyrosine respectively.

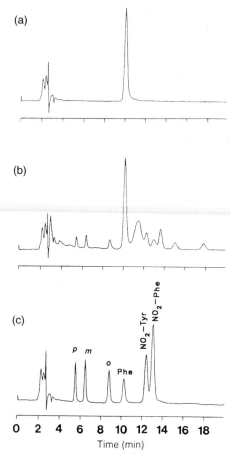

Figure 5. Hydroxylation and nitration of phenylalanine (Phe) by peroxynitrite analysed by HPLC, namely: Phe on its own (a) or after incubation with 1.0 mM peroxynitrite (b), and a mixture of Phe, 3-nitrotyrosine (NO$_2$-Tyr), 4-nitrophenylalanine (NO$_2$-Phe), and *p*-, *m*-, and *o*-tyrosine (labelled *p*, *m*, and *o*, respectively) (c).

of ˙OH, measured in terms of levels of *o*- and *m*-tyrosines. The amount of ˙OH detected (as measured by tyrosines) is decreased markedly by super-oxide dismutase, catalase, desferrioxamine, and mercaptopropionylglycine.

As mentioned earlier, peroxynitrite (ONOO$^-$) has been suggested to decompose to ˙OH, though Koppenol *et al.* (25) claimed that thermodynamic calculations preclude this process from taking place. In order to test the initial postulate, phenylalanine was successfully used to trap the ˙OH with the added advantage that products of nitration can also be measured using the method outlined in *Protocol 4*, as shown in *Figure 5* (26). Chromatograms of extracts of 5 mM phenylalanine alone (*Figure 5a*) and after incubation with 1.0 mM peroxynitrite (*Figure 5b*) are shown. *Figure 5c* is a chromatogram of a mixture of 40 μM each of *p*-tyrosine (*p*), *m*-tyrosine (*m*), *o*-tyrosine (*o*), 2 mM phenylalanine (Phe), 20 μM 3-nitrotyrosine (NO_2-Tyr), and 20 μM 4-nitrophenylalanine (NO_2-Phe) and demonstrates the formation of both hydroxylated products and nitro-adducts from phenylalanine.

This result has now been repeated using salicylic acid instead of phenylalanine in the authors' laboratory (unpublished work). The detection of nitrated products is very useful in investigating the role of $ONOO^-$ and more importantly the implication of nitric oxide (NO·) in disease states, e.g. RA (27) and neurodegenerative disease such as Parkinson's.

3. Conclusions

In conclusion, aromatic hydroxylation using trapping molecules is highly sensitive when developed *in vitro* but its full potential for *in vivo* studies has yet to be developed.

Acknowledgements

The authors thank the Arthritis and Rheumatism Council and MAFF for research support. The late Mr Rodney Boardman and Miss Julie Lumley are thanked for their invaluable advice with column selection during method development.

References

1. Halliwell, B. (1995). *Ann. Rheum. Dis.*, **54**, 505.
2. von Sonntag, C. (ed.) (1987). In *The chemical basis of radiation biology*. Taylor and Francis, London.
3. Halliwell, B. and Gutteridge, J. M. C. (1990). In *Methods in Enzymology* (ed. L. Packer and A. N. Glazer), Vol. 186, p. 1. Academic Press, London.
4. Halliwell, B. and Gutteridge, J. M. C. (1992). *FEBS Lett.*, **307**, 108.
5. Beauchamp, C. and Fridovich, I. (1970). *J. Biol. Chem.*, **245**, 4646.
6. Beckman, J. S., Beckman, T. W., Chen, J., Marshall, P. A., and Freeman, B. A. (1990). *Proc. Natl Acad. Sci. USA*, **87**, 1620.
7. McCord, J. M. and Day, E. D. (1978). *FEBS Lett.*, **86**, 139.
8. Babbs, C. F. and Griffin, D. W. (1989). *Free Radic. Res. Commun.*, **6**, 493.
9. Sagone, A. L., Decker, M. A., Wells, R. M., and de Mocko, C. (1980). *Biochem. Biophys. Acta*, **628**, 90.
10. Halliwell, B. and Gutteridge, J. M. C. (1981). *FEBS Lett.*, **128**, 347.
11. Yamazaki, I. and Piette, L. H. (1990). *J. Biol. Chem.*, **265**, 13589.
12. Raghavan, N. V. and Steenken, S. (1980). *J. Am. Chem. Soc.*, **102**, 3495.
13. Halliwell, B., Grootveld, M., and Gutteridge, J. M. C. (1988). *Methods Biochem. Anal.*, **33**, 59.
14. Richmond, R., Halliwell, B., Chauhan, J., and Darbre, A. (1981). *Anal. Biochem.*, **118**, 228.
15. Ingelman-Sundberg, M., Kaur, H., Terelis, Y., Perrson, J. O., and Halliwell, B. (1991). *Biochem. J.*, **276**, 753.

16. Grootveld, M. and Halliwell, B. (1986). *Biochem. J.*, **237**, 499.
17. Floyd, R.A., Henderson, R., Watson, J.J., and Wong, P.K. (1986). *Free Radic. Biol. Med.*, **2**, 13.
18. O'Connell, M. J. and Webster, N. R. (1990). *J. Pharm. Pharmacol.*, **42**, 205.
19. Ghiselli, A., Laurenti, O., Mattia, G. D., Mariani, G., and Ferro-Luzzi, A. (1992). *Free Radic. Biol. Med.*, **13**, 621.
20. Gelvan, D., Moreno, V., Gassman, W., Hegenauer, J., and Saltman, P. (1992). *Biochim. Biophys. Acta*, **1116**, 183.
21. Sun, Z. Z., Kaur, H., Halliwell, B., Li, X. Y., and Bolli, R. (1993). *Circ. Res.*, **73**, 534.
22. Powell, S. R. (1994). *Free Radic. Res.*, **21**, 355.
23. Kaur, H. and Halliwell, B. (1994). *Anal. Biochem.*, **220**, 11.
24. Kaur, H., Fagerheim, I., Grootveld, M., Puppo, A., and Halliwell, B. (1988). *Anal. Biochem.*, **172**, 360.
25. Koppenol, W. H., Moreno, J. J., Pryor, W. A., Ischiropoulos, H., and Beckman, J. S. (1992). *Chem. Res. Toxicol.*, **5**, 834.
26. van Der Vliet, A., O'Neill, C. A., Halliwell, B., Cross, C. E., and Kaur, H. (1994). *FEBS Lett.*, **339**, 89.
27. Kaur, H. and Halliwell, B. (1994). *FEBS Lett.*, **350**, 9.

Part III
Measurement of free radical products

8

Peroxides and other products

R. K. BROWN and F. J. KELLY

1. Introduction

The only definitive way to demonstrate excessive free radical activity *in vivo* is by electron spin resonance as already discussed in Chapter 2, but clearly this is currently inapplicable in clinical practice due to the nature of the spin traps that must be used. Instead investigators must rely upon the measurement of established markers of free radical activity in biological fluids and tissues. Traditionally, most markers of oxidative injury utilized reflect free radical attack on polyunsaturated fatty acids, with the classical route of attack involving lipid peroxidation, generating hydroperoxides, endoperoxides, long-lived aldehydes and the end-products malondialdehyde, ethanc, and pentane.

Oxidation of lipids was first reported by Farmer *et al.* (1) and was one of the earliest recognized consequences of free radical attack, leading to rancidity in fats and oils, the development of characteristic unpleasant tastes and odours, and changes in colour and viscosity. To improve understanding of these processes, a variety of methods have been developed to measure the numerous products formed by lipid peroxidation both *in vivo* and *in vitro*. Although investigation of lipid peroxidation *in vitro* is relatively straightforward, *in vivo* systems present additional problems in the determination of markers of lipid peroxidation due to their intrinsic heterogeneity. This chapter will concentrate on measurement of products of lipid peroxidation in body fluids and tissues.

1.1 Initiation of lipid peroxidation

Lipid peroxidation is initiated by the attack of an acyl side-chain on a fatty acid by any chemical species that has sufficient reactivity to abstract a hydrogen atom from a methylene carbon in the side-chain. The greater the number of double bonds in a fatty acid side-chain, the easier is the removal of a hydrogen atom, which is why the polyunsaturated fats, arachidonic acid (20:4) and docosahexaenoic acid (22:6), are more susceptible to lipid peroxidation than linoleic acid (18:2). The carbon-centred radical resulting from hydrogen abstraction from a fatty acid can have several fates, but the most likely one in an aerobic environment is to undergo a molecular rearrangement, followed by reaction with oxygen to give a peroxyl radical (ROO$^{\bullet}$). Peroxyl radicals are capable of abstracting hydrogen from an adjacent fatty acid side-chain in a membrane, and hence propagating a chain reaction of lipid peroxidation resulting in the conversion of hundreds of fatty acid side-chains into lipid hydroperoxides. In biological membranes this can lead to impairment of membrane function and decreased fluidity and has been associated with a number of disease states from atherosclerosis (2) and ischaemia–reperfusion injury (3) to hyperoxic lung injury (4).

1.2 Detection and measurement of lipid peroxidation

The extent of lipid peroxidation can be determined by a number of methods which can be divided between those that measure primary products of lipid peroxidation such as hydroperoxides; and those that measure the secondary breakdown products of lipid hydroperoxides such as malondialdehyde, 4-hydroxynonenal (see Chapter 9) and volatile hydrocarbons (breath ethane and pentane).

The chemical composition of the end-products of peroxidation will depend both on the fatty acid composition of the target lipid and on what (if any) metal ions are free to participate in the reaction. For the most accurate measure of lipid peroxidation it is preferable to determine the primary products of oxidative attack such as lipid hydroperoxides. The measurement of lipid hydroperoxides by high-performance liquid chromatography (HPLC) will be described here together with two methods for determining the most widely studied secondary product of lipid peroxidation, namely malondialdehyde. Determination of volatile hydrocarbons will not be covered as it is cumbersome, requires specialist equipment, and is subject to a number of artefacts; however, for further information the reader is referred to ref. 5.

2. HPLC determination of lipid hydroperoxides

The method described is based on the luminol (5-amino-2,3-dihydro-1,4-phthalazinedione) chemiluminescence assay for the detection of hydrogen peroxide and is well known for its picomole sensitivity. Microperoxide has been shown to be the most effective catalyst for this assay. The reaction mechanism of the chemiluminescent reactions of isoluminol and lipid hydroperoxides is assumed to be as follows:

$$\text{Lipid hydroperoxides (LOOH)} + \text{microperoxidase} \longrightarrow \text{LO}^{\bullet} \quad (1)$$
$$\text{LO}^{\bullet} + \text{isoluminol (QH}^{-}) \longrightarrow \text{LOH} + \text{semiquinone radical (Q}^{\bullet}) \quad (2)$$
$$Q^{\bullet} + O_2 \longrightarrow Q + O_2^{\bullet} \quad (3)$$
$$Q^{\bullet} + O_2^{\bullet} \longrightarrow \text{isoluminol endoperoxide} \longrightarrow \text{light } (\lambda_{max} = 430 \text{ nm}) \quad (4)$$

Problems that arise from the use of the isoluminol chemiluminescence assay in biological samples stem from the fact that any antioxidants present quench the chemiluminescence signal by scavenging the oxygen radicals. This problem has been partially circumvented by the development of HPLC separation techniques and different extraction methods have been employed to separate antioxidants from lipid hydroperoxides. The method described here is based on that of Holley and Slater (6) and utilizes a different extraction procedure from that described in the original method of Yamamoto *et al.* (7) in an attempt to prevent artefactual peroxidation. In our laboratory, this approach has been utilized to examine a variety of body fluids and tissue samples for lipid hydroperoxides.

Protocol 1. Chemiluminescent detection of lipid hydroperoxides

Equipment and reagents
- Heparinized blood collection tubes
- Butylated hydroxytoluene (BHT), 2 mM
- Desferrioxamine mesylate (Desferal), 2 mM
- Citric acid, 0.2 M
- Sodium borate buffer, 0.1 M (adjusted to pH 10.0 with 10 M NaOH)
- 6-Amino-2,3-dihydro-1,4-phthalazinedione (isoluminol) (Sigma)
- Microperoxidase (MP-11), sodium salt (Sigma)
- HPLC-grade methanol, hexane, water, and triethylamine
- Millipore vacuum filtration equipment
- Glass stoppered tubes, 10 ml
- Microcentrifuge

A. *Plasma samples*
1. Collect 2.0 ml of blood into a heparinized tube on ice containing 10 μl of 2 mM BHT and 10 μl of 2 mM Desferal per ml of blood.
2. Centrifuge the blood immediately (1000 *g*, 10 min, 4°C) to obtain plasma.
3. Extract plasma samples immediately as described below in 'C'[a].

B. *Tissue samples*
1. Homogenize the sample in 100 mM phosphate buffer, pH 7.4, containing 10 μl 2 mM BHT and 10 μl Desferal per ml to give a final concentration of 100 mg tissue/ml buffer.
2. Extract samples immediately as described below. Storage of the homogenate will result in degradation of the lipid hydroperoxides within 24 h.

C. *Extraction of lipid hydroperoxides from plasma or tissue samples*
1. Carry out all extractions on ice and protected from strong light.
2. Add 0.25 ml of 0.2 M citric acid to 0.5 ml of plasma or homogenate in a 10 ml glass tube on ice.

Protocol 1. *Continued*

3. Mix for 20 sec.

4. Add 6 ml of hexane and mix for 30 sec.

5. Centrifuge the solutions at 2000 g for 5 min at 4°C.

6. Remove the hexane layer into a clean amber HPLC tube and evaporate off under a stream of nitrogen, protected from light.

7. Store desiccated samples under nitrogen at −70°C until analysis. The samples stored in this way are stable for up to 1 month.

D. HPLC running conditions

1. Column: Apex II ODS 5 μm C_{18} column (Jones Chromatography, Hongoed, Wales), dimensions 25 cm × 5 mm connected in series with an Apex II ODS 8 μm C_{18}, 3 cm × 5 mm guard column.

2. Mobile phase: methanol containing 0.02% (v/v) triethylamine.

3. Flow rate: 1 ml/min with a sample run-time of 35 min.

4. Post-column chemiluminescence reagent: set the post-column pump to deliver an aqueous mixture of isoluminol/microperoxidase buffer to the column eluent at 1.75 ml/min. The isoluminol/microperoxidase reagent consists of: microperoxidase (12.5 mg/ml microperoxidase solution in 0.1 M Tris buffer pH 7.4[b]) dissolved to a final concentration of 1 μg/ml in the isoluminol reagent (1 mM 6-amino-2,3-dihydro-1,4-phthalazinedione (isoluminol) in methanol/0.1 M sodium borate buffer (30:70, v/v)[c]) and filtered through a 0.22 μm Millipore membrane.[d]

5. Detector: dual wavelength UV detector (232 and 254 nm) in series with a programmable fluorescence detector fitted with a 10 μl flow cell. The fluorimeter is used in this instance as a photon detector with the excitation source turned off and the UV detector is present to monitor the absorption of conjugated dienes in the samples at 232 nm.

[a] Holley and Slater (6) have shown that lipid hydroperoxides are stable in plasma stored at −70°C in the presence of antioxidants for up to 2 weeks.
[b] The concentration of the microperoxidase solution can be varied to increase or decrease the chemiluminescence signal; however, with increasing concentration there is a concomitant rise in signal/background noise. This solution can be stored for up to 4 days at 4°C.
[c] The order of addition is important when making up the isoluminol/microperoxidase chemiluminescence reagent. To ensure complete mixing, the isoluminol must be added to the microperoxidase solution.
[d] This solution must be made up at least 48 h before use to allow the inherent chemiluminescence to subside. All solutions should be kept in brown bottles to minimize light-induced generation of chemiluminescence-producing material.

A diagram of the HPLC system described is shown in *Figure 1*. The desiccated samples are resuspended in 100 μl of HPLC-grade methanol and, after

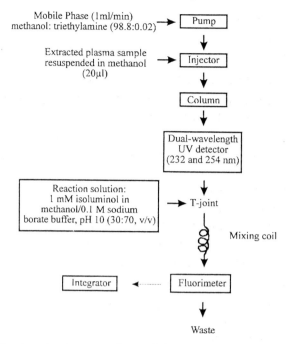

Figure 1. HPLC set-up for the separation and detection by chemiluminescence of lipid hydroperoxides.

injection of 20 μl of sample via a valve injector (20 μl loop), the sample passes through the C_{18} reversed phase column and UV detector after which the column eluent is combined with the chemiluminescence reagent in a T-joint and incubated in a 15 sec delay coil (approximately 43 cm in length).

2.1 Peak identification and quantification

The advantage of the HPLC technique described, is that it allows the separation and quantification of a variety of different products of lipid peroxidation. These include:

- free fatty acid hydroperoxides
- cholesterol ester hydroperoxides
- cholesterol hydroperoxides
- phosphatidylcholine hydroperoxides

The limiting factor is usually the availability of standards of these oxidation products. The majority of standards of compounds resulting from oxidative attack on arachidonic acid (5(S),12(S),15(S)-HPETE) can be obtained from Cascade Biochem. Identification of an unknown peak is accomplished by comparison of the run time with the standards (see Figure 2). Standard

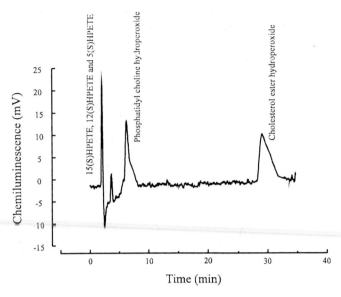

Figure 2. Sample chromatogram showing the retention times of several lipid hydroperoxide standards.

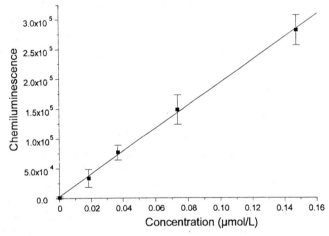

Figure 3. A typical standard curve for 15(S)HPETE as run for determination of lipid hydroperoxides in plasma.

curves are built by chromatographing known amounts of 15(*S*)-HPETE, phosphatidylcholine hydroperoxides, or other standards and integrating chemiluminescence counts of the peak area after subtraction of background noise. An example of a typical standard curve is given in *Figure 3*. All samples and standards are run in duplicate and the mean value is taken.

Table 1. Concentrations of lipid hydroperoxides in plasma measured by chemiluminescence–HPLC detection

Patient group	Lipid hydroperoxide concentration (μM)	Reference
Control patients	< 0.03	20, 5
Adult respiratory distress syndrome (ARDS) patients	0.02–3.30	5
Angioplasty patients	0.02–0.92	5
Cystic fibrosis patients	0.023–0.30	11
Various hospitalized patients (suffering from diabetes, liver injuries, ischaemic heart disease and hyperlipidaemia)	0.5–9.0	21

2.2 Plasma concentrations of lipid hydroperoxides

Table 1 summarizes reported concentrations of lipid hydroperoxides in the plasma of healthy controls and patients with various diseases. In controls, no evidence of lipid hydroperoxides has been found; however, detectable concentrations of lipid peroxidation have been found in numerous disease states implicating oxidative damage in their aetiology. There are no reports of lipid hydroperoxides being found in tissue samples using this method. This may be due to the presence of the intracellular antioxidant enzymes, catalase and superoxide dismutase, which may inhibit their formation, or the efficient breakdown of the hydroperoxides once they are formed in the tissue.

3. Determination of malondialdehyde

3.1 Introduction

Aldehydes are always produced when lipid hydroperoxides are metabolized in biological systems (8) and identification and quantification of these compounds gives an indirect index of oxidative injury which results in lipid peroxidation. Malondialdehyde (MDA) is in many instances the most abundant aldehyde arising from lipid peroxidation and its determination, by measurement of the coloured product formed upon reaction with thiobarbituric acid (TBA), is one of the most common assays used in lipid peroxidation studies. The TBA test works well when applied to defined membrane systems (e.g. microsomes and liposomes) but its application to body fluids and tissue extracts has produced a host of problems. These stem mainly from the fact that the TBA reaction is notoriously non-specific because many non-lipid compounds present in biological samples, including carbohydrates, pyrimidines, and haemoglobin, also react with TBA forming colours that have spectra overlapping that of authentic MDA–TBA.

Unfortunately, these problems associated with the TBA assay have been largely ignored by many workers who have developed their own versions of the method. This makes it virtually impossible to interpret and compare many of the publications on plasma lipid peroxides in health and disease states. With care, many of the artefacts that occur in the TBA test can be avoided and the method described below incorporates some of these.

As mentioned above, the TBA reaction has the inherent advantage of simplicity. However, in view of the problems associated with it, it is prudent in each particular case to cross-check its relationship with lipid peroxidation by reference to another method. Moreover, if the results of the TBA reaction are to be equated with malondialdehyde content then a direct measurement of malondialdehyde should be carried out using an HPLC technique such as the one described in section 3.3.

3.2 Determination of malondialdehyde

The TBA test is probably the single most widely used single assay for the measurement of lipid peroxidation. The sample under test is treated with TBA at low pH, and a pink chromogen is measured. In the TBA reaction, one molecule of MDA reacts with two molecules of TBA with the production of a pink pigment with an absorption maximum at 532–535 nm (see *Figure 4*). Amplification of peroxidation during the assay is prevented by the addition of the chain-breaking antioxidant BHT and the iron chelator desferrioxamine to the sample at the time of collection.

Numerous methods have been developed for the determination of TBA reactive substances (TBARS) in biological samples, though most rely on spectrophotometric detection at 532 nm following reaction in an acid environment and incubation at 95 °C. The method described in *Protocol 2* is a variation on the method first reported by Yagi *et al.* (9) and involves fluorimetric detection of TBARS. This method is more sensitive and specific than spectrophotometric methods for the measurement of malondialdehyde (10), as few of the contaminants that interfere with the spectrophotometric determination fluoresce at the same wavelength as MDA.

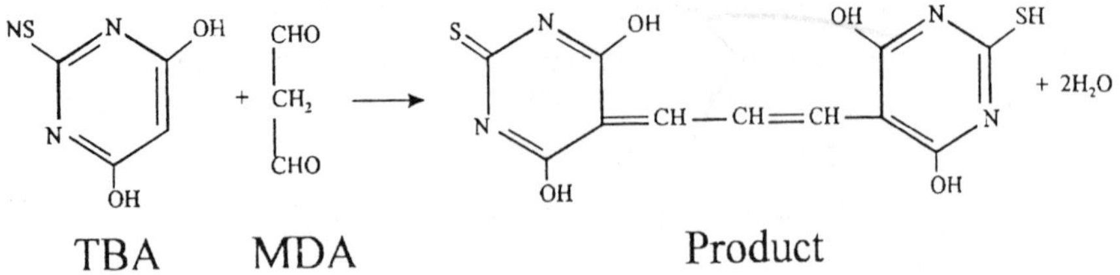

TBA   MDA   Product

Figure 4. The reaction of thiobarbituric acid with malondialdehyde to produce a coloured product with absorbance at 532 nm.

Protocol 2. The TBA test

Equipment and reagents

- Tetraethoxypropane (TEP) or tetramethoxypropane (TMP), 0.125 μM
- Phosphotungstic acid, 10% (w/v)
- TBA reagent: 0.67% (w/v) TBA in water and equal volume glacial acetic acid
- *n*-Butanol
- H_2SO_4, 4 M
- Sorvall tubes, 15 ml

- Microcentrifuge
- Bench-top water bath
- Fluorimeter
- Fluorescence cuvettes (with four clear sides)
- Butylatedl hydroxytoluene (BHT)
- Desferal

A. *Collection of plasma samples*

1. Collect blood into a heparinized tube on ice containing 10 μl of 2 mM BHT and 10 μl of 2 mM Desferal per ml of blood.

2. Centrifuge the blood immediately (1000 *g*, 10 min, 4°C) to obtain plasma.

3. Aliquot plasma into 1.0 ml Eppendorf tubes and store at −70°C until analysis. Samples should be analysed within two weeks of collection as content of TBARS has been shown to increase with storage.

B. *Sample preparation and incubation*

1. Add 2 ml of 4 M H_2SO_4 and 0.25 ml of 10% (w/v) phosphotungstic acid to 10 μl of plasma or serum.

2. Mix and leave to stand at room temperature for 5 min.

3. Centrifuge at 2000 *g* for 10 min at room temperature.

4. Aspirate and discard the supernatant. To the remaining pellet add 1 ml of 4 M H_2SO_4 and 0.15 ml of 10% (w/v) phosphotungstic acid.

5. Mix and leave to stand at room temperature for 5 min.

6. Centrifuge at 2000 *g* for 10 min.

7. Remove the supernatant and resuspend the pellet in 2.0 ml of water.

8. Add 0.5 ml of TBA reagent and mix.

9. Heat the capped tubes for 60 min at 95°C in a water bath.

10. Cool the tubes on ice, add 2.5 ml of butanol and mix vigorously for 30 sec.

11. Centrifuge at 2000 *g* for 10 min.

12. Transfer the butanol layer to a fluorescence cuvette.[a]

C. *Preparation of standard curve*

1. Prepare MDA standards from the 0.125 μM stock TEP (range

Protocol 2. *Continued*

    0.125–0.004 μM) which yields equimolar amounts of MDA under the conditions of the reaction.

  2. Add 0.5 ml of TBA reagent to 2.0 ml of diluted standards.

  3. Heat for 60 min at 95°C in a water bath in sealed glass tubes.

  4. Cool the tubes on ice, add 2.5 ml butanol, and mix vigorously for 30 sec.

  5. Centrifuge at 2000 *g* for 10 min.

  6. Transfer the butanol layer to a fluorescence cuvette.

D. *Sample analysis*

  1. Set excitation at 515 nm and emission at 553 nm [b]

  2. Autozero the detector using a control containing butanol incubated with TBA reagent.

  3. Read the fluorescence of the standards and the samples.

  4. Calculate the sample concentration of thiobarbituric acid reactive substances by plotting a standard curve of fluorescence against MDA concentration obtained from the TEP standards.

[a] Butanol will begin to degrade most plastic cuvettes after 1 h, so transfer into the cuvettes should take place immediately before determination of the fluorescence.
[b] To increase the sensitivity, the slit width of the excitation source can be decreased and that of the emission source increased where possible. The settings we use are 5 and 15 mm for excitation and emission, respectively.

All the samples and standards are run in duplicate and the mean reading is taken for each. Values obtained for plasma lipid hydroperoxides using this method range from 3–5 μM for healthy controls to 5–9 μM in diabetic patients (11, 12) and we have measured concentrations of 7–15 μM in patients with cystic fibrosis (13).

3.3 Detection of malondialdehyde by HPLC

To minimize the problems of interference that affect the spectrophotochemical determination of MDA, a number of HPLC methods for measuring the MDA–TBA adduct have been developed (14, 15). Such techniques give significantly lower values for tissue and serum MDA levels than traditional spectrophotometric methods due to separation of the MDA–TBA adduct from contaminants that fluoresce at the same wavelength. Most methods require complex sample preparation to remove contaminants, or sample extraction into organic solvents to improve sensitivity and peak separation. These methods are not applicable for the determination of MDA in body fluids due to contaminating peaks that elute near to, or at the position of MDA. A number of workers (15–17) have used an aminophase column with isocratic elution with acetonitrile/0.03 M Tris buffer (pH 7.4, 1:9 v/v) and UV detection at 270 nm to determine free MDA in cells, cell fractions, liver microsomes, and autoxidized fatty acids, but the method was found to be unsuitable for plasma samples. The method described in *Protocol 3*, developed by Young and Trimble (18), combines HPLC separation with fluorimetric detection of MDA. It does not require complex preparation or extraction into organic solvents before injection on to the column and can be applied to biological fluids such as plasma and serum with relative ease and good reproducibility.

Protocol 3. Measurement of malondialdehyde by HPLC with fluorimetric detection

Reagents and equipment

- Dipotassium EDTA blood collection tubes
- TEP or TMP standard, 0.125 μM
- Phosphoric acid, 1.22 M
- TBA reagent: 0.67% (w/v) TBA in water and an equal volume of glacial acetic acid
- HPLC-grade methanol
- NAOH, 1 M
- Phosphate buffer, 25 mM, pH 6.5
- Sorvall tubes, 15 ml
- Microcentrifuge
- Bench-top water bath

A. *TBA reaction*

1. Collect venous blood samples into dipotassium EDTA and prepare plasma as described in *Protocol 2A*, steps 2 and 3.

2. To 50 μl of sample/standard add 250 μl of 1.22 M phosphoric acid, 450 μl of HPLC water, and 250 μl of TBA reagent.

3. Mix for 30 sec.

4. Prepare MDA standards (0, 0.24, 0.48, 1.2, 2.4, and 4.8 μM) using TEP which yields equimolar amounts of MDA under the conditions of the reaction.

5. Incubate the reaction mixtures at 95°C in a water bath for 60 min in sealed glass tubes.

6. Cool the samples/standards to 4°C in ice.

7. Add 360 μl of HPLC-grade methanol and 40 μl of 1 M NaOH to 200 μl of sample or standard to neutralize the solutions and precipitate the proteins before injection on to the column.

B. *HPLC running conditions*

1. Column: μBondapak C$_{18}$, dimensions 30 cm $\times$ 3.9 mm (i.d.) (10 μm silica bonded with ODS groups) with a μBondapak C$_{18}$ guard column (3.9 $\times$ 23 mm) (Waters-Millipore).

Protocol 3. *Continued*

2. Mobile phase: 50% (v/v) methanol/25 mM phosphate buffer, pH 6.5, filtered through a 0.22 μm Millipore membrane and made up fresh every day.

3. Flow rate: 0.8 ml/min.

4. Detector: programmable fluorescence detector with excitation set at 532 nm and emission at 553 nm with gain × 1000.

All samples and standards should be analysed in duplicate and the yield of MDA from TEP can be calculated using the Beer–Lambert Law (Equation 5) with the molar extinction coefficient (ϵ_o) for the MDA–TBA adduct at 532 nm (153 000):

$$\text{Absorbance} = \epsilon_o \times \text{concentration} \tag{5}$$

Using the above method, Young and Trimble (18) have reported plasma concentrations of MDA of 0.59 μM in control subjects, which compare favourably with similar concentrations reported by Knight *et al.* (19).

4. Conclusions and future directions

Investigations into the mechanisms of lipid peroxidation *in vivo* require increasingly sensitive techniques to provide specific chemical information about the products that are formed. Thus, HPLC techniques such as those described in *Protocols 1* and *3* are extremely useful in that they identify specifically the products formed and hence provide clearer information on the source of the lipid hydroperoxides and the type of oxidative attack occurring. Direct determination of lipid hydroperoxides and MDA takes on a greater significance with the publication of evidence implicating these products of lipid peroxidation as secondary mediators of oxidative stress both extracellularly and intracellularly (20) and specifically to DNA (21).

One method not discussed here is gas chromatography–mass spectroscopy analysis of lipid peroxidation which has the inherent advantages of superior sensitivity and specificity and as such may be used to probe in depth the various products of lipid peroxidation produced by oxidative reactions. However, the disadvantages of cost and time efficiency for processing large numbers of samples mean that until these can be reduced, improvements in the sensitivity and specificity of existing techniques will be of great benefit.

Acknowledgements

The authors acknowledge the support of the CF Trust (UK) and the Leopard Müller Trust in their research.

References

1. Farmer, E. H., Koch, H. P., and Sutton, D. A. (1943). *J. Chem. Soc.*, **14**, 541.
2. Munro, J. M. and Cotran, R. S. (1988). *Lab Invest.*, **59**, 249.
3. Repine, J. E., Cheronis, J. C., Rodell, T. C., Linas, S. L., and Patt, A. (1987). *Am. Rev. Respir. Dis.*, **136**, 483.
4. Smith, C. V., Hansen, T. N., Martin, N. E., McMicken, H. W., and Elliot, S. J. (1993). *Pediatr. Res.*, **34**, 360.
5. Kneepkens, C. M. F., Ferreira, C., Lepage, G., and Roy, C. C. (1992). *Clin. Invest. Med.*, **15**, 163.
6. Holley, A. E. and Slater, T. F. (1991). *Free Radic. Res. Commun.*, **15**, 51.
7. Yamamoto, Y., Brodsky, M. H., Baker, J. C., and Ames, B. N. (1987). *Anal. Biochem.*, **160**, 7.
8. Kagan, V. E. (1988). In *Lipid peroxidation in biomembranes* (ed. V. E. Kagan), pp. 13–54. CRC Press, Boca Raton, FL.
9. Yagi, K. (1984). In *Methods in enzymology* (ed. L. Packer), Vol. 105, pp. 328–331. Academic Press, London.
10. Draper, H. H., Squires, E. J., Mahmoodi, H., Agarwal, S., Wu, J., and Hadley, M. (1993). *Free Radic. Biol. Med.*, **15**, 353.
11. Sato, Y., Hotta, N., Saxamoto, N., Matsuoka, S., Ohnishi, N., and Yagi, K. (1979). *Biochem. Med.*, **21**, 104.
12. Yagi, K. (1987). *Chem. Phys. Lipids*, **45**, 337.
13. Brown, R. K. and Kelly, F. J. (1994). *Pediatr. Res.*, **36**, 487.
14. Bird, R. P., Hung, S. S., Hadley, M., and Draper, H. H. (1983). *Anal. Biochem.*, **128**, 240.
15. Esterbauer, H., Lang, J., Zadravec, S. and Slater, T. F. (1984). In *Methods in enzymology* (ed. L. Packer), Vol. 105, pp. 319–327. Academic Press, London.
16. Wade, C. R., Jackson, P. G., and van Rij, A. (1985). *Biochem. Med.*, **33**, 219.
17. Largilliere, C. and Melancon, S. B. (1988). *Anal. Biochem.*, **170**, 123.
18. Young, I. S. and Trimble, E. R. (1991). *Ann. Clin. Biochem.*, **28**, 504.
19. Knight, J. A., Smith, S. E., Kinder, V. E., and Anstall, H. B. (1987). *Clin. Chem.*, **33**, 2289.
20. Esterbauer, H. (1993). *Am. J. Clin. Nutr.*, **57**, 779S.
21. Park, J. and Floyd, R. A. (1991). *Free Radic. Biol. Med.*, **12**, 245.

Further reading

Frei, B., Stocker, R., and Ames, B. N. (1988). *Proc. Natl Acad. Sci. USA*, **85**, 9748.
Pompella, A., Maellaro, E., Casini, A., Ferri, M., Ciccoli, L., and Comporti, M. (1987). *Lipids*, **22**, 206.
Slater, T. F. (1984). In *Methods in enzymology* (ed. L. Packer), Vol. 105, pp. 283–293. Academic Press, London.
Smith, C. V. and Anderson, R. E. (1987). *Free Radic. Biol. Med.*, **3**, 341.
Wendel, A. (1987). *Free Radic. Biol. Med.*, **3**, 355.

Quantitative analysis of 4-hydroxy-2-nonenal

MICHAEL KINTER

1. Introduction

Tissue damage resulting from the production of reactive oxygen species has been implicated in a diverse series of pathologies. One component of tissue damage by reactive oxygen species is the oxidation of cell membrane poly-unsaturated fatty acids. Polyunsaturated fatty acid oxidation can proceed by a chain reaction mechanism referred to as autooxidation in which the termination reactions form a variety of products including a series of aldehydes. The aldehydes reported as products of polyunsaturated fatty acid oxidation are listed in *Table 1*, which shows that one class of aldehydes formed is the 4-hydroxy-2-alkenals, exemplified by 4-hydroxy-2-nonenal (4HNE).

4HNE is produced by the oxidation of the two most abundant poly-unsaturated fatty acids in mammalian tissues, linoleic and arachidonic acids, although the chemical mechanism of formation has not been determined. Relative to the other aldehydes, 4HNE is formed in significant abundance and possesses an array of biological effects, many of which have been associated with the damaging effects of oxidative stress (1). Cytostatic effects of 4HNE were under investigation in the 1960s (2). Later, 4HNE was identified as the diffusible cytotoxin formed by the peroxidation of rat liver microsomes (3). 4HNE has since been linked to atherosclerosis (4, 5), oxygen toxicity (6), liver injury (7, 8), and so on. As a whole, the literature strongly supports the hypothesis that 4HNE contributes to tissue damage which results from oxidative stress.

Based on this literature, the quantitative analysis of 4HNE can be viewed on two levels. On the first level, 4HNE quantification can be used as an index to monitor the overall progression of polyunsaturated fatty acid oxidation reactions since 4HNE is a known product of these reactions. The quantitative analysis of malondialdehyde using the thiobarbituric acid test has been used for this purpose for many years. Experiments in which the progression of polyunsaturated fatty acid oxidation reactions is monitored are useful for investigations of factors which control the oxidation reactions, links between

Table 1 Aldehydes formed by polyunsaturated fatty acid oxidation

Aldchyde	Fatty acid[a]	Reference
Malondialdehyde	20:4, 20:5, 22:6	8
Propanal	18:2, 20:4	9
Butanal	18:2, 20:4	9
Pentanal	18:2, 20:4	9
Hexanal	18:2, 20:4	9
2,4-Heptadienal	18:2, 20:4	9
4-Hydroxy-2-hexenal	22:6	10
4-Hydroxy-2-octenal	18:2, 20:4	9
4-Hydroxy-2-nonenal	18:2, 20:4	9
4-Hydroxy-nona-2,6-dienal	20:5, 22:6	11
4,5-Dihydroxydecenal	20:4	12

[a]Abbreviations: 18:2, linoleic acid; 20:4, arachidonic acid; 20:5, eicosapentaenoic acid; 22:6, docosahexaenoic acid.

oxidative stress and a given tissue damage, and the efficacy of intervention strategies. On the second level, the biological activities of 4HNE and the strong link between many of these activities and cellular damage indicates that 4HNE quantification may be a direct measure of an important toxic species. In this regard, quantitation of 4HNE is superior to the quantification of malondialdehyde.

It is important to carefully consider the metabolism and disposition of 4HNE when evaluating the utility of its quantification in a given system. An overview of the current understanding of 4HNE metabolism is shown in *Figure 1*. As would be expected for an active xenobiotic, the metabolism of 4HNE can be complex. The assays described in this chapter detect two components of this metabolic scheme: free 4HNE (*Figure 1*, compound I) and 4HNE bound to proteins through Schiff base formation (*Figure 1*, compound II). These are important dispositions of 4HNE, because they represent either unmetabolized xenobiotic or xenobiotic which has reacted with protein in a injurious manner, whereas a number of the other metabolites are detoxified species and, therefore, would be of less interest in investigations of oxidative tissue damage. However, it is clear that in systems where components of this metabolic scheme are active, free and Schiff-base bound 4HNE represent only a fraction of the total 4HNE produced.

2. Determination by UV spectroscopy

2.1 Principle of analysis

Aldehydes react with 2,4-dinitrophenylhydrazine (DPNH) to form 2,4-dinitrophenylhydrazones which have a UV absorbance band at approximately 350 nm with molar extinction coefficients (ϵ) greater than $20\,000\ \mathrm{M}^{-1}\ \mathrm{cm}^{-1}$. When the aldehyde–2,4-dinitrophenylhydrazone products are isolated from

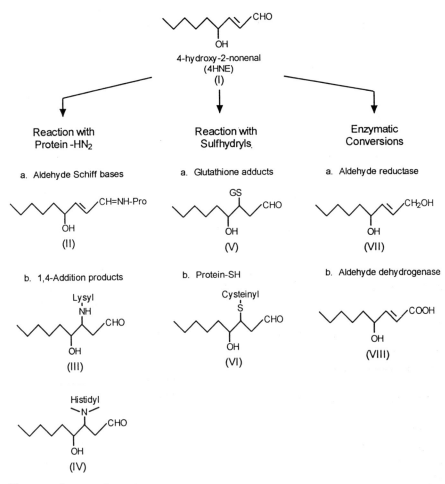

Figure 1. An overview of the metabolism of 4HNE compiled through a survey of the literature.

the DPNH reagent, A_{350} can be related to concentration of aldehyde using Beer's law. The UV-absorbance band for 4HNE at approximately 220 nm is generally not useful for quantitative analysis because so many unrelated compounds also absorb in this region. It should be noted that all aldehydes can react with DPNH under the conditions used in *Protocol 1*. Therefore, the quantitative results obtained using this protocol are best considered as a measure of total aldehyde production. Additional specificity for 4HNE can be obtained using HPLC separation of the products as described in *Protocol 2*.

   For best results, the 4HNE primary standard is diluted into the same buffer/solvent system used for the samples to generate the standard curve. The volume of standards should be adjusted by dilution with less buffer/solvent or

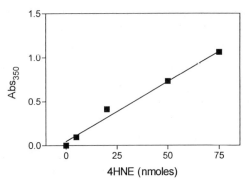

Figure 2. A calibration curve for the quantification of 4HNE by UV spectroscopy.

addition of more buffer/solvent to match the sample volumes. In the following dilution scheme 1.0 ml samples are used. Design the standard curve to
bracket the concentration range encountered in the samples. In general a
three or four point calibration curve is sufficient. A typical calibration curve
is shown in *Figure 2*. Aqueous samples containing up to 50% methanol have
been analysed using this method. If the protein content of the samples is
greater than 1 mg/ml, precipitation and removal of the proteins following
derivatization (*Protocol 1*, step 3) and before extraction (*Protocol 1*, step 4)
may be required.

Protocol 1. Quantification of aldehyde–2,4-
dinitrophenylhydrazones by UV absorbance

Equipment and reagents

- A UV/visible spectrophotometer able to measure absorbance at wavelengths ranging 220 nm and 350 nm, i.e. a Beckman DU® Series 60 UV/visible spectrophotometer
- Hexane
- Methanol
- Hydrochloric acid, 0.5 M: dilute the concentrated hydrochloric acid by adding 21 ml of acid to 479 ml of water.
- 2,4-Dinitrophenylhydrazine (DNPH) reagent, 5 mM: add 100 mg of DNPH to 100 ml of 0.5 M hydrochloric acid, stir for approximately 1 h at room temperature, filter to remove undissolved DNPH, and store in an amber bottle for up to one month

- 4HNE primary standard: dissolve 1.0 mg of 4HNE (6.4 μmol) in 6.4 ml of methanol to give a 1.0 mM solution. Standardize the solution by diluting 100 μl to 5.0 ml of methanol in a volumetric flask and measuring the absorbance at 220 nm in a 1 cm quartz cuvette versus a methanol blank. Calculate the concentration of 4HNE in the primary standard using eqn 1. Store the 4HNE primary standard at −20°C for up to 3 months.

$$[4HNE] = (A_{220}/13\,750) \times 50 \quad (1)$$

where A_{220} is the absorbance at 220 nm relative to a methanol blank, $13\,750$ = the molar extinction coefficient for 4HNE in methanol, and 50 = the dilution factor.

Method

1. Prepare the standard curve as follows:

   0 nmol standard, 0 μl primary standard, and 1000 μl buffer/solvent
   5 nmol standard, 5 μl primary standard, and 995 μl buffer/solvent

20 nmol standard, 20 µl primary standard, and 980 µl buffer/solvent
50 nmol standard, 50 µl primary standard, and 950 µl buffer/solvent
75 nmol standard, 75 µl primary standard, and 925 µl buffer/solvent

2. Pipette samples up to 2 ml into a 13 × 100 mm glass tube with cap.

3. Add 1 ml of DNPH reagent, then vortex mix the sample and allow it to react at room temperature for 1 h.

4. Extract the sample with three, 2 ml aliquots of hexane combining the extracts in a 13 × 100 mm glass tube.

5. Evaporate the extract to dryness under argon at 40 °C and reconstitute the residue in 1.0 ml of methanol, adding the methanol with a volumetric pipette.

6. Read the absorbance at 350 nm using the 0 µM standard as the blank.

7. Calculate the results as follows:

   (a) Plot the absorbance readings for the standards versus quantity of 4HNE.

   (b) Perform a linear regression analysis to determine the slope and intercept of the calibration line.

   (c) Calculate the quantity of 4HNE in the sample based on the linear regression analysis using eqn 2.

$$\text{nmol of 4HNE} = (A_{350} \times \text{slope}) - \text{intercept} \qquad (2)$$

   where A_{350} is the absorbance at 350 nm versus the 0 µM blank, 'slope' is the calculated slope of the linear regression line, and 'intercept' is the calculated y-intercept of the linear regression line.

2.2 Adaptability to other aldehydes of interest

As described in Section 2.1, this method is not specific for 4HNE and the results are best treated as a total aldehyde determination.

3. Determination by HPLC

3.1 Principle of analysis

As described in Section 2.1, aldehyde–dinitrophenylhydrazone derivatives are formed by reaction of the aldehyde with DNPH. The derivatives are isolated by extraction, separated by reversed phase HPLC and detected by monitoring absorbance at 350 nm. Individual aldehydes are quantified by integration of the appropriate chromatographic peak.

It is recommended that the 4HNE primary standard is diluted into the same buffer/solvent system used for the samples to generate the standard curve. The volume of standards is not critical for this analysis but accurate reconstitution prior to the HPLC analysis is important. The standard curve

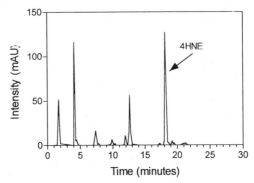

Figure 3. A typical chromatogram for the quantification of 4HNE by HPLC. The sample is 1 ml of a 5 μM 4HNE standard analysed as described in *Protocol 2*.

should be designed to bracket the concentration range encountered in the samples. In general a three or four point calibration curve is sufficient. Aqueous samples containing up to 50% methanol have been analysed using this method. If the protein content of the samples is greater than 1 mg/ml, precipitation and removal of the proteins following derivatization (*Protocol 2*, step 3) and before extraction (*Protocol 2*, step 4) may be required. A typical chromatogram is shown in *Figure 3*.

Protocol 2. Quantification of aldehyde–dinitrophenylhydrazones by HPLC (7)

Equipment and reagents

- Any HPLC system capable of performing gradient elutions with absorbance detection at 350 nm equipped with an ODS column, 200 mm × 2.1 mm with 5 μm packing. For example a Hewlett–Packard 1090 series II HPLC system equipped with a Hewlett–Packard ODS Hypersil column, 200 mm × 2.1 mm with 5 μm packing and a computer-based chromatographic data system.

- All items listed in *Protocol 1*
- HPLC-grade acetonitrile
- HPLC-grade water
- Standard prepared as in *Protocol 1*

Method

1. Pipette samples up to 2 ml into a 13 × 100 mm glass tube with cap.

2. Add 1 ml of the DNPH reagent, vortex mix the sample, and allow it to react at room temperature for 1 h.

3. Extract the sample with three 2 ml aliquots of hexane combining the extracts in a 13 × 100 mm glass tube.

4. Evaporate the extract to dryness under argon at 40 °C.

5. Add 100 μl of acetonitrile and vortex mix to reconstitute the residue then add 100 μl of water to adjust the solvent to 50% (v/v) acetonitrile in water.

6. Perform HPLC analysis as follows:

   (a) Use a flow rate of 250 μl/min with a column temperature of 40 °C. This flow rate is based on the requirements of narrow bore columns; use higher flow rates for standard dimension columns.

   (b) Set the detector to monitor absorbance at 355 nm with a 10 nm bandwidth.

   (c) Inject 25 μl aliquots. Larger sample volumes can be injected depending on the quantities of 4HNE, the nature of the sample, and the dimensions of the HPLC column.

   (d) Programme the mobile phase from 50% acetonitrile/water to 85% acetonitrile/water in 20 min.

7. Calculate the results as follows:

   (a) Plot the integrated peak areas for the standards versus amount of 4HNE.

   (b) Perform a linear regression analysis to determine the slope and intercept of the calibration line.

   (c) Calculate the quantity of 4HNE in the sample based on the results of the linear regression analysis using eqn 3.

   $$\text{nmol of 4HNE} = (\text{4HNE peak area} \times \text{slope}) - \text{intercept} \qquad (3)$$

   where 4HNE peak area is the integrated area of the 4HNE chromatographic peak, 'slope' is the calculated slope of the linear regression line, and 'intercept' is the calculated y-intercept of the linear regression line.

3.2 Adaptability to other aldehydes of interest

As described in Section 2.1, all aldehydes can react with the DNPH reagent under these conditions. The method can be adapted to other aldehydes by developing effective chromatographic conditions so that the aldehyde of interest can be separated from any interfering species.

4. Determination by gas chromatography–mass spectrometry (GC–MS)

4.1 Principle of analysis

The principle of any assay which uses GC–MS is the same as that of an HPLC assay, namely the relation of detector response, determined by integrating the chromatographic peak area, to quantity of analyte. The primary advantages of using a mass spectrometer as the chromatographic detector are increased sensitivity due to the high sensitivity of mass spectrometers and increased selectivity (often referred to as specificity) due to the high information content of a mass spectrum. In the case of 4HNE analysis, increased sensitivity and selectivity allows quantification of as little as 1–5 pmol of 4HNE in complex samples such as cultured cells, tissues, and blood. The disadvantages of GC–MS analyses are the need for more extensive sample processing to prepare the samples for GC–MS analysis and the relative expense and difficulty of maintaining and operating mass spectrometer systems. The most important feature of this system is the high sensitivity of a 'research-grade' mass spectrometer. Successful analyses have also been performed using a research-grade quadrupole mass spectrometer while attempts to use a less sensitive 'table-top' mass spectrometer system have not been successful.

Several types of biological samples have been analysed using this method; cultured cells, precipitated proteins, tissues, whole blood, plasma, and serum. A typical chromatogram for the analysis of 4HNE by GC–MS is shown in *Figure 4*. A feature of the derivatization scheme used is the formation of both *syn*- and *anti*-isomers of the oximation product. These isomers can be separated by capillary GC columns to give a characteristic doublet of chromatographic peaks.

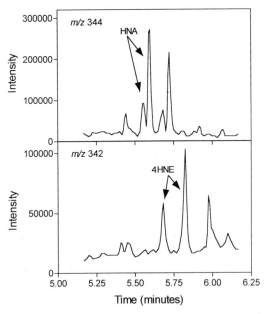

Figure 4. A typical chromatogram for the quantification of 4HNE by GC–MS. The sample, 0.5 ml of human whole blood, was found to contain 46 pmol of 4HNE using the method described in *Protocol 3*. Ions at *m/z* 342 and *m/z* 344 are monitored and plotted independently. The doublet of chromatographic peaks seen for both HNA and 4HNE is due to the separation of the *syn*- and *anti*-isomers formed in the oximation reaction; the more abundant peak in each doublet is integrated for quantification.

Protocol 3. Quantitative analysis of 4HNE by GC–MS

Equipment and reagents

- GC–MS system: a Finnigan-MAT Model 8230 double focusing, reversed geometry mass spectrometer is used. The instrument is equipped with a Varian 3700 gas chromatograph with an Alltech Associates SE-30 Econocap 30 m × 0.32 mm capillary column with a 0.25 μm film thickness bonded polydimethylsiloxane stationary phase.
- Octadecylsilyl solid-phase extraction columns
- Polypropylene tubes (17 × 100 mm) with caps
- Temperature-controlled heating block
- HPLC-grade methanol
- HPLC-grade methylene chloride
- Anhydrous dimethylformamide (DMF)
- Hydroxynonanal (HNA) is synthesized by the reaction of mercuric acetate with 2-nonenal (13)
- Acetate buffer: 0.1 M sodium acetate with 1 mM diethylenetriaminepenta-acetic acid (DTPA), pH 5. Dissolve 8.2 g sodium acetate and 0.4 g DTPA in 1 litre of water. Adjust the pH to 5 with acetic acid. Store at room temperature for up to 0 months.
- Oximation reagent: 0.3 M hydroxylamine hydrochloride (oxime) in acetate buffer. Dissolve 2.0 g of hydroxylamine hydrochloride in 100 ml of the acetate buffer.

- *N*-Methyl-*N*-*tert*-butyldimethylsilyl trifluoroacetamide (MTBSTFA)
- Butylated hydroxytoluene (BHT), 50 mM in methanol. Dissolve 1.1 g BHT in 100 ml of methanol. Store at room temperature for up to 1 month.
- Bovine serum albumin (BSA), essentially fatty acid free, 1 mg/ml: dissolve 10 mg of BSA in 10 ml of water. Store at 4°C for up to 1 month.
- 4HNE primary standard: dissolve 1.0 mg of 4HNE (6.4 μmol) in 6.4 ml of methanol to give a 1.0 mM solution. Store the 4HNE primary standard at −20°C for up to 6 months.
- 4HNE intermediate standard: dilute 100 μl of the 4HNE primary standard to 10.0 ml of methanol in a volumetric flask to give a 10 μM standard. Standardize the solution as in *Protocol 4*. Store at −20°C for 1 month.
- HNA internal standard stock solution: dissolve 4.7 mg of HNA (30 μmol) in 10.0 ml of methanol to give a 3.0 mM solution. Store this stock standard at −20°C for up to 6 months.
- HNA internal standard intermediate dilution: dilute 100 μl of the HNA stock solution to 10.0 ml of methanol in a volumetric flask to give a 30 μM standard. Store this solution at −20°C for 6 months.

Method

1. Prepare the standards as follows.

   (a) Prepare a 4HNE working standard solution by diluting 1.0 ml of the 4HNE intermediate standard using a volumetric pipette into 10 ml of methanol in a volumetric flask. Prepare this solution daily.

   (b) Prepare an HNA working internal standard solution by diluting 1 ml of the HNA intermediate dilution to 10 ml of methanol in a volumetric flask. Prepare this solution daily.

   (c) Transfer 0, 10, 50, and 100 μl of the 4HNE working standard solution into a series of sample tubes. These standards contain 0, 10, 50, and 100 pmol of 4HNE, respectively.

   (d) Add 1.0 ml of the BSA solution to each standard and gently mix.

2. Prepare the samples as follows.

   (a) For analysis of cultured cells, pour off the medium and collect the cells by scraping into the cold saline. Pellet the cells by centrifugation, discard the supernatant, and lyse the cells in 2.0 ml of

Protocol 3. *Continued*

deionized water. Take small aliquots for protein analysis and use the remainder of the sample for the 4HNE analysis. Express 4HNE concentration on a per mg cell protein basis.

(b) For analysis of tissue samples, homogenize the tissue in a 50 mM phosphate buffer containing 1 mM DETAPAC and 1 mM BHT, pH 7, to give homogenates with protein concentrations of approximately 10 mg/ml. Take 0.5 ml aliquots of the homogenate for 4HNE analysis. Measure total protein concentration on separate aliquots and express the 4HNE concentration on a per mg protein basis.

(c) For analysis of aliquots of whole blood, serum or plasma, take aliquots ranging from 0.2 ml to 1 ml for analysis. Express the 4HNE concentrations on a per volume basis.

3. Add 4 ml of the oximation reagent, 50 µl of 50 mM BHT in methanol, and 100 µl of the HNA internal standard to each sample.

4. Thoroughly mix the samples with gentle rocking, react for 1 h at 70°C, and allow to cool to room temperature.

5. If needed, pellet precipitated proteins by centrifugation.

6. Recover the aldehyde–oximes by octadecylsilyl solid phase extraction as follows.

(a) Condition the columns with 1 ml of methylene chloride, 2 ml of methanol, and 6 ml of 0.1 M acetic acid.

(b) Apply the samples to the columns, passing the sample through the column in a dropwise manner.

(c) Wash the column with two, 4 ml aliquots of 0.1 M acetic acid followed by 1 ml of methylene chloride. Discard all washings.

(d) Elute the 4HNE–oxime from the column with two, 2 ml aliquots of methanol, passing the methanol through the column dropwise and collecting the eluent in a glass tube.

7. Transfer the eluent to a 5 ml reaction vial and evaporate the sample to dryness under argon at 40°C. Be aware that, because of water in the methanol, this evaporation step can take approximately 1 h. Additional effort to dry the column out after the washing step (step 6c) and before the elution step (step 6d) can speed this evaporation step but the sample must be taken to dryness.

8. Reconstitute the residue in 200 µl of DMF, add 50 µl of MTBSTFA, and react the sample at 70°C for 1 h.

9. Allow the sample to cool to room temperature.

10. Add 250 µl of saturated NaCl and extract the mixture with two, 600 µl aliquots of petroleum ether, combining the petroleum ether in a 1 ml reaction vial.

11. Evaporate the extract to dryness and reconstitute in 50 μl of isooctane for analysis.

12. Gas chromatography–mass spectrometry:

    (a) Operate the mass spectrometer with 70 eV electron ionization in the selected ion monitoring mode monitoring m/z 342.2 (4HNE, M-57$^+$), m/z 344.2 (HNA, M-57$^+$).

    (b) Make injections on-column and program the gas chromatograph from 100°C to 290°C at 15°C/min.

13. Perform the calculations as follows.

    (a) Calculate the ratio of the integrated 4HNE chromatographic peak area to the integrated HNA chromatographic peak area and plot this ratio versus quantity of 4HNE. The respective chromatographic peaks are indicated in *Figure 4*.

    (b) Perform a linear regression analysis to determine the slope and intercept of the calibration line.

    (c) Calculate the amount of 4HNE in the sample based on the linear regression analysis using Equation 4.

    pmol of 4HNE = (4HNE:HNA ratio × slope) − intercept   (4)

    where 4HNE:HNA ratio is the ratio of the integrated 4HNE chromatographic peak area to the integrated HNA chromatographic peak area, slope is the calculated slope of the linear regression line, and intercept is the calculated *y*-intercept of the linear regression line.

    (d) Normalize the amount of 4HNE as described for the different types of samples in step 2.

4.2 Adaptability to other aldehydes of interest

All aldehydes and aldehyde-Schiff bases should react with the oximation reagent under these conditions. As a result, the method can be adapted to other aldehydes by determining what ion in the mass spectrum is best used for detection and developing effective chromatographic conditions to separate the aldehyde of interest from any interfering species. In some cases it may also be necessary to develop additional internal standards for optimum assay performance.

5. Summary

The protocols presented here cover a range of specificity, sensitivity and ease of use. *Protocol 1* offers the advantage of a rapid spectroscopic assay which can detect amounts of 4HNE above approximately 5 nmol in the sample, remembering that the protocol is not specific for 4HNE. *Protocol 2* extends the limit-of-detection for 4HNE down to approximately 1 nmol in the sample with ability to differentiate 4HNE formation from the formation of other aldehydes. Readers should be aware that, due to recycling and equilibration times associated with a gradient elution, sample throughput for this analysis is approximately 1 sample per hour. *Protocol 3* offers the highest sensitivity, highest specificity assay and, as a result, limits-of-detection for 4HNE down to approximately 5 pmol in the sample. However, this protocol requires GC–MS instrumentation.

References

1. Schaur, R. J., Zollner, H., and Esterbauer, H. (1991). In *Membrane lipid oxidation* (ed. C. Vigo-Pelfry), Vol. III, pp. 141–163. CRC Press, Boston, FL.
2. Schauenstein, E., Esterbauer, H., Jaag, G., and Taufer, M. (1965). *Monatschr. Chem.*, **95**, 180.
3. Benedetti, A., Comporti, M., and Esterbauer, H. (1980). *Biochim. Biophys. Acta*, **620**, 281.
4. Hoff, H. F., O'Neil, J., Chisolm, G. M., III, Cole, T. B., Quehenberger, O., Esterbauer, H., and Jurgens, G. (1989). *Arteriosclerosis*, **9**, 538.
5. Kinter, M., Robinson, C. S., Grimminger, L. C., Gillies, P. J., Shimshick, E. J., and Ayers, C. (1994). *Biochem. Biophys. Res. Commun.*, **199**, 671.
6. Sullivan, S. J., Roberts, R. J., and Spitz, D. R. (1991). *J. Cell. Physiol.*, **147**, 427.
7. Kanimura, S., Gaal, K., Britton, R. S., Bacon, B. R., Triadafilopoulos, G., and Tsukamoto, H. (1992). *Hepatology*, **16**, 448.
8. Esterbauer, H., Schaur, R. J., and Zollner, H. (1991). *Free Radic. Biol. Med.*, **11**, 81.
9. Esterbauer, H., Jurgens, G., Quehenberger, O., and Kolner, E. (1987). *J. Lipid Res.*, **28**, 495.
10. Van Kuijk, F. J. G. M., Holte, L. L., and Dratz, E. A. (1990). *Biochim. Biophys. Acta*, **1043**, 116.
11. Beckman, J. K., Howard, M. J., and Greene, H. L. (1990). *Biochem. Biophys. Res. Commun.*, **169**, 75.
12. Benedetti, A., Comporti, M., Fulceri, R., and Esterbauer, H. (1984). *Biochim. Biophys. Acta*, **792**, 172.
13. Kinter, M., Sullivan, S., Roberts, R. J., and Spitz, D. R. (1992). *J. Chromatogr.*, **578**, 9.

Further reading

Blanchflower, W. J., Walsh, D. M., Kennedy, S., and Kennedy, D. G. (1993). *Lipids*, **28**, 261.
Bringmann, G., Gassen, M., and Schneider, S. (1994). *J. Chromatogr. A*, **670**, 153.
Des Rosiers, C., Rivest, M. J., Boily, M. J., Jette, M., Carrobe-Cohen, A. and Kumar, A. (1993). *Anal. Biochem.*, **208**, 161.
Goldring, C., Casini, A. F., Maellaro, E., Del Bello, B., and Comporti, M. (1993). *Lipids*, **28**, 141.

Kautiainen, A. (1992). *Chem. Biol. Int.*, **83**, 55.
Tamura, H. and Shibamoto, T. (1991). *Lipids*, **26**, 170.
Uchida, K. and Stadtman, E. R. (1994). In *Methods in enzymology* (ed., Packer), Vol. 233, pp. 371–380. Academic Press, London.
Uchida, K., Szweda, L. I., Chae, H.-Z., and Stadtman, E. R. (1993). *Proc. Natl Acad. Sci. USA*, **90**, 8742.

F$_2$-Isoprostanes: prostaglandin-like products of lipid peroxidation

JASON D. MORROW and L. JACKSON ROBERTS

1. Introduction

Free radicals are believed to be involved in a variety of disease processes (1). Much of the evidence for this, however, is indirect and circumstantial because of limitations in the techniques currently available to quantify free radicals or their products in biological systems. This is a particular problem when assessing oxidant injury *in vivo* in humans (2).

Measures of lipid peroxidation are often employed to implicate free radicals in pathophysiological process. These measures include quantification of short-chain alkanes, malondialdehyde, or conjugated dienes. Each of these assays, however, suffers from inherent problems related to sensitivity and specificity, again primarily when applied to *in vivo* situations. In addition, artefactual generation of these products can occur *ex vivo*, and endogenous factors such as metabolism can greatly affect product levels (2).

Recently, we described a series of prostaglandin (PG)-F$_2$ like compounds, termed F$_2$-isoprostanes, that are produced *in vivo* in humans by a non-cyclo-oxygenase free radical catalysed mechanism involving the peroxidation of arachidonic acid (3). Formation of these compounds initially involves formation of four positional peroxyl radical isomers which endocyclize to form PGG$_2$-like intermediates which are then reduced to PGF$_2$-like compounds. Theoretically, four F$_2$-isoprostane regioisomers are formed (*Figure 1*), each consisting of eight racemic diasteromers (4).

Subsequently, we have found that quantification of F$_2$-isoprostanes represents a reliable and useful approach to assess lipid peroxidation and oxidant stress *in vivo*. For example, we have shown that the formation of isoprostanes increases dramatically in animal models of lipid peroxidation and correlates with the degree of tissue damage (4, 5). Further, F$_2$-isoprostanes are present in easily detectable concentrations in normal human biological fluids, such as urine and plasma, allowing for a normal range to be defined and also for mild increases in these compounds to be measured in settings of minimal oxidant

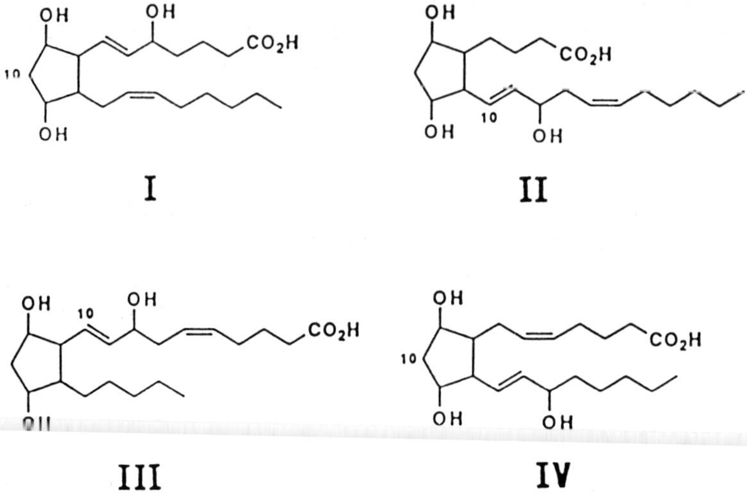

Figure 1. Structures of the four F_2-isoprostane regioisomers. Stereochemistry is not indicated although each regioisomer consists of eight racemic diastereomers.

stress (6). In addition, F_2-isoprostanes can be detected in all types of biological fluids and tissues we have examined thus far, providing an opportunity to assess the formation of these compounds at local sites of oxidant injury (6).

As noted, the precursor of the isoprostanes is arachidonic acid. The majority of arachidonic acid present *in vivo* exists esterified to phospholipids. Recently, we reported that the F_2-isoprostanes are initially formed *in situ* from arachidonate esterified to phospholipids and then subsequently released preformed by phospholipases (7). From this discovery emerges an important concept regarding the assessment of isoprostane formation *in vivo*. Depending on the studies that are undertaken, total isoprostane production may be more accurately assessed by measuring levels of both free and esterified isoprostanes. Further, the fact that F_2-isoprostanes are formed *in situ* on phospholipids can be utilized in an advantageous way to assess oxidant injury in specific human organs by analysing levels of isoprostanes esterified to phospholipids in tissue biopsy specimens that may be obtained for diagnostic purposes.

The purpose of this chapter is to discuss methods for the analysis of F_2-isoprostanes from biological sources. Procedures are outlined for the analysis of both free and esterified F_2-isoprostanes. F_2-isoprostanes from biological sources can only be quantified as free compounds using gas chromatography (GC)–mass spectrometry (MS). Therefore, to measure levels of isoprostanes esterified to phospholipids, the phospholipids are first extracted from a tissue or fluid sample and then subjected to alkaline hydrolysis to liberate the F_2-isoprostanes from the phospholipids. Following hydrolysis, the free F_2-isoprostanes are measured using the same techniques as free compounds in biological fluids.

Thus, the following methods first outline the extraction and hydrolysis of F_2-isoprostane-containing phospholipids from tissue or fluid samples. Subsequently, the methods for the analysis of free compounds are discussed.

2. Method of assay

2.1 Handling and storage of biological fluids and tissues for quantification of F_2-isoprostanes

As noted, we have detected measurable levels of F_2-isoprostanes in virtually all biological fluids examined thus far (6). One potential drawback of quantifying F_2-isoprostanes as a measure of lipid peroxidation, however, is that these compounds can be artefactually generated *ex vivo* in biological fluids, such as plasma, in which arachidonyl-containing lipids are present (4). Importantly we have shown that:

(a) F_2-isoprostanes can be generated *ex vivo* in biological fluids which are allowed to remain at room temperature or which are frozen at $-20\,^{\circ}\mathrm{C}$ (4).

(b) Artefactual generation does not occur, however, when fluids or tissues are stored at $-70\,^{\circ}\mathrm{C}$ for up to 6 months (6).

(c) Further, formation of isoprostanes does not occur if biological fluids are processed immediately after procurement and if the free radical scavenger, butylated hydroxytoluene (BHT), and/or the reducing agent, triphenylophosphine (TPP) is added to the organic solvent during extraction of lipids (6).

Therefore, samples obtained for analysis should be processed either immediately or stored at $-70\,^{\circ}\mathrm{C}$. In addition, samples of tissue or biological fluids that are stored should be rapidly cooled to $-70\,^{\circ}\mathrm{C}$ or lower. This can be accomplished by snap freezing the samples in liquid nitrogen prior to placing them at $-70\,^{\circ}\mathrm{C}$.

2.2 Extraction and hydrolysis of F_2-isoprostane-containing phospholipids

Esterified F_2-isoprostanes can be analysed in either biological fluids or tissues (6). The method of lipid extraction for tissue varies somewhat from that employed for fluids and thus each is discussed separately. Nonetheless, perhaps the most important aspect of the extraction procedure for phospholipids from either biological source is the fact that utmost care must be taken during the extraction procedure to prevent *ex vivo* generation of F_2-isoprostanes during sample handling (4). This is more likely to occur if reagents contain significant trace metal contamination. Thus, it is imperative that the reagents are of extremely high purity. The quality of water used during lipid extraction, in particular, is very important. We use water which has been triply distilled (or its equivalent) and subsequently passed over a column containing the ion-exchange resin Chelex (100 mesh; Bio-Rad Laboratories) to remove trace metals. In addition, agents such as BHT or TPP are routinely employed to suppress autoxidation of lipids during extraction. Finally, all glass and plastic-ware employed are hand washed by us and rinsed with ultrapure water prior to use.

2.2.1 Extraction of tissue lipids for the analysis of F_2-isoprostanes

The method outlined in this section is applicable to essentially all animal and human tissue samples.

Protocol 1. Extraction of tissue lipids for F_2-isoprostane analysis

Equipment and reagents

- Ultrapure water (see above)
- High purity organic reagents: methanol and chloroform (ethanol as preservative) (Burdick and Jackson brand, Baxter Diagnostics Inc.)
- Butylated hydroxytoluene (BHT) (Aldrich Chemical Co.)
- Sodium chloride

- Blade homogenizer-PTA 10S generator (Brinkmann Instruments)
- Table-top centrifuge
- Rotavap unit (Brinkmann)
- Conical glass centrifuge tubes, 40 ml
- Conical bottom flasks, 100 ml
- Tank of nitrogen gas

Method

1. Weigh out 0.05–1 g of either fresh tissue or tissue frozen at −70°C.

2. Add tissue to 20 ml of ice-cold Folch solution (chloroform/methanol; 2:1 v/v) containing 0.005% BHT in a 40 ml glass centrifuge tube.

3. Homogenize tissue with blade homogenizer at full speed for 30 sec.

4. Purge airspace in centrifuge tube with nitrogen and cap. Let solution stand at room temperature for 1 h to effect maximum extraction of lipids from ground tissue. Shake tube occasionally for several seconds during this time period.

5. Add 4 ml aqueous NaCl (0.9%). Vortex vigorously for 30 sec.

6. Centrifuge 800 *g* for 10 min at room temperature.

7. After centrifugation, carefully pipette off top aqueous layer and discard. Remove the lower organic layer carefully from under the intermediate semisolid proteinaceous layer and transfer to a 100 conical bottom flask.

8. Evaporate under vacuum on rotavap unit to dryness. When dry, immediately proceed to hydrolysis step (*Protocol 3*).

2.2.2 Extraction of F₂-isoprostane-containing lipids from biological fluids

The technique used to extract F_2-isoprostane-containing phospholipids from biological fluids differs slightly from the methods used for tissues. These modifications, however, are necessary because we have found that the precursor of the F_2-isoprostanes, arachidonic acid, is much more susceptible to autoxidation during the extraction of lipids, e.g. from plasma, as opposed to tissue phospholipids (unpublished data, Roberts and Morrow). The primary modifications involve the use of TPP in the Folch solution to reduce arachidonyl peroxides to alcohols and evaporation of solvent under nitrogen.

Protocol 2. Extraction of lipids from biological fluids

Equipment and reagents

- Ultrapure water (see Section 2.2)
- Organic reagents: chloroform and methanol (see *Protocol 1*)
- BHT
- TPP (triphenylphosphine) (Aldrich Chemical Co.)
- MgCl₂

- Table-top centrifuge
- Conical glass centrifuge tubes, 40 ml
- Conical bottom flasks, 100 ml
- Tank of nitrogen
- Analytical evaporation unit (such as Meyer N-Evap, Organomation)

Method

1. Add 1 ml of a biological fluid to 20 ml cold Folch solution (*Protocol 1*) containing 0.005% BHT and 5 mg TPP in a 40 ml glass conical centrifuge tube.
2. Shake mixture vigorously for 2 min.
3. Add 10 ml 0.043% MgCl₂ and shake vigorously for 2 min.
4. Centrifuge 800 *g* for 10 min at room temperature.
5. Separate lower organic layer from other layers (*Protocol 1*) and transfer to conical bottom flask.
6. Dry organic layer at room temperature under nitrogen.
7. Proceed immediately to hydrolysis procedure (*Protocol 3*).

2.3 Hydrolyis of lipid extracts

Hydrolysis of tissue or fluid lipids should be performed immediately after lipid extraction to avoid potential autoxidation of arachidonic acid contained in the phospholipids. The technique of hydrolysis employs chemical saponification to yield free (unesterified) F_2-isoprostanes for subsequent purification, derivatization and mass spectrometric analysis (6).

Protocol 3. Hydrolysis of lipids

Equipment and reagents

- Ultrapure water (Section 2.2)
- Methanol (high purity) (*Protocol 1*)
- Potassium hydroxide pellets
- BHT

- HCl (American Chemical Society (ACS) certified or equivalent grade)
- Water bath at 37°C
- Tank of nitrogen

Method

1. To lipid residue in conical flask, add 4 ml methanol containing 0.005% BHT and then add 1 ml aqueous KOH (15%). Purge flask with nitrogen and cap.

2. Incubate mixture at 37°C for 30 min.

3. After incubation, acidify mixture to pH 3 with 1 M HCl. Then dilute mixture to 80 ml with H_2O (pH 3) in preparation for extraction of free F_2-isoprostanes. Note: dilution of methanol in the solution with water to 5% or less is necessary to ensure proper column extraction of F_2-isoprostanes in the subsequent purification procedure (*Protocol 4*).

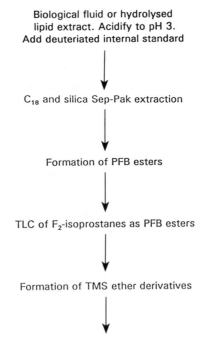

Biological fluid or hydrolysed
lipid extract. Acidify to pH 3.
Add deuteriated internal standard

C_{18} and silica Sep-Pak extraction

Formation of PFB esters

TLC of F_2-isoprostanes as PFB esters

Formation of TMS ether derivatives

Quantification by selected ion monitoring GC–NICI MS

Figure 2. Outline of the procedures used for the extraction, purification, derivatization, and mass spectrometric analysis of F_2-isoprostanes from biological sources.

2.4 Purification and derivatization of F₂-isoprostanes

Quantification of F₂-isoprostanes by GC–negative ion chemical ionization (NICI) MS is extremely sensitive with lower limits of detection in the 3–15 fmol range using a deuterated standard with a blank of less than 5 parts per thousand (6). Thus, it is not necessary to assay more than 1–3 ml of a fluid such as plasma, 0.1 ml of urine, or 50–1000 mg of tissue. The assay method for purification and derivatization of free F₂-isoprostanes in biological fluids or in hydrolysed lipid extracts is outlined in *Figure 2*. For purposes of discussion, the method used to purify and derivatize F₂-isoprostanes in plasma is detailed in *Protocol 4* but this method is equally adaptable to other biological fluids or tissue extracts. We had previously employed a second thin-layer chromatography (TLC) step in the analysis of some fluids, such as urine, but recently have found this unnecessary if small amounts of urine (e.g. 0.1 ml) are used (6). The aim of this assay method is to separate F₂-isoprostanes from other biological impurities and convert these molecules into derivatives suitable for mass spectrometric analysis.

Protocol 4. Purification and derivatization of F₂-isoprostanes

Equipment and reagents

- Ultrapure water (Section 2.2)
- HCl (reagent ACS)
- Organic reagents (high purity): methanol, ethyl acetate, heptane, acetonitrile, chloroform, ethanol
- Pentafluorobenzylbromide (PFBB) (Aldrich Chemical Co.)
- *N,N'*-Diisopropylethylamine (DIPE) (Aldrich)
- Dimethylformamide (Aldrich)
- *N,O*-bis(trimethylsilyl)trifluoroacetamide (BSTFA) (Supelco Inc.)
- Undecane (Aldrich)
- Phosphomolybdic acid (Sigma Chemical Co.)
- Na₂SO₄ (anhydrous)
- [²H₄]PGF₂α internal standard (Cayman Chemical Co.)
- Sep Pak cartidges: C₁₈ and silica (Millipore Corp.)
- Disposable plastic syringes, 10 ml
- Glass scintillation vials
- Reactivials (5 ml)
- TLC plates: LK6D silica (Whatman)
- Microcentrifuge tubes
- Analytical evaporator (*Protocol 2*)
- Tank of nitrogen
- Microcentrifuge

A. *Purification of F₂-isoprostanes*

1. Acidify 1–3 ml of plasma to pH 3 with 1 M HCl.

2. Add 1–5 pmol of deuterated [²H₄]PGF₂α internal standard and vortex.

3. Apply mixture to a C₁₈ Sep Pak column connected to a 10 ml syringe preconditioned with 5 ml methanol and 5 ml pH 3 water.

4. Wash column with 10 ml pH 3 water and then 10 ml heptane.

5. Elute isoprostanes from column with 10 ml ethyl acetate/heptane (50:50, v/v) into a glass scintillation vial.

Protocol 4. *Continued*

6. Add 5 g anhydrous Na_2SO_4 to vial and swirl gently. Note: this removes residual water from the eluate.

7. Apply eluate to silica Sep Pak preconditioned with 5 ml methanol and 5 ml ethyl acetate.

8. Wash cartridge with 5 ml ethyl acetate.

9. Elute isoprostanes from silica Sep Pak with ethyl acetate/methanol (50:50, v/v) into reactivial.

10. Evaporate eluate under nitrogen.

11. To convert F_2 isoprostanes into pentafluorobenzyl esters add 40 μl 10% (v/v) PFBB in acetonitrile and 20 μl 10% (v/v) DIPE in acetonitrile to residue for 30 min at room temperature.

12. Dry reagents under nitrogen and repeat step 11 to ensure quantitative esterification.

13. Dry reagents under nitrogen and resuspend in 50 μl methanol.

14. Apply mixture to a lane on a silica TLC plate (LK6B) that has been prewashed with methanol. Chromatograph to 13 cm in a solvent system of chloroform/ethanol (93:7, v/v). For a TLC standard, apply approximately 2–5 μg of the methyl ester of $PGF_{2\alpha}$ to another TLC lane. Visualize the TLC standard by spraying the lane with a 10% solution of phosphomolybdic acid in ethanol and heating.

15. Scrape silica from the TLC plate in a region of the methyl ester of the $PGF_{2\alpha}$ standard (R_f 0.15) and adjacent areas 1 cm above and below.

16. Place silica in microcentrifuge tube and add 1 ml ethyl acetate. Vortex vigorously for 20 sec.

17. Pour off ethyl acetate into another microcentrifuge tube taking care not to disrupt silica pellet.

18. Dry organic layer under nitrogen.

B. *Conversion of F_2-isoprostanes into trimethylsilyether ether derivatives*

1. Add 20 μl BSTFA and 6 μl DMF to residue.

2. Vortex well and incubate tube at 37°C for 20 min.

3. Dry reagents under nitrogen.

4. Add 10 μl undecane (which has been stored over calcium hydride to prevent water accumulation). Sample is now ready for mass spectrometric analysis.

2.5 Quantification of F$_2$-isoprostanes using GC–MS

Quantification of F$_2$-isoprostanes in biological fluids using the methods out-lined in this chapter requires mass spectrometry with NICI capabilities. The use of NICI MS enhances the sensitivity of the assay by orders of magnitude compared with standard techniques employing electron ionization MS. For the analysis of isoprostanes, we routinely use a Nermag R10–10C mass spectro-meter interfaced with a DEC-PDP 11/23 Plus computer system although any other mass spectrometer with NICI capabilities can be utilized. The iso-prostanes are chromatographed on a 15 m DB1701 fused silica capillary col-umn (J and W Scientific). We have found that this GC phase gives superior separation and resolution of the individual F$_2$-isoprostanes compared with other columns (6). The column temperature is programmed from 190°C to 300°C at 20°C/min. Methane is used as the carrier gas for NICI at a flow rate of 1 ml/min. The ion source temperature is 250°C, the electron energy is 70 eV, and the filament current is 0.25 mA. The ion monitored for endo-genous F$_2$-isoprostanes is the carboxylate anion m/z 569 (M-181, loss of $CH_2C_6F_5$). The corresponding carboxylate anion for the deuterated internal standard is m/z 573.

3. Application of the assay to biological samples

As mentioned, we have successfully used this assay to quantify F$_2$-iso-prostanes in a number of diverse biological fluids (including urine, plasma, joint fluid, and cerebrospinal fluid) and tissues (such as muscle, heart, liver, brain, and kidney) (6). *Figure 3* shows selected ion current chromatograms obtained from the analysis of F$_2$-isoprostanes in plasma of a rat following treatment with CCl$_4$ to induce endogenous lipid peroxidation. The series of peaks in the upper m/z 569 chromatogram represent endogenous F$_2$-iso-prostanes. This pattern of peaks is virtually identical to that obtained from all other biological fluids and tissues examined thus far. The peak in the lower m/z 573 chromatogram in *Figure 3* represents the [^{2}H$_4$]PGF$_{2\alpha}$ internal standard.

By using a variety of chemical and mass spectrometric approaches, we have found that all the peaks present in the m/z 569 chromatogram in *Figure 3* represent different F$_2$-isoprostanes (3, 4). However, the peak denoted by an asterisk (*), which elutes approximately 13 sec before the [^{3}H$_4$]PGF$_{2\alpha}$ internal standard is routinely used for quantification purposes. Using the ratio of the intensity of this peak to that of the internal standard, the concentration of F$_2$-isoprostanes in the plasma sample in *Figure 3* was calculated to be 2340 pmol/l, which is approximately 25-fold above normal. Normal plasma levels of F$_2$-isoprostanes in rats and in humans are in the range 60–170 pmol/l. Levels in other tissues and fluids have been reported elsewhere (3, 5). The assay for F$_2$-isoprostanes in biological fluids is highly precise and accurate. The precision is within ±6% and the accuracy is 96%.

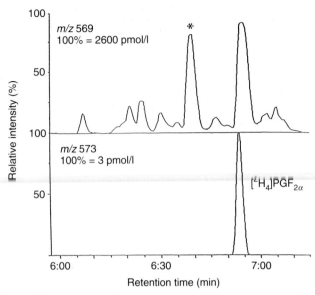

Figure 3. Analysis of F_2-isoprostanes in plasma obtained from a rat several hours follow-ing treatment with CCl_4 (2 ml/kg orogastrically) to induce lipid peroxidation. Plasma was assayed using $[^2H_4]PGF_{2\alpha}$ as the internal standard. The *m/z* 569 chromatographic peaks represent endogenous F_2-isoprostanes whereas the peak in the *m/z* 573 chromatogram represents the deuterated $PGF_{2\alpha}$ internal standard. In the *m/z* 569 chromatogram, the asterisk marks the peaks routinely used for quantification of the F_2-isoprostanes.

In summary, this chapter outlines methods to assess lipid peroxidation associated with oxidant injury *in vivo* by quantifying concentrations of free and esterified F_2-isoprostanes in biological fluids and tissues. The assay described herein is highly precise ($\pm$ 6%) and accurate (96%) (6). Measure-ment of F_2-isoprostanes appears to overcome many of the shortcomings associated with other methods to assess oxidant status, especially when applied to the *in vivo* situation in humans. Thus, measurement of F_2-isoprostanes may represent an important advance in our ability to assess the role of oxidant stress and lipid peroxidation in human disease.

Acknowledgements

This work was supported by NIH grants GM 42056, DK 48831, and GM 15431. J. D. Morrow is a Howard Hughes Medical Institute Physician Research Fel-low and recipient of a career development award from the International Life Science Institute.

References

1. Southorn, P. A. (1988). *Mayo Clin. Proc.*, **63**, 390.

2. Halliwell, B. and Grootveld, M. (1987). *FEBS Lett.*, **213**, 9.
3. Morrow, J. D., Hill, K. E., Burk, R. F., Nammour, T. M., Badr, K. F., and Roberts, L. J. (1990). *Proc. Natl Acad. Sci. USA*, **87**, 9383.
4. Morrow, J. D., Harris, T. M., and Roberts, L. J. (1990). *Anal. Biochem.*, **184**, 1.
5. Morrow, J. D., Awad, J. A., Kato, T., Takahashi, K., Badr, K. F., Roberts, L. J., and Burk, R. F. (1992). *J. Clin. Invest.*, **90**, 2502.
6. Morrow, J. D. and Roberts, L. J. (1994). In *Methods in enzymology* (ed. L. Packer), Vol. 233, pp. 163–174. Academic Press, San Diego.
7. Morrow, J. D., Awad, J. A., Boss, H. J., Blair, I. A., and Roberts, L. J. (1992). *Proc. Natl Acad. Sci. USA*, **89**, 10721.

11

Determination of carbonyl groups in oxidized proteins

EMILY SHACTER, JOY A. WILLIAMS, EARL R. STADTMAN,
and RODNEY L. LEVINE

1. Introduction

There is now a bewildering array of biological processes in which free radicals have been implicated (2), and we assume that enzymes and structural proteins may be attacked whenever free radicals are generated. As a consequence, oxidative modification of proteins may occur in a variety of physiological and pathological processes. Although the distinction is sometimes arbitrary, these modifications may be primary or secondary. Primary modifications occur in metal-catalysed oxidation, radiation-mediated oxidation, and in oxidation by ozone or oxides of nitrogen. Secondary modifications occur when proteins are modified by molecules generated by oxidation of other molecules. One important example is the covalent modification of proteins by hydroxynonenal produced by oxidation of lipids (3).

Carbonyl groups (aldehydes and ketones) may be introduced into proteins by any of these reactions, and the appearance of such carbonyl groups is taken as presumptive evidence of oxidative modification. The word *presumptive* is important because the appearance of carbonyl groups is certainly not specific for oxidative modification. For example, glycation of proteins may add carbonyl groups on to amino acid residues. Despite this caveat, assay of carbonyl groups in proteins provides a convenient technique for detecting and quantifying oxidative modification of proteins.

Methods for determination of carbonyl content were discussed in reference 4. The methodology and discussion in that article remain useful and should be read in conjunction with this chapter, which presents newer methods based on the reaction of carbonyl groups with 2,4-dinitrophenylhydrazine (DNPH) to form a 2,4-dinitrophenylhydrazone. The assays provide substantial improvements in both sensitivity and specificity. They employ high-performance liquid chromatographic (HPLC) gel filtration or electrophoresis for removal of excess reagent and introduce Western blotting for sensitive and specific detection of the 2,4-dinitrophenyl group (5). A similar technique was developed independently by Keller *et al.* (6).

2. Reaction with 2,4-dinitrophenylhydrazine in guanidine

2,4-Dinitrophenylhydrazine is a classical carbonyl reagent (7), and it has emerged as the most commonly used reagent in the assay of oxidatively modified proteins. However, quantitative derivatization requires a large excess of reagent to be present, and that reagent must be removed to allow spectrophotometric determination of the protein-bound hydrazone. Removal is required because the DNPH reagent has significant absorbance at 370 nm so that residual reagent can cause an artefactual increase in the apparent carbonyl content of the sample. Recent investigations suggest that this may occasionally be a problem with the filter paper method, especially with samples containing low amounts of protein (4). Previous publications described several methods for removal of reagent including extraction by ethanol/ethyl acetate, reversed phase chromatography, and gel filtration (4, 8). After extraction, the hydrazone can be determined spectrophotometrically by its absorbance at 370 nm.

HPLCs are widely available, and gel filtration by HPLC has proven to be a convenient and efficient technique for removal of excess reagent. Derivatized proteins are also separated by molecular weight, allowing a more specific analysis of carbonyl content. HPLC spectrophotometric detectors are also far more sensitive than stand-alone spectrophotometers, so much less sample is required for quantification of carbonyl content. The HPLC method has sufficient sensitivity for analysis of cells from a single tissue culture dish ($\sim 500\,000$ cells) or for analysis of the small amounts of tissue available from biopsy and autopsy samples. Most HPLC detectors provide chromatograms at two or more wavelengths, allowing carbonyl content to be followed at 370 nm and protein content at around 276 nm, obviating the need for a separate protein assay. If a diode-array detector is available, then full spectra of the peaks are obtained. These can be useful in checking for contaminants which might artefactually affect either carbonyl or protein determination. As noted earlier, nucleic acid contamination could potentially interfere with the assay (4); such contamination would be suspected if the protein spectrum were skewed from the 276 nm protein peak towards the 260 nm nucleic acid peak. Also, any chromophore which absorbs at 370 nm would interfere with quantification of the hydrazone. Whereas total background can be determined on a blank sample not treated with DNPH, availability of spectra may assist in identifying the chromophore and in devising methods for minimizing the background. Examples of such chromophores include heme from contaminating haemoglobin and retinoids from tissues such as liver (L. Szweda and R. L. Levine, unpublished observations).

Derivatizations with DNPH are classically performed in solutions of strong acids such as 2 M HCl. There are two disadvantages to the use of 2 M HCl in preparing samples for HPLC analysis. First, very few HPLC systems or columns can tolerate the HCl. Second, many proteins are insoluble in HCl and must be solubilized before injection into the HPLC system. These problems are dealt with by derivatization in 6 M guanidine at pH 2.5. The guanidine effectively solubilizes most proteins whereas pH 2.5 is compatible with most HPLC systems.

The rate of reaction of proteins studied thus far is much faster than suggested earlier (4), being essentially complete within 5 min. The time course may vary for different proteins and should be checked if this is a concern. Moreover, samples should not be allowed to stand in the derivatization solution for longer than about 15 min because a slow reaction leads to the introduction of the 2,4-dinitrophenyl moiety in a non-hydrazone linkage. When either glutamine synthetase or bovine serum albumin was incubated for 120 min, the apparent carbonyl content was artefactually increased by ~0.2 mol/subunit. The nature of the reaction is not yet understood, but it can be avoided simply by injecting the sample onto the HPLC at a fixed reaction time, such as 10 min. Autosamplers sold by various manufacturers are capable of performing the derivatization and then injecting the sample at fixed reaction times.

Protocol 1. Reaction with 2,4-dinitrophenylhydrazine in 6 M guanidine

Equipment and reagents

- Buffer for gel filtration and for derivatization blank: 6.0 M guanidine-HCl, 0.5 M potassium phosphate, pH 2.5
- Column: Zorbax GF450, 92 × 250 mm, at 2 ml/min flow rate (Mac-Mod Analytical or Dionex)[a]

- Derivatization solution: 10 mM DNPH, 6.0 M guanidine HCl, 0.5 M potassium phosphate, pH 2.5. (Solutions of DNPH develop precipitates during storage, and these can be removed by centrifugation before use. Stock solutions are usable for months.)

A. *Preparation of buffer for gel filtration and for derivatization blank (1 litre)*

1. Dissolve 573 g guanidine-HCl in water. Use only about 200 ml water initially to avoid overdilution after the guanidine dissolves.

2. Bring the volume to about 850 ml, warming as needed to dissolve the guanidine.

3. Add 33.3 ml concentrated phosphoric acid (85%) slowly with stirring.

4. Adjust the pH to 2.5 with 10 M KOH.

5. Adjust the volume to 1000 ml and filter through a 0.45 μm filter.

B. *Preparation of derivatization solution (100 ml)*

1. Dissolve the solid DNPH in 3.33 ml concentrated phosphoric acid (85%); 100 ml derivatization solution requires 198 mg DNPH. You

Protocol 1. *Continued*

must take into account the actual content of DNPH in your supply, which is typically 60–70%, with the remainder being water.

2. Dissolve 57.3 g guanidine-HCl in water to give a volume of about 80 ml. This will require adding only about 25 ml water.

3. With stirring, add the DNPH solution dropwise to the guanidine solution.

4. Use 10 M KOH, added dropwise, to bring the solution to pH 2.5.

5. Adjust the volume to 100 ml.

C. *Method*[b]

1. Prepare the sample as desired. For example, concentrate by precipitation with trichloroacetic acid or ammonium sulphate.

2. Split the sample in half if a derivatization blank is also being prepared.

3. Add 3 vol. of derivatization solution and mix to dissolve the sample.

4. If a blank is being prepared, treat it with the buffer without DNPH.

5. Allow to stand 10 min at room temperature.

6. The in-line filter will remove particulates, but it is possible to lengthen the time between filter changes by centrifuging the sample for 3 min in a tabletop microcentrifuge (11 000 g).

7. Inject the sample and follow the chromatograms at 276 and 370 nm.[c] Inject the next sample when the reagent has washed through and the baseline has restabilized. A typical chromatogram and peak spectrum are shown in *Figure 1*.

[a] The flow rate of 2 ml/min allows sample to be injected at least every 15 min. It is important to protect the column from fouling by particulates which are often present in crude samples, and this is easily accomplished by placing an in-line filter just before the column (Item no. A314, Upchurch Scientific). The GF450 (or GF250) column has proven quite rugged and easily tolerates the back pressure generated when pumping the guanidine/phosphate buffer at 2.0 ml/min. Other columns could be substituted, but those which we have evaluated have generally not worked as well, either because of poorer separations or because they could not tolerate the 2 ml/min flow rate. Some columns, such as the Fast Desalting Column (Pharmacia), cannot tolerate the 6 M guanidine.

[b] Derivatization is conveniently carried out in 1.5 ml screw-topped or snap-top plastic tubes. Sensitivity is comparable to the borotritide method so that one can use 10 μg of protein containing 1 mol carbonyl/mol protein or 100 μg of protein containing 0.10 mol carbonyl/mol protein. As with any gel filtration method, larger sample volumes will cause peak broadening. If one wishes to estimate the molecular weights of the labelled proteins then the sample volume before derivatization should be 75 μl or less; with larger volumes it will take longer for the later-eluting reagent to come off the column, thus requiring the injection of the next sample to be delayed by a few minutes to permit the absorbance to return to baseline.

[c] A 10 nm bandwidth is appropriate for detectors with adjustable bandwidths. If the detector can only follow one wavelength, use 370 nm to detect the hydrazone and determine protein content by a separate assay. In most HPLC systems, the proteins will begin to elute after 3 min and the reagent will elute after 6 min.

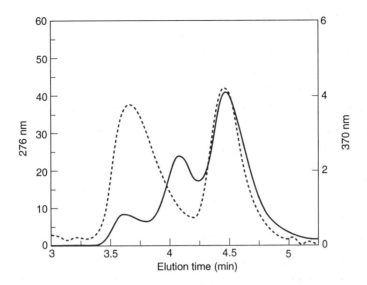

Figure 1. HPLC chromatograms of mouse peritoneal fluid after treatment with the inflammatory oil pristane (14). The tracing at 276 nm (solid line) tracks protein, while that at 370 nm (dashed line) tracks 2,4-dinitrophenylhydrazones. Note that the first eluting protein peak ($M_r > 200\,000$ Da) appears relatively enriched in carbonyl-containing protein. The second peak is primarily IgG and the third peak is primarily albumin.

2.1 Calculations

Carbonyl content may be expressed either as nmol carbonyl/mg protein or as mol carbonyl/mol protein. In either case, integrate the chromatograms to obtain the area of the protein peaks at 276 and 370 nm. Stop integration before reagent begins to elute, preferably at a specific molecular weight determined by calibration of the column. We typically exclude material eluting below 10 000 Da.

If a blank was run, subtract its area at 370 nm from that of the sample. For the molar absorbtivity of the hydrazone use $\epsilon_{M370\ nm} = 22\,000$ and $\epsilon_{M276\ nm} = 9460$ (43% of that at 370 nm). If the molar absorbtivity of the protein is known, use it. If not, use 50 000 which is a good estimate of the molar absorbtivity of a protein of average amino acid composition (9) and a molecular weight of 50 000. Individual proteins can deviate substantially from this average value. For example, glutamine synthetase from *Escherichia coli* has an ϵ_{M276} of 33 000. To determine the mol of carbonyl/mol of protein:

$$\text{Mol carbonyl/mol protein} = \frac{(\epsilon_{\text{protein}_{276}}) \times (\text{Area}_{370})}{(22\,000) \times (\text{area}_{276} - 0.43 \times \text{area}_{370})}$$

Because the result is a ratio, this calculation is independent of the amount of sample injected. However, when desired, the actual mass in a given peak can be determined. For a typical injection one will have nanogram quantities of protein (pmol) which will contain pmol of carbonyl. The general equation for determination of the amount of material in a peak is:

$$\text{mol} = \frac{\text{Area} \times \text{Flow}}{\epsilon_M \times \text{Path length}}$$

Note that the path lengths of HPLC detectors are usually not 1.0 cm. In the case of the Hewlett–Packard 1040 diode array detector, the path length is 0.6 cm and the area units are mAU-sec. Switching to micromolar absorbtivity and pmol, the equation becomes

$$\text{pmol} = \frac{\text{Area} \times \text{Flow}}{\epsilon_{\mu M} \times 36}$$

Given a flow rate of 2 ml/min and $\epsilon_{M370\ nm} = 22\,000$, the calculation for carbonyl is simply

$$\text{pmol carbonyl} = 2.53 \times \text{Area}_{370nm}$$

For protein determination with $\epsilon_{M276\ nm} = 50\,000$ the equation would be

$$\text{pmol protein} = 1.11 \times (\text{Area}_{276nm} - 0.43 \times \text{Area}_{370nm})$$

with the area at 276 nm being corrected for any contribution from the hydrazone as noted above. For a molecular weight of 50 000 then,

$$\text{ng protein} = 55.5 \times (\text{Area}_{276nm} - 0.43 \times \text{Area}_{370nm})$$

2.2 Comments

This HPLC gel filtration method has been used for a large number of analyses in our laboratory. It is especially convenient if an autosampler is available, obviating the need to inject samples manually. However, 6 M guanidine-HCl is not kind to pump seals, injector rotors, and other components of the HPLC system. Even small leaks will deposit substantial amounts of guanidine which may cause corrosion. We make it a practice to flush with water and to inspect the system at the end of each day's use, taking care to correct any small leaks that may be noted. Pump seals should be changed at the earliest sign of wear. With most solvent systems a delay in changing the seals is of no consequence, but with 6 M guanidine serious damage can result.

The guanidine solution is relatively viscous, leading to rather high back pressures at the 2 ml/min flow rate. Typical back pressures are on the order of 200 bar, but systems with long runs of microbore tubing may be substantially higher. Only HPLC systems can generate the required pumping pressure, so this method cannot be used with low pressure pumping systems such as Pharmacia's FPLC.

When an HPLC is not available or not convenient, reagent can also be removed by gel filtration under gravity flow on single use columns such as the PD-10 from Pharmacia. Of course, the sample will be diluted more than in the HPLC method. Sample can be prepared as above or in 2 M HCl as described earlier (4). In the latter case, the sample can be precipitated with trichloroacetic acid, then taken up in the guanidine buffer. The Sephadex beads tend to compact in 6 M guanidine causing flow to become quite slow. To minimize this problem, the column should be washed with the smallest volume of guanidine buffer required for equilibration. This can be done by washing with 15 ml of water, then 10 ml of buffer. The sample is usually applied in a total volume of 0.5 ml, then eluted with the guanidine buffer. One millilitre fractions are collected, with the protein emerging in the third millilitre (including the sample volume). The carbonyl is quantified spectrophotometrically as described (4). Reagent begins to elute in the fifth millilitre, but Sephadex tends to bind or react with DNPH, so some of the reagent will remain bound to the column.

3. Reaction with DNPH in sodium dodecyl sulphate

The comments above point out that use of the pH 2.5, 6 M guanidine buffer can be detrimental to the HPLC system. Moreover, the guanidine must be removed if the derivatized proteins are to be analysed by sodium dodecyl sulphate (SDS) polyacrylamide gel electrophoresis. To obviate these two problems a method has been developed to derivatize proteins in SDS instead of guanidine (10). *Figure 2* provides a flow chart which summarizes the preparation of samples for analysis by either HPLC or gel electrophoresis and Western blotting.

We have less experience with the SDS system than with the guanidine system, and more will be required to determine whether the SDS or guanidine systems may be preferred for particular samples. When checked with several samples of oxidized glutamine synthetase, the analytical results were the same with both methods. There are at least two advantages of the SDS system over the guanidine system: Back-pressure is substantially lower with the SDS system, and the problem of salt corrosion is essentially eliminated. One disadvantage of SDS is that resolution of proteins from each other and from reagent may not be as good as with guanidine, presumably due to the relatively large size of the SDS micelles (11).

Protocol 2. Reaction with DNPH in SDS[a]

Equipment and reagents

- Buffer for gel filtration: 200 mM sodium phosphate, pH 6.5, 1% SDS
- Derivatization solution: 20 mM DNPH in 10% trifluoroacetic acid (v/v)[b]
- Derivatization blank solution: 10% trifluoroacetic acid (v/v)

- Neutralization solution: 2 M Tris (free base, not the HCl salt), 30% glycerol[c]
- 12% SDS[d]
- Column: Zorbax GF450, 92 × 250 mm, at 2 ml/min flow rate (see *Protocol 1*)

Protocol 2. *Continued*

Method

1. Prepare the sample as desired. Samples with high protein concentrations (>10mg/ml) usually require dilution to assure solubility after addition of SDS and acid. Potassium dodecyl sulphate is less soluble than the sodium salt so that the sample should not contain more than about 50 mM potassium.

2. Split the sample in half if a derivatization blank is also being prepared. The volumes mentioned below always refer to the volume of the sample at this point, that is, before the addition of derivatizing reagents.

3. Add 1 volume of 12% SDS with mixing. Do this even if the sample contains some SDS, because it is important to have at least 6% SDS before adding the trifluoroacetic acid.

4. Add 2 vol. DNPH solution and mix.

5. If a blank is desired, treat it with 2 vol. 10% trifluoroacetic acid alone.

6. Allow to stand for 10 min at room temperature.

7. Bring the solution approximately to neutrality by adding 2 M Tris/30% glycerol.[e]

8. For HPLC analysis, inject all or part of the sample and analyse the chromatograms as described above for the guanidine system, except that the hydrazones are monitored at 360 nm in SDS instead of the 370 nm used in guanidine.

9. Calculate the results as described in Section 2.2.

[a] The procedure for derivatization in SDS is very similar to that for guanidine, but has been designed to facilitate preparation of the sample for analysis by both HPLC gel filtration and Western blotting, if desired.

[b] As noted in *Protocol 1*, take into account the actual content of DNPH in the reagent as most manufacturers supply DNPH with at least 30% water content. For 100 ml stock, dissolve the 396 mg DNPH in 10 ml pure trifluoroacetic acid (15.4 g), then dilute to 100 ml with water. Precipitates which appear during storage can be removed by centrifugation.

[c] Glycerol is included to facilitate analysis by SDS gel electrophoresis. It may be omitted if only the HPLC analysis is being carried out.

[d] Warming the water will speed dissolution of the solid.

[e] You will need about 1.5 times the original sample volume. When neutralized there will be a noticeable colour change from light yellow to red–orange. For analysis by HPLC, the exact pH is not critical. You must simply neutralize the bulk of the acid, taking care to keep the pH below 8 to avoid alkali-induced changes in the 2,4-dinitrophenylhydrazones. For analysis by SDS gel electrophoresis, the pH needs to be controlled more closely. Inclusion of bromophenol blue in the neutralization solution (0.06%) provides a tracking dye for following the electrophoretic front.

Protocol 3. Immunodetection of protein-bound 2,4-dinitrophenylhydrazones (Western blotting)

Equipment and reagents

- Derivatization reagents as listed in *Protocol 2*
- Apparatus for gel electrophoresis
- Apparatus for electroelution (Western blotting)
- Antibodies to the 2,4-dinitrophenyl moiety[a]
- Labelled secondary antibody to the species chosen for the antibody to the 2,4-dinitrophenyl moiety[b]

Method

1. Derivatize the samples as described in *Protocol 2*, steps 1–7.[c]

2. Load the samples and perform SDS gel electrophoresis.[d]

3. Transfer to nitrocellulose and immunodetect the labelled proteins (Western blotting) (12).[e]

[a] These are available from several suppliers as monoclonal and polyclonal antibodies. We have most experience with the mouse monoclonal IgE from Sigma (cat. no. D-8406) and the rabbit polyclonal from Dako (cat. no. V401). Typically a 1:1000 dilution is used, but it is best to check the optimal dilution yourself.

[b] When the first antibody is rabbit, we have used the goat anti-rabbit IgG conjugated to alkaline phosphatase (part of the Western-Light chemiluminescent kit WL10RC from Tropix), typically at a 1:10 000 dilution. When the first antibody is the mouse monoclonal IgE, we have used biotinylated rat anti-mouse IgE from Southern Biotechnology (cat. no. 1130–08), typically at a 1:4000 dilution. A biotin–avidin amplification system provides increased sensitivity and works well with purified proteins. However, false positive bands are often observed in cruder mixtures which contain biotin-binding proteins. In our hands, it is not possible reproducibly to prevent this false-positive by blocking, so we simply use a second antibody conjugated to horseradish peroxidase.

[c] If desired, one portion of the derivatized sample can be analysed by HPLC and another by SDS gel electrophoresis. Also, fractions from the HPLC gel filtration may subsequently be analysed by SDS gel electrophoresis.

[d] After derivatization and neutralization with the 2 M Tris/30% glycerol the samples may be loaded directly on to the gel and electrophoresed as usual because the sample is now in a solution similar to that used for electrophoresis (10). When desired, β-mercaptoethanol can be added to the sample solution. If all samples are to be treated with β-mercaptoethanol, the neutralization solution can be changed to 2 M Tris/30% glycerol/19% β-mercaptoethanol, yielding 5% β-mercaptoethanol in the neutralized sample. Note that samples are not heated before analysis because heating is generally not required for reduction of disulphide bonds, and we do not know the stability of protein-bound hydrazones to heating. Also, inclusion of β-mercaptoethanol generally intensifies the bands in the chemiluminescent detection system so that it is best to either include or omit the thiol from all samples which are to be directly compared.

[e] Either colorimetric or chemiluminescent detection protocols may be used. With chemiluminescence, the lower limit of detection is about 30 ng of oxidized glutamine synthetase containing 0.5 carbonyl groups per subunit (that is, about 0.3 pmol carbonyl). If chemiluminescence is chosen, it is convenient to make exposures of varying lengths. Times of 0.5–5 min have worked well when using the kit from Tropix. Shorter times, 2–30 sec, were used with Amersham's kit (cat. no. RPN 2109).

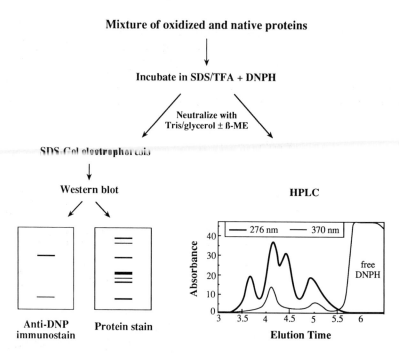

Figure 2. Schematic overview of the methods for separating and detecting proteins derivatized with DNPH in SDS.

3.1 Comments

A typical Western blot is shown in *Figure 3*. The excess reagent from the sample does not interfere with gel electrophoresis or with Western blotting. Lack of interference may result from poor accessibility of the antibody to reagent on nitrocellulose, a suggestion which follows from the observation that the monoclonal antibody did not detect reagent spotted directly on to the nitrocellulose. As with all Western blot techniques, this method is not quantitative, but serial dilutions of the sample should provide an estimate of the amount of labelled material. In addition, it is important to include standard proteins of known carbonyl content in each gel. These facilitate selection of development times which allow distinction between carbonyl-positive and carbonyl-negative proteins. Specificity for derivatized carbonyl groups can be checked by testing samples which have been pretreated with sodium borohydride to reduce carbonyl groups and Schiff bases (4). The technique also works with isoelectric focusing gels, so it should be possible to extend the method to two-dimensional gels. Use of these antibodies for immunoaffinity purification and for immunocytochemical localization of oxidized proteins also appears feasible (13).

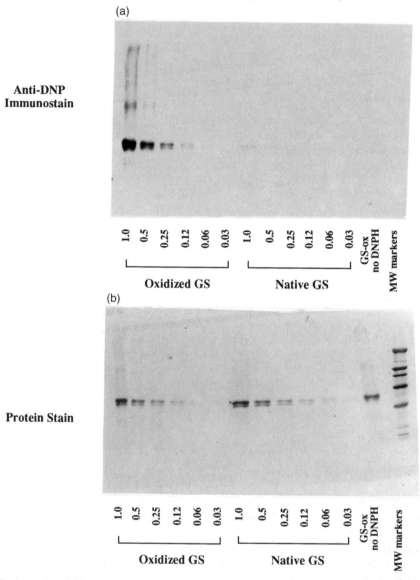

Figure 3. Western blot immunoassay for protein-bound carbonyl groups in glutamine synthetase. Glutamine synthetase (GS) from *E. coli* was oxidized with an iron/ascorbate/oxygen system (15) to yield a carbonyl content of ~0.5 mol/subunit. The untreated or native GS contained ~0.5 mol carbonyl/subunit. Both native and oxidized proteins were derivatized for 15 min with DNPH in SDS as described in the text. After neutralization, serial dilutions were applied to the gel for electrophoresis and Western blotting (a). A replicate blot was stained for protein with Amido Black (b). The amount of protein applied ranged from 30 ng to 1 μg, which in the case of oxidized GS corresponds to 0.3–10 pmol of carbonyl. A negative control (GS-ox no DNPH) contained oxidized GS which was not treated with DNPH.

Acknowledgement

The chapter has been modified from ref. 1.

References

1. Levine, R. L., Williams, J. A., Stadtman, E. R., and Shacter, E. (1994). In *Methods in enzymology* (ed. L. Packer), Vol. 233, p. 346. Academic Press, San Diego.
2. Halliwell, B. and Gutteridge, J. M. C. (1990). In *Methods in Enzymology* (ed. L. Packer and A. N. Glazer), Vol. 186, p. 1. Academic Press, London.
3. Levine, R. L., Oliver, C. N., Fulks, R. M., and Stadtman, E. R. (1981). *Proc. Natl Acad. Sci. USA*, **78**, 2120.
4. Levine, R. L., Garland, D., Oliver, C. N., Amici, A., Climent, I., Lenz, A. G., Ahn, B. W., Shaltiel, S., and Stadtman, E. R. (1990). In *Methods in Enzymology* (ed. L. Packer and A. N. Glazer), Vol. 186, p. 464. Academic Press, London.
5. Shacter, E., Williams, J. A., and Levine, R. L. (1995). *Free Radic. Biol. Med.*, **18**, 815.
6. Keller, R. J., Halmes, N. C., Hinson, J. A., and Pumford, N. R. (1993). *Chem. Res. Toxicol.*, **6**, 430.
7. Jones, L. A., Holmes, J. C., and Seligman, R. B. (1956). *Anal. Chem.*, **28**, 191.
8. Levine, R. L. (1984). In *Methods in Enzymology* (ed. L. Packer), Vol. 105, p. 370. Academic Press, London.
9. *Protein Identification Resource*, Release 26.0. (1990). National Biomedical Research Foundation, Washington, DC.
10. Laemmli, U. K. (1970). *Nature*, **227**, 680.
11. Helenius, A., McCaslin, D. R., Fries, E., and Tanford, C. (1979). In *Methods in Enzymology* (ed. S. Fleischer and L. Packer), Vol. 56, p. 734. Academic Press, London.
12. Towbin, H., Staehelin, T., and Gordon, J. (1979). *Proc. Natl Acad. Sci. USA*, **76**, 4350.
13. Pompella, A. and Comporti, M. (1991). *Histochemistry*, **95**, 255.
14. Shacter, E., Arzadon, G. K., and Williams, J. (1992). *Blood*, **80**, 194.
15. Rivett, A. J. and Levine, R. L. (1990). *Arch. Biochem. Biophys.*, **278**, 26.

12

Protein hydroperoxides, protein hydroxides, and protein-bound DOPA

ROGER T. DEAN, SHANLIN FU, STEVEN GIESEG, and
SHARYN G. ARMSTRONG

1. General background

Other chapters in this volume describe the determination of stable, relatively unreactive products of protein oxidation, such as protein carbonyls. However, we have recently shown that radical attack on proteins can generate two kinds of reactive moieties that are fairly long-lived under appropriate conditions: protein hydroperoxides (oxidizing species) and protein-bound DOPA (3,4-dihydroxyphenylalanine) (a reducing species (1, 2); reviewed in ref. 3).

We have argued that these species may be particularly important in pathophysiology. The hydroperoxides probably have a particular relevance to the chain reaction of protein oxidation (4), since they can participate in reaction chains in which propagation is by the generation of further radicals from the hydroperoxide, and does not necessarily involve the repeated consumption of oxygen. We have demonstrated the formation of radicals by reaction of protein and amino acid hydroperoxides with transition metals (5). We have identified the nature of many of these radicals, and their potential to eliminate chain-carrying radicals, such as the hydroperoxy radical, has been revealed. The hydroperoxides can also consume biological reductants, such as ascorbate and glutathione, and thus transfer oxidative damage to other targets and locations. In these reductions, and during transition metal and glutathione peroxidase attack, the major product is the hydroxide (6) (Fu *et al.*, unpublished work), and hence below we describe methods for determining these hydroxides, in the hope that their estimation will be valuable in assessing their role *in vivo*.

The reducing species, protein bound DOPA, may also be of particular interest because of its ability to initiate secondary reactions, which again may result in the transfer of oxidative damage to other targets and locations (2). This moiety can reduce free and chelated transition metals, and so facilitate secondary reactions of these species. For example, reduced transition metals may interact with oxygen to generate radical fluxes, and can cleave hydroperoxides much more rapidly than their higher-valency state counterparts, thus propagating peroxidation processes, such as lipid peroxidation. Again, the investigation of the possible damaging roles of this protein oxidation product requires assays for its occurrence *in vivo*, and we present such methods here.

2. General methods involved in determination of protein hydroperoxides, hydroxides, and protein-bound DOPA

When it is necessary to determine specific amino acid hydroxides or DOPA from protein samples, protein hydrolysis is a necessary first step. When tissue samples are studied, delipidation is desirable prior to hydrolysis (see *Protocol 2*). The hydrolysis method (*Protocol 1*) indicates some of the precautions appropriate to avoid losses of the DOPA. Valine hydroperoxides under such hydrolysis conditions are only partially recovered as valine hydroxides (Val.OH), ranging from 20 to 80% according to samples. Thus to determine protein-bound valine hydroperoxides as hydroxides it is necessary to reduce these species fully with borohydride prior to protein hydrolysis. Methods for preparing tissue samples which permit the recovery of protein hydroxides and protein-DOPA are also summarized in *Procotol 2*. For satisfactory hydrolysis and subsequent analysis, it is necessary to remove some of the contaminating materials, particularly lipids. The procedure listed in *Protocol 2* is suitable. For amino acid hydroxide determinations, complete borohydride reduction (see below) is required before all other steps. With certain tissue homogenates, the success of this reduction may require careful investigation. The methods described here are developments from previously published methods (2, 6).

Protocol 1. Protein hydrolysis

Equipment and reagents

- L-DOPA (Sigma)
- 6 M hydrochloric acid containing 1% phenol (v/v)
- Mercaptoacetic acid (95%, BDH)
- 700 μl brown glass autosampler vials (All-tech)
- Pico-Tag reaction vessel (Millipore Waters)
- Oven
- Nitrogen or argon gas cylinder
- Water aspirator
- Freeze-dryer (Speedy-Vac)

Method

1. Add appropriate internal standards, such as DOPA or Val.OH, to some protein samples (for recovery determination), and then freeze-dry the samples in glass autosampler vials.

---

2. Place sample vials (maximum of eight) in Pico-Tag reaction vessel, to which 1 ml of 6 M HCl containing 1% (v/v) phenol and 50 μl mercaptoacetic acid have already been added.

3. Evacuate air from reaction vessel using water aspirator hose over inlet hole.

4. Fill vessel with nitrogen or argon gas (also via inlet hole).

5. Repeat steps 3 and 4 once, followed by step 3 again.

6. Place reaction vessel in an oven at 110°C for 16 h.

7. Remove reaction vessel from oven and allow to cool.

8. Remove sample vials from reaction vessel and evaporate any residual acid using Speedy-Vac. Redissolve samples as described below for the different protocols

---

Protocol 2. Preparation of biological samples for protein hydrolysis

Equipment and reagents

- Phosphate-buffered saline (PBS): 0.2 g potassium phosphate monobasic, 0.2 g potassium chloride, 8 g sodium chloride, and 1.15 g sodium phosphate dibasic per litre of nanopure water, preferably Chelex-pretreated to minimize metal contamination
- Butylated hydroxytoluene (BHT)
- EDTA

- 0.30% NaDOC (w/v), 50% trichloroacetic acid (TCA) (w/v)
- Acetone, diethyl ether
- Round-bottomed brown glass autosampler vials (1 ml, Alltech)
- Centrifuge
- Speedy-Vac freeze dryer

Method

1. Homogenize fresh rinsed tissue in PBS containing 100 μM BHT (as antioxidant), and 1 mM EDTA (as metal chelator), at 4°C.

2. Put 700 μl of homogenate (with a protein concentration of about 1 mg/ml) in a 1 ml brown glass autosampler vial. Add into the vial 50 μl of 0.3% NaDOC and 100 μl of 50% TCA.

3. Mix the solution, centrifuge at 9000 g for 10 min at −10°C, and then discard the surpernatant.

4. Add 800 μl of acetone to the precipitate.

5. Repeat steps 3 and 4.

6. Add 800 μl of diethyl ether.

7. Repeat steps 3 and 4.

8. Dry the precipitate in a Speedy-vac freeze dryer.

3. Protein hydroperoxides

3.1 Introduction

In our laboratory, two analytical methods are often used for determination of total protein hydroperoxides. One is the iodometric assay (7), the principle of which is that iodine (I_2) is generated from iodide (I^-) by the reaction with peroxides in acidic solution, and is then converted into tri-iodide ion (I_3^-) in the presence of an excess of I^-. The amount of I_3^- formed can be measured spectrophotometrically at 360 nm, allowing quantitative estimation of the formation of peroxides as the reaction between I^- and the hydroperoxy (-OOH) group normally follows 1:1 stoichiometry. The method is simple and relatively sensitive but has some disadvantages. For example, protein at a high concentration (>30 mg/ml) precipitates in the assay system and thus interferes with the accuracy. Protein also tends to bind I_2 generated during the assay, which can result in falsely low I_3^- values.

The other method used is isoluminol chemiluminescence detection. Chemi-luminescence based HPLC assays have been widely used to study lipid hydroperoxides (8, 9). The method is based on light emission during the hydroperoxide-induced oxidation of isoluminol, catalysed by microperoxi-dase and metal ions. The assay is very sensitive, with the lower limit of detec-tion being about 0.1 pmol or less. The method can be adapted for detecting protein hydroperoxides (6). When protein samples are free from other biomolecules, especially antioxidants, the HPLC separation step can be omit-ted. Hydroperoxy groups (-OOH) on different molecules have been found to respond differently in the chemiluminescence system. Therefore, different hydroperoxides require different calibration curves, which makes it difficult to use for quantitative measurement of protein hydroperoxides, especially in complex samples.

3.2 Determination of stable derivatives of protein hydroperoxides

A new approach has been developed for studying protein hydroperoxides. The hydroperoxy group on oxidized protein molecules does not tolerate the drastic conditions used for protein hydrolysis, and is lost as unknown prod-ucts. A reduction step prior to hydrolysis is therefore necessary to convert the -OOH group into the hydroxy group (-OH): sodium borohydride is used as the reductant. Valine, the amino acid from which the peroxide yields are greatest (10), has been found to produce three valine hydroperoxides, i.e. β-hydroperoxyvaline (Val.OOH1), (2S, 3S)-γ-hydroperoxyvaline (Val.OOH2), and (2S, 3R)-γ-hydroperoxyvaline (Val.OOH3) after exposure to hydroxyl radicals (•OH) in oxygen saturated systems (6). These hydroperoxyvalines can be determined after conversion into the corresponding hydroxides, i.e. Val.OH1, Val.OH2, Val.OH3. Among them, β-hydroxyvaline (Val.OH1) has been chosen as the most suitable for routine assay, due to: (a) its sole origin from a hydroperoxide precursor (the 2 γ-hydroxyvalines arise additionally from NaBH$_4$ reduction of a valine carbonyl (probably the aldehyde), another oxidation product of valine (6)); (b) its higher recovery from protein hydrolysis (85% in the case of oxidized bovine serum albumin, compared with Val.OH2 25%, and Val.OH3 17%); and (c) its better separation from other amino acids in HPLC analysis of the *o*-phthaldialdehyde (OPA) derivatives.

The distribution of protein hydroperoxides in tissue samples between the reduced (hydroxide) and unreduced form is not yet known. However, it has been shown (Fu *et al.*, unpublished) that transition metals, glutathione peroxidase, reducing agents, and cells, can all cause rapid reduction of such hydroperoxides. In processing tissue samples, the reduction step using borohydride is still necessary to ensure complete conversion into hydroxides. Thereafter, the hydrolysis procedure is as described above.

3.3 Preparation of β-hydroxyvaline standard

A solution of L-valine (1 mM) in water is irradiated by gamma rays from a ^{60}Co facility to a dose of 1200 Gy while gassing continuously with O$_2$ Three valine hydroperoxides were generated and then reduced by NaBH$_4$ (1 mg/ml) to the corresponding hydroxides.

Since detection of β-hydroxyvaline from the hydrolysate of oxidized protein requires two steps of HPLC, as will be described later, its elution profile on an LC-NH$_2$ (aminopropyl derivatized silica) column, and on a reversed phase LC-18 (octadecyl-bonded silica) column (as the OPA derivative) needs to be established.

Due to the weak absorbance of β-hydroxyvaline at 210 nm, the oxidized valine solution must be concentrated at least 10-fold by using a rotary evaporator prior to HPLC using UV detector. A typical chromatogram of β-hydroxyvaline on an LC-NH$_2$ column by following *Protocol 3* is shown in *Figure 1*, with a retention time of around 18 min.

Derivatization of amino acids by OPA followed by HPLC on a reversed phase LC-18 column provides a very sensitive and accurate method for quantitative amino acid analysis (11) and has also been used in our study. Since this method is sensitive (0.1–1.0 pmol) (12), the above oxidized valine solution needs to be diluted at least 20-fold. Separation of the *OPA* derivative of β-hydroxyvaline on an LC-18 column was achieved by following *Protocol 4*. The chromatogram is shown in *Figure 2a*.

3.4 Detection of β-hydroxyvaline on oxidized protein molecules

The concentration of β-hydroxyvaline, if any, in hydrolysates of oxidized protein is usually very low, which does not facilitate its direct measurement using the OPA-HPLC method (see *Protocol 4*). A solution to this is the introduction of

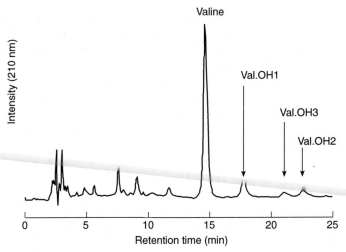

Figure 1. HPLC of gamma-radiolysed valine after NaBH₄ reduction on an LC-NH₂ column by following *Protocol 3*.

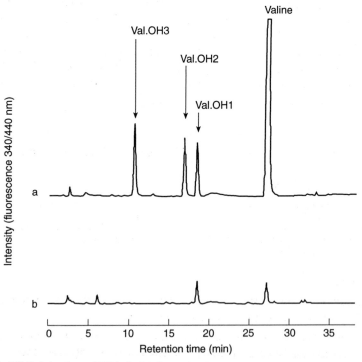

Figure 2. HPLC detection of OPA derivatives of valine hydroxides on oxidized valine and BSA. (a) Gamma-radiolysed valine solution (1 mM) was treated with NaBH₄ (1 mg/ml) and diluted 20 times before being subjected to OPA-HPLC analysis (see *Protocol 4*). (b) Gamma-radiolysed BSA solution (2 mg/ml) was first reacted with NaBH₄ and then hydrolysed before being subjected to OPA-HPLC analysis (see *Protocol 5*).

Protocol 5. An additional purification and concentration step prior to OPA derivatization was implemented (see *Protocol 5,* Part B). The hydrolysate was chromatographed first by HPLC on the analytical LC-NH$_2$ column, from which only the fraction which corresponds to the elution window of Val.OH1 was collected (no visible β-hydroxyvaline peak in the UV trace can be observed due to the low concentration). OPA measurement of this purified fraction was then carried out. *Figure 2b* shows that Val.OH1 can be detected in the hydrolysate of reduced, gamma-radiolysed bovine serum albumin (BSA) using *Protocol 5.* About 0.1% of valines in the oxidized BSA molecules were converted into β-hydroxyvaline in this particular case. Sensitivity for detecting β-hydroxyvaline in protein samples by using this method is very good, with a detection limit as low as 0.5 pmol.

Caution needs to be taken when this method is used. First, the retention time for β-hydroxyvaline chromatographed on the LC-NH$_2$ column (see *Protocol 3*) decreases as the column becomes older. Therefore a standard elution profile needs to be obtained before each experiment, to guide the fraction collection step in *Protocol 5,* Part B, step 3. Slight modifications of the mobile phase are often necessary to obtain a satisfactory separation, and can usually be achieved by increasing the acetonitrile composition in the mobile phase to 85% (instead of 82% as in *Protocol 3*). Secondly, if derivatization of amino acids by OPA is performed manually rather than by using an autosampler as described in *Protocol 4,* the time interval between the onset of derivatization to column injection needs to be consistent for each sample, as the decomposition of OPA derivatives is time-dependent (13).

Protocol 3. HPLC of β-hydroxyvaline on an LC-NH$_2$ column

Equipment and reagents

- LC-NH$_2$ column (25 cm × 4.6 mm, 5 μm particle size, Supelco) with a Pelliguard NH$_2$ guard column (2 cm)
- Mobile phase: 82% acetonitrile in 10 mM NaH$_2$PO$_4$ buffer, pH 4.3
- HPLC system (Shimadzu) with a manual injection port
- UV detector (Shimadzu SPD-10A)
- 25 μl HPLC syringe

Method

1. Equilibrate the column with the mobile phase eluting at 1.5 ml/min for 30 min (or until a steady baseline is reached).

2. Inject 20 μl of the oxidized valine sample on to the HPLC column. The gamma-radiolysed and NaBH$_4$-treated valine sample needs to be concentrated at least 20-fold by using a rotarary evaporator or freeze dryer.

3. Elute at 1.5 ml/min and monitor the eluent at 210 nm.

Protocol 4. HPLC of the OPA derivatives of amino acids on a reversed phase LC-18 column

Equipment and reagents

- *o*-Phthaldialdehyde (OPA) reagent solution (incomplete) (Sigma)
- 2-Mercaptoethanol (Merck)
- Screw-topped vial (2 ml) with 8 mm screw (Activon, Australia), 8 mm Telflon septa (Shimadzu), and 200 μl flat bottom insert (Edwards Instruments, Australia)
- LC-18 column (15 cm × 4.6 mm, 3 μm particle size, Supelco) with a Pelliguard C18 guard column (2 cm)
- Mobile phase: solvent A, methanol/tetrahydrofuran/20 mM sodium acetate pH 5.2, 20:2.5:77.5 (by vol.); solvent B, methanol/tetrahydrofuran/20 mM sodium acetate pH 5.2, 80:2.5:17.5 (by vol.)
- Autosampler (Shimadzu)
- Gradient HPLC system with two pumps and a high pressure mixer (Shimadzu)
- Fluorescence detector (Hitachi F-1050)

A. Preparation of OPA derivatives of amino acids

1. Cool the autosampler to 5°C.

2. Constitute 1 ml of the OPA solution with 3 μl 2-mercaptoethanol in a vial and put it in the autosampler.

3. Place 40 μl aliquots of amino acid sample in H_2O (50 nM to 100 μM) into the insert of a vial and put it in the autosampler.

4. Execute a pretreatment file to take 20 μl of the OPA reagent (from step 2) and mix with the amino acid sample (in step 3) for 2 min precisely.

B. HPLC of the OPA derivatives

1. Equilibrate the column with solvent A for 30 min at 1 ml/min.

2. Inject automatically 15 μl of the reaction solution (from Part A, step 4) to the HPLC column.

3. Elute at 1 ml/min with a gradient of solvent A and B as follows: 0 min, 0% B; to 50% B in 20 min; then to 100% B in 5 min; isocratic elution for 5 min at 100% B; then isocratic elution for 5 min at 0% B.

4. Monitor the eluent using a fluorescence detector with excitation wavelength at 340 nm and emission wavelength at 440 nm

---

Protocol 5. Detection of β-hydroxyvaline from hydrolysates of oxidized proteins

Equipment and reagents
- Same equipments and reagents as listed in *Protocols 1, 3,* and *4*

A. Hydrolysis of oxidized proteins

1. Add $NaBH_4$ (1 mg/ml) to the oxidized protein solution.

2. Hydrolyse the protein following *Protocol 1*.

B. *Purification of the hydrolysates by HPLC on an LC-NH$_2$ column*

1. Redissolve the hydrolysate from Part A in 25 μl H$_2$O.

2. Inject 20 μl of the hydrolysate to an LC-NH$_2$ column (see *Protocol 3*).

3. Collect the fraction corresponding to the elution window of β-hydroxy-valine.[a]

4. Lyophilize the collected fraction by freeze drying.

C. *Determination of β-hydroxyvaline by HPLC of the OPA derivative on an LC-18 column*

1. Redissolve the fraction collected from Part B, step 4 in 50 μl H$_2$O.

2. Take aliquots of 40 μl for OPA-HPLC analysis by using *Protocol 4*.

3. Use a calibration curve of valine for calculation of the quantity of β-hydroxyvaline.[b]

---

[a] The retention time for standard β-hydroxyvaline on the LC-NH$_2$ column needs to be established at the beginning of each experiment.
[b] The OPA derivative of β-hydroxyvaline and valine have the same fluorescent yield, as demonstrated previously (6). We observe a linear relationship between 0.05 μM and 100 μM for valine in the OPA-HPLC assay.

4. Protein-bound DOPA

4.1 Introduction

Many proteins that function as varnishes, adhesives, and eggshell components contain protein-bound DOPA (14, 15). These have been detected by the colorimetric Arnow assay (16) and by borate–hydrochloric acid difference spectroscopy (17). Although these methods provide simple means of detection of DOPA residues in intact protein, they lack adequate sensitivity; the Arnow assay detects down to 0.5–1.0 μg DOPA/ml of sample (representing 2.5–5.0 nmol/assay) and the borate–hydrochloric acid method detecting only down to 2 μg/ml. The latter also suffers from numerous interferences, including those from tyrosine and tryptophan.

Quinoproteins can be detected and quantified using the nitroblue tetrazolium/glycine assay (18), which has been shown to be sensitive down to 2–10 pmol of quinone. However, it relies on redox cycling of protein-bound catechols/quinones and is thus not specific to protein-bound DOPA. It has also failed to detect quinones in a number of known quinoproteins (18), and can thus only be used reliably as a way to confirm the presence of protein-bound quinone.

Waite (19) has developed a modification of a conventional amino acid analysis method using HPLC detection of precolumn derivatized amino acids. The method described here has been developed for the determination of protein-bound DOPA by acid hydrolysis of proteins and subsequent HPLC analysis with fluorimetric detection. It is appropriate for use on protein samples containing as little as 5 pmol DOPA/50 μl injection (or 20 ng DOPA/ml hydrolysate). The sensitivity is likely to be enhanced, however, on newer fluorescence detectors than the Hitachi F-1050 we used.

Although it is fairly stable to acidic conditions, care must be taken to avoid oxidative loss of DOPA during hydrolysis of the protein. (We envisage this oxidation to occur via a pathway generating a number of bicyclic *o*-diphenol and *o*-benzoquinone intermediates, including 5,6-dihydroxyindole (2).) Precautions (20) were taken, including the addition of mercaptoacetic acid to the reaction vessel, as well as flushing with an inert gas, followed by evacuation (see *Protocol 1*). Phenol (1%) was also included, to act as an antioxidant. Gas phase hydrolysis was used to aid in DOPA recovery (21). Maximum DOPA recovery from oxidized proteins was achieved when the hydrolysis was carried out at 110°C for 16 h. Recovery of authentic L-DOPA used to spike protein samples was determined to be 98%, whereas that for tyrosine was 100%.

The HPLC gradient conditions have been slightly altered from those of our original method (2), after a small peak was found to co-elute with DOPA in some (especially stored) samples. GC–MS of trifluoroethanol/pentafluoropropionic acid-derivatized hydrolysates was used to confirm the identity of the DOPA peak as well as to validate the quantification by fluorescence measurement.

Peak purity for this HPLC method was confirmed by comparison of UV spectra taken at three positions within a DOPA peak from an oxidized protein's hydrolysate, using diode array detection.

Protocol 6. HPLC determination of DOPA in protein hydrolysates

Equipment and reagents

- L-DOPA (Sigma)
- Trifluoroacetic acid 0.1% (v/v)
- Solvent A: 0.1% trifluoroacetic acid adjusted to pH 2.5
- Solvent B: 40% methanol, 0.1% trifluoroacetic acid, adjusted to pH 2.5
- Gradient HPLC system with appropriate integration facility

- C-18 HPLC column (such as the Supelcosil LC-18 250 × 4.6 mm, 5 μm column (Supelco) used here, supplied by Activon Scientific Products Co. Ltd, Australia) fitted with a C-18 guard column
- Fluorescence detector, set at 280 nm excitation and 320 nm emission (Hitachi F-1050 used here)

Method

1. Dissolve each sample in freshly prepared (daily) 0.1% trifluoroacetic acid (200 μl). Inject hydrolysed sample (20–50 μl) into HPLC and run at a flow rate of 1.0 ml/min.

2. Elute using the following gradient: 100% solvent A for 15 min, then a linear slope of up to 25% solvent B in the next 10 min. DOPA elutes

during the first phase, and tyrosine during the linear gradient. This is followed by a wash phase of up to 100% solvent B, then a gradual return to 100% solvent A and an equilibration for 10 min ready for the next injection.

3. Quantification of DOPA is performed by comparing peak area with a standard curve developed on the same system using known concentrations of authentic DOPA (also prepared in 0.1% trifluoroacetic acid).

4.2 DOPA: general comments

DOPA (only) as a product of oxidative attack on protein is measured by the procedure given here. Under our routine conditions for protein-bound DOPA generation (γ-irradiation of aqueous solutions with O_2 gassing) dityrosine crosslinks are not formed. If dityrosine production is to be measured, however, the hydrolysis conditions (with minor changes) are appropriate (22). For detection of dityrosine, an extended solvent gradient is required to elute this more hydrophobic compound. The eluted dityrosine is detected by fluorescence at 280/410 nm (excitation/emission). Its identity can be confirmed by collecting the peak, buffering it to pH 8 or above and examining it for the characteristic dityrosine fluorescence at 323 nm/410 nm (excitation/emission).

It is possible that some of the DOPA in oxidized proteins may form cysteinyl-DOPA adducts. It has been found that free DOPA can bind to proteins through cysteine residues, forming the adducts 2-, 5-, and 6-cysteinyl-DOPA (23–26). We found that authentic cys-DOPA standards (kindly supplied by Dr Ito) could be separated by our HPLC conditions (with UV detection at 210 nm). In our system 2- and 6-cysDOPA elute well before free DOPA, whereas 5-cysDOPA (the predominant adduct) elutes after tyrosine. We determined a recovery of 80% of added free 5-cysDOPA (expected to be the predominant form) through the acid hydrolysis procedure. However, these products were not detected in hydrolysates of our radiolysed proteins.

The borate–hydrochloric acid difference spectroscopy method (17) has also been used on oxidized proteins, yielding DOPA concentrations in agreement with those found by the HPLC–fluorescence method given here. The oxidation product 3,4-dihydroxy-α,β-dehydrodopa, which can also be detected by this method (17, 27, 28), was not found in our γ-irradiated proteins.

Giulivi and Davies (29) suggested the presence of dopamine and dopamine quinone in pronase digests of proteins treated with H_2O_2. This would require a decarboxylation step, which has not been well established under these oxidative conditions. The HPLC conditions of this paper were duplicated by us in an attempt to find dopamine in our samples. On injection of authentic dopamine, however, we found that it co-eluted with tyrosine. It was also found that dopamine does not fluoresce at 308/395 nm, as was claimed, but shares a fluorescence maximum with tyrosine (280/320 nm). Dopamine (if

Roger T. Dean et al.

present) also co-elutes with tyrosine under our HPLC conditions. It must be mentioned, however, that our column was 5 cm shorter than that of Giulivi and Davies, which could possibly account for our inability to resolve the two compounds. A further concern is that the retention time for the dopamine peak shown by these authors was approximately 1.1 min, which at the quoted flow rate of 1.5 ml/min indicates that it eluted within the first 1.7 ml of mobile phase. Such a small volume would be less than the void volume for the 30 cm column used. These observations place in serious doubt the identification of the peak in question as dopamine.

References

1. Simpson, J., Narita, S., Gieseg, S., Gebicki, S., Gebicki, J. M., and Dean, R. T. (1992). *Biochem. J.*, **282**, 621.
2. Gieseg, S. P., Simpson, J. A., Charlton, T. S., Duncan, M. W., and Dean, R. T. (1993). *Biochemistry*, **32**, 4780.
3. Dean, R. T., Gieseg, S., and Davies, M. J. (1993). *Trends Biochem. Sci.*, **18**, 437.
4. Neuzil, J., Gebicki, J. M., and Stocker, R. (1993). *Biochem. J.*, **293**, 601.
5. Davies, M. J., Fu, S., and Dean, R. T. (1995). *Biochem. J.*, **305** 643.
6. Fu, S., Hick, L. A., Sheil, M. M., and Dean, R. T. (1995). *Free Radic. Biol. Med.*, **19**, 281.
7. Jessup, W., Dean, R. T., and Gebicki, J. M. (1994). In *Methods in enzymology* (ed. L. Packer), Vol. 233, p. 289. Academic Press, San Diego.
8. Yamamoto, Y., Brodsky, M. H., Baker, J. C., and Ames, B. N. (1987). *Anal. Biochem.*, **160**, 7.
9. Sattler, W., Mohr, D., and Stocker, R. (1994). In *Methods in enzymology* (ed. L. Packer), Vol. 233, p. 469. Academic Press, San Diego.
10. Gebicki, S. and Gebicki, J. M. (1993). *Biochem. J.*, **289**, 743.
11. Turnell, D. C. and Cooper, J. D. H. (1982). *Clin. Chem.*, **28**, 527.
12. Umagat, H. and Kucera, P. (1982). *J. Chromatog.*, **239**, 463.
13. Simons, S. S. and Johnson, D. G. (1976). *J. Am. Chem. Soc.*, **98**, 7098.
14. Waite, J. H. (1990). *Comp. Biochem. Physiol.*, **97B**, 19.
15. Huggins, L. G. and Waite, J. H. (1993). *J. Exp. Zool.*, **265**, 549.
16. Waite, J. H. and Tanzer, M. L. (1981). *Anal. Biochem.*, **11**, 131.
17. Waite, J. H. (1984). *Anal. Chem.*, **56**, 1935.
18. Paz, M. A., Flukiger, R., Boak, A., Kagan, H. M., and Gallop, P. A. (1991). *J. Biol. Chem.*, **266**, 689.
19. Waite, J. H. (1991). *Anal. Biochem.*, **192**, 429.
20. Ito, S., Kato, T., Shinpo, K., and Fujita, K. (1984). *Biochem. J.*, **222**, 407.
21. Meltzer, N. M., Torus, G. I., Gruber, S., and Stein, S. (1987). *Anal. Biochem.*, **160**, 356.
22. Heinecke, J. W., Li, W., Francis, G. A., and Goldstein, J. A. (1993). *J. Clin. Invest.*, **91**, 2866.
23. Ito, S. and Fujita, K. (1982). *Biochem. Pharmacol.*, **31**, 2887.
24. Ito, S. and Fujita, K. (1984). *Biochem. Pharmacol.*, **33**, 2193.
25. Kato, T., Ito, S., and Fujita, K. (1984) *Biochem. Biophys. Acta*, **881**, 415.

26. Ito, S., Kato, T., and Fujita, K. (1988). *Biochem. Pharmacol.*, **37** 1707.
27. Rzepecki, L. M. and Waite, J. H. (1991). *Arch. Biochem. Biophys.*, **285**, 27.
28. Rzepecki, L. M., Nagafuchi, T., and Waite, J. H. (1991). *Arch. Biochem. Biophys.*, **285**, 17.
29. Giulivi, C. and Davies, K. J. A. (1993). *J. Biol. Chem.*, **268**, 8752.

13

Investigating the effects of oxygen free radicals on carbohydrates in biological systems

H. R. GRIFFITHS and J. LUNEC

1. Introduction

Carbohydrates are a large family of related molecules defined as polyhydroxy aldehydes or ketones, and their derivatives. They perform a diverse range of biological functions, both in their own right, in the derivation of energy for cellular maintenance and survival, or as conjugates to larger macromolecules such as proteins or lipids. In addition, the 5-C sugar deoxyribose is an integral component of the DNA backbone and its oxygenated form is critical to the integrity of RNA.

In common with other biological structures carbohydrates are susceptible to the effects of oxygen free radicals (OFRs). Their rate of reaction is dependent on structure but, in general, OFRs are 10 times less reactive with sugars than they are with either the nucleotide bases or aromatic or sulphur-containing amino acids. Specific rate constants of reaction for the hydroxyl radical ($^{\bullet}$OH) with hexose and pentose sugars are reported by Anbar and Neta (1).

An examination of rates of reactivity allows a general picture of the degree of denaturation afforded by OFRs to be constructed. However, what is also critical in determining the degree of damage in biological systems is the relative concentration of different molecules and the subcellular location of catalytic metal ions such as Fe^{2+} and Cu^{+} (2). These ions can catalyse the production of $^{\bullet}$OH in their immediate microenvironment, and since $^{\bullet}$OH rarely diffuse further than 1 nm before reacting, this site-specific form of oxidative damage may be the most important mechanism in the denaturation of carbohydrates by OFRs.

There are comparatively few studies reporting the effects of OFRs on carbohydrates owing to the complex methods of analysis. Elegant studies have been reported (3) using radiolysis techniques and also gas chromatography–mass spectrometry (GC–MS). Although these can provide a detailed picture of the complex array of derivatives formed, such techniques are largely inaccessible for *ex vivo* and *in vitro* measurements. Therefore, herein we will focus on selected techniques that can be applied in most laboratories for the simple determination of OFR damage to carbohydrates. This chapter considers methods that are suitable to *in vitro* and/or *ex vivo* systems and describes their application to several pathological states.

1.1 Method of choice

As implied above, isolated sugar molecules required for the production of cellular energy, such as glucose and fructose, are susceptible to the effects of OFRs. Such *in vitro* effects can be studied by high performance liquid chromatography (HPLC) techniques (Section 2.2) or by more specialist GC–MS techniques. In addition, certain hexose and pentose sugars degrade following OFR attack to release thiobarbituric acid (TBA)-reactive products, where the resultant chromogen can be determined spectrophotometrically (Section 2.2). However, since proteins and lipids can also produce TBA-reactive products following OFR attack this technique is only of use in ultrapure systems.

In vivo, most cell surface and serum-derived proteins have branched oligosaccharides covalently linked to asparagine (Asn) or threonine (Thr) in the primary sequence, where they account for up to 10% of the molecular weight. Abnormalities in the glycoform of immunoglobulin G (IgG) during rheumatoid arthritis (RA) and inflammatory osteoarthritis (OA) have been recognized for almost two decades, and *in vitro* studies have shown that OFRs can mimic these effects (4). A lectin-based enzyme-linked absorbent assay (ELISA) technique is described in Section 2.3 for the rapid determination of modifications to the carbohydrate moiety of isolated IgG.

The related molecules, proteoglycans, have a much larger percentage of carbohydrate relative to protein and are important components of the extracellular matrix. The major extracellular matrix biomolecule in synovial fluid is the mucopolysaccharide, hyaluronic acid. In inflammatory conditions characterized by an infiltrate of monocytes and macrophages capable of eliciting the respiratory burst, hyaluronic acid is degraded. The latter can be measured by a change in viscosity (see Section 3.1).

The deoxyribose moiety of DNA degrades following OFR attack with the induction of strand breaks. In ultrapure systems, deoxyribose degradation can be monitored by the measurement of a TBA-reactive chromophore (Section 2.2). In cellular systems, the consequences of OFR attack on sugars in DNA can be more clearly assessed using a technique for the end-point analysis of strand breaks, such as the fluorescence-activated DNA unwinding (FADU) technique or the comet assay (see Section 2.4).

2. *In vitro* analyses

3.1 Methods of carbohydrate damage by *in vitro* OFR generation

Several different systems can be used to generate OFRs *in vitro* that differ in their complexity, the purity of the species obtained, and their relevance to biological systems. Although the use of steady-state gamma radiolysis bears little resemblance to ongoing free radical events *in vivo*, it does allow the generation of pure solutions of radical species (*Protocol 1*).

However, in normoxic conditions the addition of an Fe^{2+} stress can alone catalyse the induction of free radical damage. The generation of $^{\bullet}OH$ via Fenton chemistry represents one of the most important mechanisms in biological systems. Denaturation of carbohydrates in this way is reported to have effects on both isolated sugars in addition to the deoxyribose molecule integrated into DNA (5, 6). The method (in *Protocol 1*) for iron-catalysed OFR generation is adapted from that described by Gutteridge (5).

Both techniques are described below, to allow the generation of positive control material for the subsequent assays, and where possible, results of their application are presented.

Protocol 1. Carbohydrate damage by Fe^{2+} catalysed production of OFRs

Equipment and reagents
- 0.1 M phosphate buffer pH 7.4 in 0.15 M NaCl (PBS)
- Deoxyribose
- $Fe(NH_4)_2(SO_4)_2$ in PBS
- Thermostatically controlled water bath set at 37°C (Fisons)

Method

1. Prepare a 0.5 ml solution of 5 mM carbohydrate in PBS.

2. Initiate the reaction at 37°C by the addition of Fe^{2+} to a final concentration of 2 mM.

3. Allow reaction to proceed for 15 min.

As with all free radical reactions, it is important to have allowed the reaction to go to completion or to be able to stop the reaction proceeding at a given time, to minimize experimental variation. Analysis of the damage induced can be made using one of the protocols outlined in Section 2.2.

2.1.1 Steady-state gamma radiolysis

Exposure of solutions to high-energy ionizing radiation, whether it be pulsed

187

or continuous from a ^{60}Co gamma source, ionizes the water molecules, as described by the following equation:

$$2H_2O \longrightarrow {}^{\bullet}OH + H_3O^+ + e_{aq}$$

with yields of $G({}^{\bullet}OH) = 2.8$, $G(H_3O^+) = 2.8$, $G(e_{aq}) = 0.6$, per 100 eV.

The reducing aqueous electron (e_{aq}) and oxidizing free radical formed will then go on to react rapidly with themselves or any other solutes present depending on their relative concentrations and rate constants. In the presence of oxygen, e_{aq} reacts rapidly to produce superoxide anion ($O_2^{\bullet-}$).

$$O_2 + e_{aq} \longrightarrow O_2^{\bullet-} \tag{1}$$

By the addition of various solutes to the irradiated solution, other defined radical species can be formed. In the presence of 100 mM formate, pure $O_2^{\bullet-}$ is produced in solution;

$$2H_2O \longrightarrow {}^{\bullet}OH + H_3O^+ + e_{aq} \tag{2}$$

$$e_{aq} + O_2 \longrightarrow O_2^{\bullet-} \tag{3}$$

$$H_3O^+ + O_2 \longrightarrow HO_2^{\bullet} + H_2O \tag{4}$$

$$HO_2^{\bullet} \longrightarrow O_2^{\bullet-} + H^+ \tag{5}$$

$${}^{\bullet}OH + HCO_2 \longrightarrow H_2O + CO_2^{\bullet-} \tag{6}$$

$$CO_2^{\bullet-} + O_2 \longrightarrow O_2^{\bullet-} + CO_2 \tag{7}$$

In the presence of a UV-absorbing aliphatic or aromatic compound (e.g. phenylalanine or thymine), ${}^{\bullet}OH$ and $O_2^{\bullet-}$ are scavenged to produce the corresponding peroxy radical (ROO${}^{\bullet}$)

$${}^{\bullet}OH + R \longrightarrow R({}^{\bullet}OH) \tag{8}$$

$$R({}^{\bullet}OH) \longrightarrow R(OH)O_2^{\bullet} \tag{9}$$

Protocol 2. Carbohydrate damage by OFR generation during radiolysis

Equipment and reagents

- Water, Millipore-filtered and deionized to ensure an iron-free solution
- Carbohydrate, e.g. deoxyribose
- ^{60}Cobalt γ source

- Potassium phosphate buffer, 40 mM, pH 7.4 (the presence of any higher salt concentration leads to some quenching of the radical species generated)

Method

1. Prepare a solution of simple or complexed carbohydrate to 5 mM in air-saturated solutions of 40 mM phosphate buffer, pH 7.4.

2. Expose to radiation doses of up to 2 kGy from a ^{60}Co source at a dose rate approximating 10 Gy/min in the presence or absence of 100 mM formate or 100 mM phenylalanine.

3. Agitate tubes every 20 min to ensure complete aeration.

4. Owing to the lability of some of the products formed, specimens should be analysed immediately or stored for up to a month at $-70°C$.

2.2 Analysis of OFR damage to monosaccharides

In ultrapure systems, in the absence of contaminating lipids or proteins, the TBA test can be applied to OFR-damaged carbohydrates. Carbohydrates are susceptible to the effects of radiation and it has been claimed that they produce 'malonaldehyde' (7); however, this may be formed from precursors during the heating stage in the TBA reaction. Thus, a more detailed analysis of OFR damage to monosaccharides may be achieved by using techniques such as HPLC. The TBA reaction has been extensively applied to the detection of lipid peroxidation. Lipid peroxides break down to form carbonyls, which form a characteristic chromogen with two molecules of TBA. In a similar way, it has been demonstrated that carbohydrates react in the presence of Fe(II) to form a TBA reactive compound (5).

Protocol 3. The thiobarbituric acid reaction

Equipment and reagents

- Thiobarbituric acid
- NaOH
- OFR-carbohydrate solution
- Glacial acetic acid
- Heating block or thermostatically controlled water bath set at 100°C
- Variable wavelength fluorimeter (Perkin Elmer)

Method

1. Terminate the free radical reaction by the addition of 1 ml TBA solution (1% w/v in 0.05 M NaOH).

2. Add 1 ml glacial acetic acid.

3. Heat for 30 min at 100°C and allow to cool.

4. Read the absorbance at 532 nm of samples at 25°C against paired blanks.

5. Alternatively, measure fluorescence scan-difference spectra and express as relative fluorescence units against a block standard containing 3 μM rhodamine B (8).

Table 1. Fe(II) damage to carbohydrates (adapted from ref. 10)

Carbohydrate	As A_{532} nm	As fluoresence $E_{x,532}, E_{m,553}$
Sucrose	0.066	46
D-Glucose	0.047	15
D-Ribose	0.083	22
D-Galactose	0.068	16
D-Glucuronic acid	0.170	91
Deoxyribose	1.270	1280

Monosaccharides (2.3 mM) were denatured with 0.9 mM ferrous ammonium sulphate at pH 7.4 for 15 min at 37°C. The subsequent TBA reactivity is reported as absorbance (A) at 532 nm and fluorescence, at
excitation wavelength (E_x) 532 nm and emission wavelength (E_m) 553 nm.

2.2.1 Application of the TBA test

By using the TBA test to analyse the effects of Fe(II)-induced damage to carbohydrates, it has been demonstrated that deoxyribose and glucuronic acid
produce TBA-reactive products (*Table 1*). However, several other carbohydrates were reported to give high blank values in the absence of Fe(II), possibly due to denaturation during the heating stage of the reaction.

When the iron concentration was elevated (100–200 mM), both galactose
and sucrose were also reported to give a weak TBA chromogen which was
not detectable at lower iron concentrations.

The method described below outlines the principle of neutral sugar analysis,
and should be calibrated for the native sugars to be investigated. It relies on
measuring the loss of a native sugar moiety rather than the appearance of a
novel product. This procedure can be applied to the study of OFR-denatured
sugars or more complex glycoproteins which have been denatured by OFRs.
To study the glycoforms on the latter, contaminating protein should be
removed before continuing with the standard HPLC procedure. The oligosaccharide moiety of a glycoprotein can be removed by reducing the pH to 5
and incubating with bacterial endoglycosidase H for 72 h at 37°C in the presence of streptomycin as a bacteriostat. Hydrazine may also be used to release
the carbohydrate moiety.

Protocol 4. Neutral sugar analysis by HPLC

Equipment and reagents

- Trichloroacetic acid (TCA)
- Boric acid
- Orcinol
- Strong anion exchange column, e.g. Aminex A-28 (Alltech)

- Concentrated H_2SO_4
- HPLC (Anachem)
- Column heater set to 100°C (Anachem)
- Absorbance detector set at 425 nm (Anachem).

Method

1. Precipitate all contaminating protein using 5% (w/v) TCA.
2. Prepare borate complexes of neutral sugars by incubation with boric acid.
3. Inject the sugars on to a column of strong ion-exchange resin (e.g. Aminex A-28) at a flow rate of 1 ml/min in 0.13 M boric acid.
4. Elute the sugar–borate complexes from the column using increasing ionic strength and pH borate buffers.
5. Visualize the derivatives by mixing eluent on flow with a stream of concentrated orcinol/H_2SO_4, heating at 100°C for 15 min and subsequently detecting the chromophore at 425 nm.

2.2.2 Application of neutral sugar analysis

This HPLC procedure has been applied in our laboratory to the study of reactive oxygen species (ROS) effects on the carbohydrate moiety of immunoglobulin G (IgG) (4). This study was undertaken to investigate whether there is a role for OFRs in the denaturation of IgG commonly observed in rheumatoid arthritis. All IgG molecules are glycoproteins carrying on average 2.8 N-linked oligosaccharides, two of which are via the conserved region Asn297 residue. The structure of the glycoform is shown in *Figure 1*.

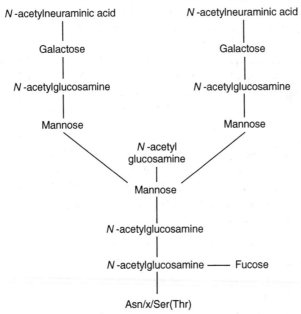

Figure 1. The IgG glycoform. A schematic representation of the carbohydrate residues present on the IgG molecule, linked through Asn297. As circulating molecules, these are rarely sialylated, and in rheumatoid arthritis, there is also reported to be a reduction in galactosylation, exposing *N*-acetylglucosamine as the new terminal residue.

Table 2. The effect of OFR on the composite neutral sugars in IgG

Radical	Dose (kGy)	% of total neutral sugars		
		Mannose	Galactose	Glucose
OH·	0	21	17	33
	0.5	21	18	32
	1	25	15	46
O$_2^-$	0	20	18	40
	1	18	19	41
	2	17	17	43
ROO·	0	18	17	33
	1	17	15	37
	2	22	10	31

IgG was denatured by defined radicals generated by steady-state radiolysis (see *Protocol 2*). Neutral sugars were released from the glycoprotein and analysed according to *Protocol 4*. The results are expressed as the means of duplicate analyses, where the coefficient of variation for the assay is 4%.

In rheumatoid arthritis, there appears to be a shift in the population of molecules towards those with a higher content of agalactosyl residues. Using neutral sugar analysis, we have demonstrated that both ·OH and ROO· radicals can cause a reduction in galactose (Gal) on IgG (see *Table 2*). This procedure can identify the loss of neutral sugars but cannot be used for derivatives such as *N*-acetylglucosamine (Glc-*N*Ac).

2.3 Complex sugar analysis using a lectin-based ELISA

In order to analyse amino sugars and other sugars commonly found in glyco-proteins, such as sugar acids, a second method was also devised using a lectin-based ELISA protocol. For a more detailed description of ELISAs and controls, see ref. 9.

Lectins are molecules that have a high specificity for defined carbohydrate residues. Some of the more important lectins that may be adopted for use in the study of glycoproteins are summarized in *Table 3*.

Table 3. Properties of lectins

Lectin	Mol. wt.	Sugar specificity
Concanavalin A	102	D-mannose, D-glucose
Erythina crystagalli	56.5	D-gal (1-4)-D-Glc-*N*Ac
Glycine max	110	D-*N*-acetylgalacturonic acid
Lathyrus odoratus	40–43	D-mannose
Limulus polyphemus	400	*N*-acetylneuraminic acid
Ricinus communis	120	D-gal

By conjugation to enzyme labels such as peroxidase, these lectins can be used in ELISA techniques to study exposed glycoforms. The ELISA protocol described below is specifically for the determination of the relative proportion of IgG molecules with Glc-NAc exposed as the new terminal residue with respect to Gal.

Protocol 5. Modified lectin ELISA

Equipment and reagents

- Glycoprotein e.g., lyophilized polyclonal IgG
- Peroxidase-conjugated *Erythina crystagalli* lectin, specific for D-galactose
- Peroxidase-conjugated *Triticum vulgaris* lectin, specific for N-acetylglucosamine
- Hydrogen peroxide, 30% solution
- o-Phenylenediamine (OPD) tablets
- Oxidized glutathione

- Carbonate buffer, pH 9.6
- Phosphate-buffered saline (see *Protocol 1*) containing 0.05% Tween 20 (PBST) as detergent
- 0.5% (w/v) sodium dodecyl sulphate in PBS (see *Protocol 1*)
- Concentrated sulphuric acid
- 96-well Nunc immuno-plates (Gibco)
- ELISA plate reader (Denley)

Method

1. Coat the immuno-plate in quadruplicate wells with glycoprotein at 50 μg/ml in carbonate buffer, pH 9.6, for 1 h at 37°C in a humidified atmosphere.

2. Wash three times in PBST before adding 0.5% (w/v) SDS to all wells to destroy secondary and tertiary conformation to allow access of lectins to internalized carbohydrate residues. Incubate for 5 min and wash as above.

3. Block any remaining sites exposed on immuno-plate with oxidized glutathione (2 mg/ml in PBS) to prevent non-specific binding of lectins to the plate. After 30 min at 37°C, wash three times in PBST.

4. To half of sample wells add peroxidase conjugated *T. vulgaris* lectin (0.1 μg/ml) and to the remaining wells add *E. crystagalli* lectin at 2 μg/ml. Incubate at room temperature for 1 h prior to washing as before.

5. Immediately before use, dissolve one OPD tablet in 20 ml of citrate phosphate buffer, pH 5.0, followed by 8 μl of H_2O_2. Add 50 μl of this substrate to all wells and allow colour to develop for 30 min at room temperature. Stop the reaction by addition of an equal volume of 4 M H_2SO_4 and measure the absorbance of the resultant colour at 492 nm in a microtire plate reader.

6. Express results as the ratio of Glc-NAc to Gal, i.e. A_{492nm} *Triticum vulgaris*:A_{492nm} *Erythina cystagalli*.

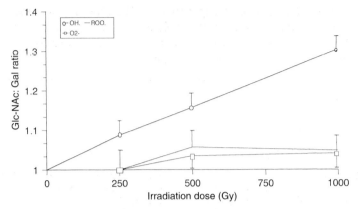

Figure 2. The effects of OFRs on the GlcNAc:Gal ratio of IgG glycoforms. The Glc-NAc:Gal ratio was measured by the binding of lectins from *Triticum vulgaris* and *Erythina crystagalli*, respectively in a modified (*Protocol 5*). Results are expressed as means and standard deviations of five separate experiments.

Using this technique, we have reported that following ˙OH attack on IgG, there is an increase in the number of terminally exposed Glc-NAc residues (4) relative to galactose (see *Figure 2*). Since the glycoform on IgG maintains its structural integrity, this may have strong implications for modifications in the antigenicity of IgG in RA.

Glycoproteins are a ubiquitous family of molecules which play important roles in cell–cell recognition. Damage to the carbohydrate moieties of such molecules may modify recognition of self in immune surveillance and lead to autoimmune phenomena. Such modifications can be studied using a lectin ELISA developed specifically for the terminal carbohydrate residues on these glycoproteins.

The majority of the techniques described above have been developed for *in vitro* analysis. However, the next method may be applied equally well *in vitro* or *ex vivo*.

2.4 Deoxyribose degradation by OFRs *in situ*

As already described, OFRs can destroy deoxyribose; within DNA this is manifested as the formation of strand breaks. The fluorescence-activated DNA unwinding (FADU) technique (10) can be used indirectly to study the effects of OFRs on deoxyribose in DNA. This technique is quantitative but complex; a recent development has been in the comet assay, for the detection of DNA strand breaks in whole cells (11). This is a simple, sensitive, visual assay for the rapid measurement of DNA strand breaks and has been usefully applied to the study of lymphocyte suspensions to examine their susceptibility to oxidative damage, UV, and ionizing radiation (12). Using the comet assay, normal cell populations can be exposed to OFR-generating systems, and subsequent strand breakage due to deoxyribose degradation can be measured. This technique can be equally well applied to *ex vivo* samples, enabling an estimation of the *in vivo* damage to the sugar moiety to be made.

Protocol 6. The comet assay

Equipment and reagents

- Single cell suspension
- Normal melting agarose (NMA)
- PBS (*Protocol 1*)
- Lysing solution (2.5 M NaCl, 100 mM Na$_2$EDTA, 10 mM Tris, pH 10, 1% (v/v), Triton X-100, 10% (v/v) dimethyl sulphoxide)
- Electrophoresis buffer (1 mM EDTA, 300 mM NaOH)
- 0.4 M Tris pH 7.5
- Ethidium bromide
- Flat bed electrophoresis system (Bio-Rad)
- Fluorescence microscope
- Callipers (Fisons)
- Dakin fully frosted microscope slides and no. 1 coverslips

Method

Note: All steps should be performed in a dark room under red light to prevent further damage to DNA

1. Layer microscope slides with 100 µl of 0.5% (w/v) NMA in PBS at 45°C, cover immediately with a no. 1 coverslip, and keep at 4°C for 10 min to allow agar to solidify.

2. Once set, gently remove coverslip and overlay 30 000–200 000 cells in 10 µl PBS mixed with 75 µl of 0.05% NMA (w/v) at 37°C. Spread using a coverslip and leave at 4°C for 10 min.

3. Remove coverslip and spread a thin layer of 0.5% NMA (w/v) as before.

4. Once solidified, remove the coverslips and immerse the slides in freshly prepared lysing solution for a minimum of 1 h at 4°C.

5. Allow lysing solution to drain away and place in a horizontal electrophoresis system with the agarose end facing the anode.

6. Fill the tank with electrophoresis solution sufficient to submerge the slides in 0.25 cm buffer.

7. Allow slides to incubate in the high pH buffer for 20 min to allow the DNA to unwind prior to electrophoresis.

8. Electrophorese samples for 20 min at 25 V, adjusted to 300 mA by altering the volume of buffer.

9. After electrophoresis, wash the slides three times in 0.4 M Tris pH 7.5 to remove alkali and detergents.

10. Stain the slides with 50–100 µl ethidium bromide in distilled water, cover, and analyse immediately.

Protocol 6. *Continued*

11. Visualize the stained nuclei using a fluorescence microscope equipped with an excitation filter of 515 560 nm from a 100 W mercury lamp and a barrier filter of 590 nm.

12. DNA migration can be determined using callipers on the negative photomicrograph, by measuring the comet length and head diameter in randomly selected cells.

3. *Ex vivo* techniques

In certain pathological conditions where a causal link exists (or is postulated) between OFR generation and disease, the measurement of OFR denaturation of biomolecules can further the understanding of both the disease process and its activity.

3.1 OFR effects on mucopolysaccharides

The mucopolysaccharides are a family of molecules which are characterized by long polymers of repeating units of acidic monosaccharides. The most abundant of these is hyaluronic acid which is present in the extracellular matrix of connective tissues. It has a repeating dimeric structure of glucuronic acid and *N*-acetylglucosamine. Hyaluronic acid is the major constituent of synovial fluid conferring the characteristic viscosity to the fluid, and was the first biological macromolecule reported to be susceptible to OFR degradation. Using an enzymic $O_2^{\bullet-}$ generating system, McCord demonstrated a decrease in the viscosity of synovial fluid which was inhibitable by superoxide dismutase (13). Halliwell furthered this work, demonstrating that the loss in viscosity was a metal ion dependent process (14), thereby implicating the Fenton reaction as an important mechanism in mucopolysaccharide degradation.

3.2 Hyaluronic acid degradation by OFRs

During inflammation in rheumatoid arthritis, the characteristic viscosity of synovial fluid is lost, probably due to the generation of OFRs *in situ*. The likely sources of OFRs in the joint are from the xanthine/xanthine oxidase system and activated polymorphs. Since these latter produce a plethora of degradative molecules, the generation of $O_2^{\bullet-}$ by xanthine/xanthine oxidase (see eqn 10 below) has been adopted as the model system in the following protocol, taken from ref. 15.

$$\text{hypoxanthine} + O_2 \longrightarrow \text{xanthine} + O_2^{\bullet-} \tag{10}$$

Protocol 7. OFR generation by xanthine/xanthine oxidase

Equipment and reagents

- Hanks' balanced salt solution (HBSS)
- EDTA
- Hepes
- $FeCl_3$

- Buttermilk xanthine oxidase
- Synovial fluid sample
- Hyaluronic acid
- Hypoxanthine

Method

1. Prepare a working buffer of $1 \times$ HBSS including 0.2 mM EDTA and 10 mM Hepes.

2. Prepare hyaluronic acid solution by allowing lyophilized powder to dissolve overnight at 4°C to a final concentration of 1.2 mg/ml in working buffer.

3. To 1.4 ml stock hyaluronic acid, add 0.2 ml hypoxanthine (5 mM in buffer), 10 μl $FeCl_3$ (4 mM in buffer), and 370 μl buffer.

4. Initiate the reaction by adding 20 μl of xanthine oxidase (1 unit per ml in buffer).

The change in viscosity induced by sugar degradation in hyaluronic acid can be measured by capillary viscometry as described in ref. 15 or, more simply, by calculating the gravitational flow rate of hyaluronic acid through a vertically mounted syringe.

Protocol 8. OFR degradation of hyaluronic acid

Equipment and reagents

- Hyaluronic acid or fluid under test
- 2 ml syringe

Method

1. Fill a 2 ml syringe, with the hyaluronic acid under test or synovial fluid.

2. At time zero, remove the syringe cap and add xanthine oxidase (as described in *Protocol 7*) to *in vitro* experiments.

3. Count the drip rate or the volume of hyaluronic acid passed through the syringe every 5 min. The greater the degradation, the lower the viscosity of hyaluronic acid and the faster the drip rate.

The use of viscometry for studying carbohydrates is clearly limited. However, the technique is easily adopted and thus analysis of hyaluronic acid degradation by novel radical generating systems may provide important information relating to OFR reactivity with carbohydrates.

There is now evidence from several different analytical procedures, that OFRs can denature carbohydrates in biological systems. Thus, these techniques may also become important in the development of new drugs aimed at minimizing the effects of OFRs on carbohydrate residues.

4. Conclusions

In common with the effects of OFRs on nucleotide bases, amino acids, and polyunsaturated fatty acids, the majority of simple analyses for OFR-induced damage of carbohydrates examines the loss of native molecules. In the last decade there has been a great deal of interest in the identification of OFR-modified DNA bases. However, relatively few studies on amino acid derivatives are reported and studies of carbohydrate products are scarce. Further characterization of these products formed from OFR attack must be achieved using pulse radiolysis, GC–MS, and NMR. Once these neo-carbohydrate derivatives are identified and their biological importance assessed, the production of specific antibodies can begin. Owing to the diversity of applications for antibodies in detection of small quantities and of novel molecules, this approach may be the way forward in establishing the frequency of OFR modifications to carbohydrates and the biological and pathological relevance of such changes.

Acknowledgement

The authors gratefully acknowledge financial support from the Arthritis and Rheumatism Council for some of the work described in this chapter.

References

1. Anbar, M. and Neta, P. (1967). *Int. J. Appl. Radiat. Isotopes*, **18**, 493.
2. Aruoma, O. A., Grootveld, M., and Halliwell, B. (1987). *J. Inorg. Biochem.*, **29**, 289.
3. Von Sonntag, C. (1980). *Adv. Carbohydr. Chem. Biochem.*, **37**, 1.
4. Griffiths, H. R. and Lunec, J. (1989). *FEBS Lett.*, **245**, 95.
5. Gutteridge, J. M. C. (1981). *FEBS Lett.*, **128**, 343.
6. Halliwell, B. and Gutteridge, J. M. C. (1981). *FEBS Lett.*, **128**, 347.
7. Gutteridge, J. M. C. (1978). *Anal. Biochem.*, **91**, 250.
8. Scherz, H. (1968). *Experientia*, **24**, 420.
9. Bampton, J. L. M., Cawston, T. E., Kyte, M. V., and Hazelman, B. L. (1985). *Ann. Rheum. Dis.*, **44**, 13.
10. Birnboim, M. C. and Jevcak, J. J. (1981). *Cancer Res.*, **41**, 1889.
11. Singh, N. P., McCoy, M. T., Tice, R. R., and Scheider, E. L. (1988). *Exp. Cell Res.*, **175**, 184.

12. Green, M. H. L., Lowe, J. E., Harcourt, S. A., Akinluyi, P., Rowe, T., Cole, J., Anstey, A. V., and Arlett, C. F. (1992). *Mutat. Res.*, **273**, 137.
13. McCord, J. M. (1974). *Science*, **189**, 529.
14. Halliwell, B. (1978). *FEBS Lett.*, **96**, 238.
15. Greenwald, R. A. and Moak, S. A. (1986). In *CRC handbook of methods for oxygen radical research* (ed. R. A. Greenwald), p. 399. CRC Press, Boca Raton, FL.

Further reading

Bampton, J. L. M., Cawston, T. E., Kyte, M. V., and Hazelman, B. L. (1985). *Ann. Rheum. Dis.*, **44**, 13.
Rademacher, T. W., Parekh, R. B., and Dwek, R. A. (1988). *Annu. Rev. Biochem.*, **57**, 785.

14

Measurement of products of free radical attack on nucleic acids

M. LINDSAY MAIDT and ROBERT A. FLOYD

1. Introduction

High performance liquid chromatography (HPLC) methods with electro-chemical methods able to measure oxygen free radical adducts to DNA became available in the 1980s (1, 2). Due to their low presence in nature, where they are found at about one adduct per 100 000 bases, they required the sensitivity afforded by HPLC coupled with electrochemical detection in order to detect and quantify them.

The most widely measured DNA oxidation product is 8-hydroxy-2'-deoxyguanosine (8-OHdG) and its RNA analogue, 8-hydroxyguanosine (8-OHG). These species can be formed through Fenton type chemistry or singlet oxygen addition (*Figure 1*). More recently, 5-hydroxy-2'-deoxycytidine (5-OHdC) has been measured as a useful biomarker (*Figure 2*).

1.1 Biological consequences

Through reverse chemical mutagenesis, the biological consequences of these altered DNA nucleotides has been worked out. Normal DNA base pairing is guanine with cytosine (G ⟶ C) and adenine with thiamine and (A ⟶ T). It has been found that 8-OHdG does not code for a C, but instead codes for

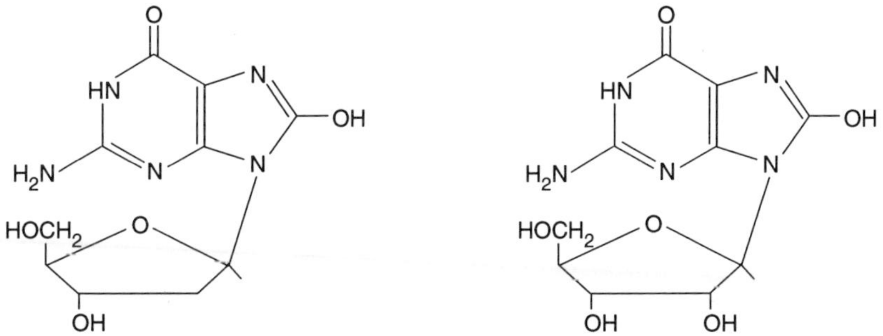

Figure 1. 8-Hydroxy-2'-deoxyguanosine (left) and 8-hydroxyguanosine (right).

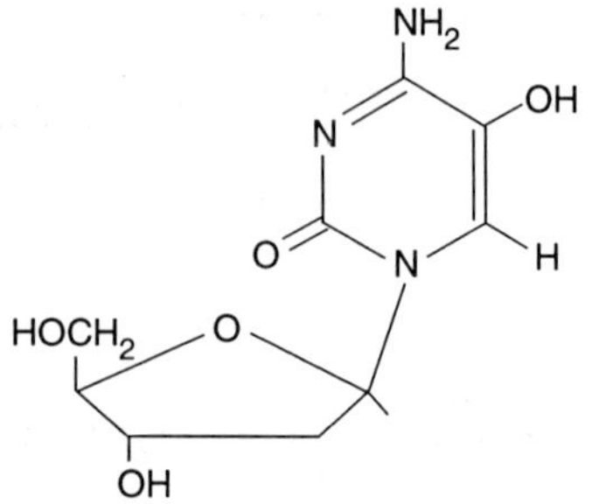

Figure 2. 5-Hydroxy-2′-deoxycytidine.

an A in the complementary strand of DNA. Upon subsequent mitotic events, this A in the complementary strand opposite the original 8-OHdG, now codes for a T. Thus one has a G ⟶ T transversion in which there is a T in the DNA where originally a G was present. Reverse chemical mutagenesis techniques show this transversion to occur about 0.6% of the time when 8-OHdG is present in DNA (3).

A similar C ⟶ T transition in which a purine replaces a purine can occur in the DNA as a result of 5-OHdC. This mutation occurs at a frequency of 2.5% (3).

Given the recent evidence of DNA transitions and transversions associated with a wide variety of cancers, the detection of these DNA adducts through the use of HPLC with electrochemical detection can give the investigator a powerful tool in understanding the aetiological processes of these diseases.

1.2 HPLC detection of DNA/RNA products

The minimum necessary equipment needed to detect oxygen free radical-damaged DNA products are shown in *Figure 3*. This includes an HPLC pump and injector, UV detector, electrochemical detector, an appropriate column for separation, and an output recording device. The nucleosides of DNA and RNA hydrolysates can be detected at 254 nm.

At the very heart of being able to quantify 8-OHdG adducts in DNA is a thorough command of electrochemical detection. Due to the extreme sensitivity of the process, many things can either help or hinder the system. Many hours of trouble-free use are possible at a detection limit of 10^{-15} M or better with proper preparation and care of electrochemical equipment.

Figure 4 is a voltammogram showing the relationship between detector response and working electrode potential. At an applied voltage of E_1, none of the analyte is oxidized and no response is seen. At E_2, the limiting current has been reached and an increase in response is not possible. At $E_{1/2}$, half the molecules have been oxidized and a response equal to half the maximum is seen.

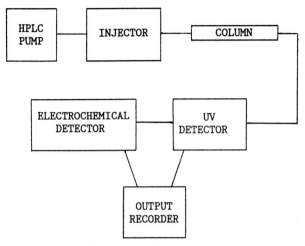

Figure 3. HPLC schematic.

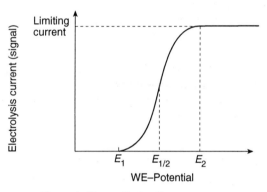

Figure 4. Theoretical voltammogram.

Figure 5 is a voltammogram for 8-OHdG in the mobile phase described. Each electrochemically active compound has a unique voltammogram under given chromatographic conditions.

Voltammograms are also affected by the pH of the mobile phase. Generally, the lower the pH, the more difficult it is to oxidize a substance, and the limits of sensitivity are not as well resolved.

A potential of 0.60 V in the oxidative mode under the conditions described is the minimum proper applied voltage for 8-OHdG/8-OHG detection.

The working electrode surface can be replaced or repolished repeatedly in order to increase the response and decrease the signal to noise ratio associated with baseline drift. Optimum electrochemical performance in practice is often a day to day trade-off involving the condition of the working electrode surface, signal to noise ratio, and selecting the proper amplification.

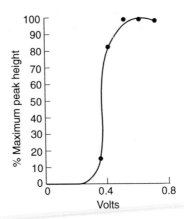

Figure 5. 8-OHdG voltammogram.

The electrochemical detector should never be turned off after achieving baseline stability. When not in use, decrease the pump rate to 0.1 ml/min. The constant flow of mobile phase through the cell will help to prevent microbial growth and maintain a good baseline.

If baseline stability cannot be achieved, the instrument cannot be used. Many factors can contribute to this dilemma. The first place to investigate problems is in the mobile phase. Microscopic contamination, barely visible to the naked eye, is the cause of 95% of all electrochemical problems. The solution is to replace the mobile phase.

Temperature variations in the laboratory can cause pressure fluctuations which show as an unstable baseline which changes with temperature. Air blowing from a ventilation outlet on to the equipment can sometimes cause problems.

Operators which have a build up of static charge can influence the electric field near the cell and cause perturbations. It is a good idea to discharge any static electricity build up before entering the laboratory.

Computer-controlled variable speed motors of central air handling units have been known to emanate frequencies affecting operation. If trouble persists, one might consider isolating the instrument in a Faraday cage.

As shown in *Figure 3*, the necessary equipment needed can be very basic and still work effectively. UV detectors that are variable in wavelength equipped with diode array abilities are available, but not necessary. Electrochemical detectors can be very simple to very complicated. Limits of resolution can often depend more on the operator than the equipment. With a knowledge of electronics, it is also possible to build your own electrochemical detector and save substantial money. A simple isocratic pump is best for electrochemical evaluations, as the introduction of gradients can influence electrochemical behaviour, leading to a changing baseline and the possibility of unreproducible results. Automated sampling equipment is also available.

The data output gives the user the most flexibility. A simple two-channel chart recorder is adequate. Computers that interface into the entire system are available, but are not really necessary.

2. DNA isolation method

Protocol 1 lists a proven technique for recovery of DNA by itself or DNA with RNA from cells or tissue.

One should be aware of the fact that the DNA repair process is an ongoing phenomenon. To obtain accurate results when working with animals the time interval between sacrifice and the start of the DNA isolation should be kept to a minimum and tissues should be kept on ice.

Protocol 1. Extraction of DNA and RNA

Reagents and equipment

- Homogenizing buffer: 0.3 M sucrose, 0.025 M Tris, 0.002 M EDTA, final pH 7.3
- RES (RNA extraction solution): 1.0 M LiCl, 2 M urea, 0.04 M sodium citrate, 0.005 M di-sodium EDTA, 2% sodium dodecyl sulphate (SDS). Final pH 6.8. Filter the solution through a 0.45 µm membrane before SDS addition and store at room temperature.
- DNase (Calbiochem, cat. no. 26095)
- RNase (Boehringer Mannheim, cat. no. 1119915)
- Proteinase K (Boehringer Mannheim, cat. no. 1373196)
- Tris/EDTA solution (TE): 0.010 M Tris, 0.001 M EDTA, final pH 7.4
- Chloroform (24 vol.)/isoamyl alcohol (1 vol.) (CIA)

Method

1. Homogenize 1 g of the tissue in 4 ml of homogenizing buffer for 10–15 sec.

2. Add an equal volume of RES.

3. For RNA-free DNA, add RNase to 100 µg/ml and incubate for 30 min at 50°C. Heat RNase for 10 min at 70°C first to inactivate DNase.

4. Add proteinase K to 100–300 µg/ml and place at 50°C for 30–120 min.

5. Extract the nucleic acids with three successive changes of CIA. Mix the aqueous/CIA phases for 15 min between changes. Spin for 5 min at 2000–3000 g to separate the phases.

6. Remove the final aqueous extraction and add 1/15 vol. of 3 M sodium acetate and 2–2.5 vol. of 95% ethanol. The pH of the sodium acetate is 7.0.

7. Place in the refrigerator for 30–60 min.

8. Spin out the precipitated nucleic acid in a microcentrifuge tube at 4000 g or at 3000 g in a floor model centrifuge.

Protocol 1. *Continued*

9. Air dry the sample then solubilize in TE if it is in the digestion tube. If a transfer is necessary, air dry the pellet and solubilize in 0.5 ml distilled water and precipitate with sodium acetate and ethanol.

10. Resuspend the final pellet after drying in 0.250 ml TE.

11. Quantify the DNA obtained spectrophotometrically with two readings at 260 and 280 nm. One absorbance unit corresponds to 50 µg/ml at 260 nm. As the ratio of absorbance at 260 and 280 nm approaches 1.8, the preparation is pure and contains no protein.

3. DNA digestion

Various methods of DNA hydrolysis are described in the literature. Some of these methods are flawed because the enzymes used contain trace metals which can catalyse the autoxidation of DNA, and therefore produce the product being measured. For this reason, protocols using DNase I should be avoided. The preferred method of DNA (or RNA) hydrolysis is the use of nuclease P_1 and alkaline phosphatase (*Protocol 2*) (4). It is very important that the DNA is in solution before attempting to digest it. Sometimes DNA can be very slow in solubilizing. Gentle warming of samples can facilitate solubilization.

Samples can be digested in clean glass tubes or plastic microcentrifuge tubes. If plastic tubes are used, they should be washed first to remove any releasing agents left during the manufacturing process which may inhibit enzymes. An ethanol wash followed by repeated water rinsing accomplishes this.

Protocol 2. DNA digestion

Reagents and equipment

- TE (see *Protocol 1*)
- 0.5 M sodium acetate, pH 5.1
- 1.0 M magnesium chloride
- Nuclease P_1 (Sigma, cat. no. N8630)
- 1.0 M Tris base
- Alkaline phosphatase: 1 unit/µl (Boehringer Mannheim, cat. no. 713023)
- 5.8 M acetic acid
- 0.2 µm HPLC filter (Gelman Acrodisk 4450)

Methods

1. Solubilize the DNA in 0.25 ml TE.

2. Add 0.025 ml 0.5 M sodium acetate, pH 5.1.

3. Add 0.00275 ml 1 M magnesium chloride.

4. Heat the samples at 100°C in water for 5 min in order to make the DNA single stranded.

5. Immediately cool the samples on ice for 5 min.

6. Add 10 μg of nuclease P_1. A stock solution of 1 mg/ml in water can be used and kept under refrigeration.

7. Incubate samples for 1 h at 37°C.

8. Adjust the pH to 7.8 with 0.008 ml 1 M Tris base.

9. Add 0.002 ml alkaline phosphatase, which is equivalent to 2 units.

10. Incubate samples for 1 h at 37°C.

11. Precipitate the enzymes by the addition of 0.004 ml 5.8 M acetic acid and filter the samples through a 0.2 μm HPLC filter. Samples are now ready for injection on to the HPLC.

4. HPLC analysis

4.1 HPLC

Since all quantitative aspects of DNA free radical damage are in reference to a known standard, it is important that the standard be accurate and kept stable at a pH of below 7. A convenient working standard of both deoxyguanosine and 8-OHdG can be made in the ratio of 500 to 1. A 250 μM solution of deoxyguanosine, made as analytically as possible, can be combined with 8-OHdG to a final concentration of 0.5 μM. A 10 ml sample of this standard can be divided into 12 portions and kept frozen. Bacterial contamination of the standard is possible and can lead to errors. By freezing the standard when not in use, this can be avoided.

8-OHdG can be purchased commercially. A molar extinction coefficient of 12 300 at 245 nm can be used to determine the concentration of both 8-OHdG and 8-OHG with a spectrophotometer.

Protocol 3. Analysis

Reagents and equipment

- HPLC pump
- HPLC injector and compatible syringe
- Column (Rainin, Microsorb 86–200-C5)
- 6× stock mobile phase (2 litres) containing: 10.6 g citric acid monohydrate, 8.3 g sodium acetate monohydrate, 4.8 g sodium hydroxide, 2.4 ml glacial acetic acid

- UV detector
- Electrochemical detector
- Output recording device
- 0.250 M 2'-deoxyguanosine (Sigma, cat. no. D7145)
- 0.5 μM 8-hydroxy-2'-deoxyguanosine (Cayman Chemical, cat. no. 89320)

Methods

1. Dilute mobile phase 5:1.

2. Add methanol to 5–10%. More methanol gives a shorter retention time.

Protocol 3. *Continued*

3. Filter the diluted mobile phase through a 0.45 μm filter and degas under vacuum.

4. Stabilize instrument at a flow rate of 1.0 ml/min.

5. Inject prepared samples.

4.2 Analysis of results

The 8-OHdG content of the DNA can be expressed as the number present per 100 000 normal deoxyguanosines. By comparing the peak height of a known standard, a number can be applied to the DNA hydrolysate and a final value can be obtained. *Table 1* shows data generated from an injection volume of 30 μl of standards from *Protocol 3* together with data from a hypothetical sample injection.

The final 8-OHdG value per 100 000 dG in a 50 μl injection of hydrolysed DNA can be calculated as follows using the data in *Table 1*.

$$8\text{-OHdG} = \frac{15 \text{ pmol}}{15.50 \text{ cm}} \times \frac{1.22 \text{ cm}}{50 \text{ μl}} = 0.0236 \text{ pmol/μl}$$

$$dg = \frac{7500 \text{ pmol}}{7.82 \text{ cm}} \times \frac{11.45 \text{ cm}}{50 \text{ μl}} = 219.63 \text{ pmol/μl}$$

$$\frac{0.0236 \text{ pmol/μl}}{219.63 \text{ pmol/μl}} \times 100\,000 = 10.75 \text{ 8-OHdG/100 000 dG}$$

5. Application

5.1 8-OHdG detection in human urine

Since 8-OHdG is now recognized as a biomarker for oxidative stress, the human output might be a relevant model for predicting the risk of cancer or rate of ageing. A modified HPLC apparatus with three columns and two pumps has been developed which allows the quantification of 8-OHdG in human urine with a minimum of sample preparation (5). The normal range of

Table 1. Data calculated using prepared standards of *Protocol 3* and hypothetical samples

	Peak height (cm)
15 pmol 8-OHdG standard	15.50
7500 pmol dG standard	7.82
8-OHdG in sample	1.22
dG in sample	11.45

8-OHdG production in humans is between 80 and 500 pmol/kg 24 h, with factors such as smoking, or lean body mass causing enhanced production.

5.2 Future developments

Advances in electrochemical detection technology will soon offer the researcher expanded ability to detect methylated DNA adducts such as 1-methylguanine, 7-methylguanine, and O^6-methylguanine. The ability to detect the human status of oxidized protein is also in development.

As medical professionals come to realize that the human body's oxidative stress may be a key parameter, these new developments coupled with existing technology offer the researcher an exciting area. The ability to measure a human's oxidative stress on an 'oxidative stress meter' might prove relevant in determining whether oxygen free radicals play an important role in the aetiology or risk of disease.

Acknowledgements

This research was supported in part by National Institutes of Health Grants CA 42854, NS 23307, and AG 09690.

References

1. Floyd, R. A., Watson, J. J., Wong, P. K., Altmiller, D. H., and Rickard, R. C. (1986). *Free Radic. Res. Commun.*, **1**, 163.
2. Floyd, R. A. (1981). *Biochem. Biophys. Res. Commun.*, **91**, 279.
3. Feig, D. I., Sowers, L. C., and Loeb, L. A. (1994). *Proc. Natl Acad. Sci. USA*, **91**, 6609.
4. Kasai, K., Nishimura, S., Kurokawa, Y., and Hayashi, Y. (1987). *Carcinogen*, **8**, 1959.
5. Loft, S., Vistisen, K., Ewertz, M., Tjonneland, A., and Powlser, H. E. (1992). *Carcinogen*, **13**, 2241.

Part IV

Measurement of antioxidants

15

Glutathione

MARY E. ANDERSON

1. Introduction

Glutathione (GSH; L-γ-glutamyl-L-cysteinylglycine) is present in most mammalian cells. It is usually the most abundant cellular thiol (up to 10 mM); cysteine is usually present in lower levels (about 10% of GSH). GSH is an antioxidant, and it participates in many biological phenomena. It is involved in the transport of certain amino acids, it is a coenzyme for various enzymes, and it protects against oxygen radicals and toxic compounds (1–6).

GSH has two interesting structural features: the γ-glutamyl peptide bond between glutamate and cysteine and the thiol moiety of cysteine. The γ-glutamyl peptide bond of GSH is not cleaved by the usual peptidases; it is only cleaved by γ-glutamyl transpeptidase. The thiol moiety is key to the antioxidant properties of GSH. Although GSH oxidizes to glutathione disulphide (GSSG), it is less easily oxidized than is cysteine or γ-glutamylcysteine. In solution, at neutral pH or higher, thiols like cysteine and glutathione are rapidly oxidized, especially if there are any metal ions such as iron or copper present. Cellular glutathione is normally present as the thiol (or reduced form); only a small (usually less than 5%) amount is present as GSSG or mixed disulphides. Larger amounts of disulphides are found after a severe oxidative stress, such as after treatment of cells with *tert*-butyl hydroperoxide or diamide, or in extracellular fluids, such as plasma. Intracellularly, GSH is maintained in the reduced form by GSSG reductase in an NADPH-dependent reaction.

GSH is synthesized intracellularly by the consecutive actions of γ-glutamylcysteine synthetase (eqn 1) and glutathione synthetase (eqn 2):

$$\text{L-glutamate} + \text{L-cysteine} + \text{ATP} \rightarrow \text{L-}\gamma\text{-glutamyl-L-cysteine} + \text{ADP} + P_i \quad (1)$$

$$\text{L-}\gamma\text{-glutamyl-L-cysteine} + \text{glycine} + \text{ATP} \rightarrow \text{GSH} + \text{ADP} + P_i \quad (2)$$

The only enzyme that degrades the γ-glutamyl peptide bond of GSH is γ-glutamyl transpeptidase.

$$\text{GSH} + \text{E} \Leftrightarrow \gamma\text{-glu-E} + H_2O \Leftrightarrow \text{glutamate} + \text{cysteinylglycine} \quad (3)$$

One of the products, cysteinylglycine, rapidly oxidizes and may catalyse the oxidation of GSH. GSSG is also a substrate of transpeptidase. It is a membrane-bound enzyme found on the external surface, usually brush border, of many cells and tissues, for example, kidney, pancreas, ciliary body, choroid plexus, intestinal epithelia, bile ductule cells, lymphoid cells, and many tumour cells. γ-Glutamyl transpeptidase is often found in plasma, especially from humans, and it is assayed in clinical chemistry laboratories as part of a liver function series. γ-Glutamyl transpeptidase plays a role in the recovery of cysteine moieties via the salvage pathway of GSH biosynthesis (1, 5–7).

2. Sample preparation

The determination of glutathione levels of biological samples requires that oxidation be minimized and that γ-glutamyl transpeptidase be inhibited. In biological samples, γ-glutamyl transpeptidase, as well as other GSH-utilizing enzymes, is usually inhibited by acidification. Additionally an inhibitor, such as acivicin or serine plus borate (at least 20 mM final concentration) may be used. Oxidation of thiols is limited by the addition of a chelating agent, such as EDTA (ethylenediamine tetra-acetic acid) or DPTA (diethylenetriaminepenta-acetic acid), and by rapid acidification. Rapid acidification also eliminates the possibility of cellular reduction of GSSG by GSSG reductase.

The choice of protein precipitant depends upon the subsequent assay method. Various methods, such as treatment of cells or tissues with perchloric acid, picric acid, *m*-phosphoric acid, 5-sulphosalicylic acid (SSA), hot ethanol, or methanesulphonic acid, have been used. Certain of these protein precipitants do not maintain the thiol–disulphide ratio. (For a more detailed discussion of protein precipitants see references 8–11.) 5-Sulphosalicylic acid is preferred for the GSH assay methods described below.

Protocol 1. Sample preparation

Reagents and equipment

- 10% SSA solution: 10% SSA (w/v) containing 0.5 mM EDTA or DPTA, prepared by dilution from a 100% (w/v) stock solution
- 5% SSA (prepared by dilution of the 10% stock solution)
- 3.33% SSA (prepared by dilution of the 10% stock solution)
- Tissue homogenizer
- Homogenizing tube (kept on ice)
- Labelled microcentrifuge tubes
- Microcentrifuge
- Balance (preferably the type that can be tared rapidly)
- Weighing paper
- 125 ml beaker with water
- Scissors
- Forceps
- Paper towels
- Dry ice
- Labelled storage containers (optional)

Method

1. Rapidly kill and exsanguinate the rodent to remove excess blood from the tissues (whole blood has about 1 mM GSH).

214

2. Rapidly remove the tissues, beginning with those containing trans-peptidase such as kidneys, pancreas, and intestine.

3. Remove excess fat from the tissues.

4. Rinse tissues in a beaker of water, lightly blot dry, and weigh.

5. Place tissue in a cold homogenizing tube and add 5% SSA solution (5 vol. per g of tissue).

6. Homogenize using a constant number of up and down strokes of the pestle (usually about 25).

7. Tissues with low transpeptidase, such as liver, heart, or muscle, can be washed and frozen in a storage container, placed first on dry ice, then at $-80\,°C$, then defrosted by swirling in water and prepared as in steps 3–6. Tissues with significant transpeptidase cannot be frozen as the activation of the enzyme during defrosting may lead to low GSH levels or high cysteine or cystine levels. Brain may be prepared immediately or the whole head frozen in dry ice/ethanol and the whole brain removed while still frozen.

8. Pour the homogenate into a labelled microcentrifuge tube and centrifuge (at least 10 000 g) for 5 min.

9. Use the supernatant for GSH determination.

Plasma GSH is notoriously difficult to determine accurately because of the rapid oxidation to GSSG and to mixed disulphides with cysteine or plasma proteins. Samples must be prepared within about 3–4 min of obtaining blood because GSH levels drop rapidly (12).

Protocol 2. Preparation of plasma samples

Reagents and equipment

- 10% SSA with chelator (see *Protocol 1*)
- Microcentrifuge tubes, 0.5 ml; tall and thin, and made of hard plastic; 1.5 ml
- Pasteur pipettes and rubber bulb
- Vortex mixer
- Razor blade (single-sided)

Method

1. Obtain blood by drawing it into an EDTA tube, a syringe, or a beaker containing EDTA (0.5 M, pH 7.00: 50 μl for mice; 100 μl for larger rodents) with gentle swirling.

2. Transfer blood carefully (without air bubbles) with a Pasteur pipette to within, at most, 0.5 cm of the top of a 0.5 ml microcentrifuge tube.

3. Centrifuge at about 10 000 g for 1.5 min.

4. Rapidly slice off the top of the tube with a razor blade.

Protocol 2. *Continued*

5. Add 100 μl of plasma to a fresh microcentrifuge tube containing 50 μl of 100% SSA solution.

6. Mix well using a vortex mixer.

7. Store the sample on ice and then centrifuge for at least 5 min.

Attached cells can either be trypsinized and then prepared or prepared *in situ*; the latter method is preferred. Cells in suspension are washed and lysed in acid.

Protocol 3. Preparation of samples from cultured cells

Reagents

- Chilled phosphate-buffered saline (PBS)
- Chilled 3.33% SSA solution (see *Protocol 1*)
- Rubber 'policemen'
- Pasteur pipettes and bulb
- Labelled microcentrifuge tubes

Method

A. *Attached cultured cells*

1. Quickly wash attached cultured cells twice with cold PBS; take care to remove all liquid on the final wash.

2. Add SSA to the culture dish and swirl until all the surface is coated. The amount of SSA depends on the size of the culture dish: a T-25 requires 1 ml, a T-75 needs 2 ml.

3. Stand the flask on ice for at least 5 min with occasional swirling (the cells look somewhat milky).

4. Scrape the cells well with a rubber 'policeman'. (The 'policeman' can be re-used by washing between samples.)

5. Transfer the particulate lysate to a labelled microcentrifuge tube with a Pasteur pipette and centrifuge as in *Protocol 1*.

B. *Cells in suspension or trypsinized cells*

1. Wash the cells at least once with cold PBS and remove the super-natant. It is important to work quickly because GSH is exported from all cells and trypsinized cells may be more leaky.

2. Add SSA solution to the cells with vortexing. The amount of SSA solution varies with the amount of cells, but typically 0.5 ml is adequate for $1-5 \times 10^6$ cells.

3. Freeze–thaw three times.

4. Centrifuge the cell lysate as in *Protocol 1*.

Other biological samples such as bile, lung, bacteria, or red blood cells need more specialized preparation (8–10, 13–15).

3. Assays for the determination of GSH

There are numerous procedures for the determination of glutathione. Simple total thiol assays using DTNB (5,5′-dithiobis-(2-nitrobenzoic acid)); Ellman's reagent (16) have been used; however, this method is non-specific and can yield misleading results, especially when GSH levels have been depleted. Some assays are specific for the thiol form of GSH, others yield information about 'total' GSH (GSH + 0.5 GSSG). 'Total' GSH is not accurate, because various forms of GSH are not measured, such as GSH S-conjugate (see ref. 9). There are a variety of enzymatic methods for the determination of GSH. Glyoxylase, GSH S-transferase, formaldehyde dehydrogenase, and maleyl acetoacetate isomerase have been used to measure the thiol form of GSH (10). A simple, sensitive assay uses GSSG reductase and can measure GSH and/or GSSG, depending on sensitivity, the assay can be automated with clinical kinetic analysers (e.g. Centrifichem or Kobas) or with a plate reader (see below). Other methods using automated column chromatography employing derivatization are also useful (see below).

3.1 DTNB–GSSG reductase recycling assay for GSH

3.1.1 GSH plus 0.5 GSSG determination

This recycling rate assay combines the selectivity of GSSG reductase with the sensitivity of DTNB using the spectrophotometric assay for GSH and GSSG (8, 10, 11, 17–19). DTNB reacts with GSH to form the highly coloured 5-thio-2-nitrobenzoic acid (TNB) anion and GSSG; GSSG is reduced by GSSG reductase coupled with NADPH to give GSH:

$$GSH + DTNB \Rightarrow GS\text{-}TNB + TNB^- + H^+ \qquad (4)$$

$$GS\text{-}TNB + GSH \Rightarrow GSSG + TNB^- + H^+ \qquad (5)$$

$$GSSG + NADPH + H^+ \Rightarrow 2GSH + NADP^+ \qquad (6)$$

Protocol 4. GSH plus 0.5 GSSG determination

Reagents and equipment

- UV/visible spectrophotometer with a water jacket sample holder (30°C); set at 412 nm (410 nm can also be used); a 0.5 OD absorbance unit scale is used and the first 0.5 absorbance unit is not usually recorded because it is not always linear
- Stock buffer: sodium phosphate (125 mM containing 6.3 mM disodium EDTA, pH 7.5); stable for months at 20°C, but can be stored at 4°C

- Daily buffer: stock buffer containing 0.3 mM NADPH (12.5 mg/50 ml); store on ice (4°C) and prepare daily (Note, phosphate occasionally precipitates and the solution must be rewarmed to solubilize it; NADPH should be stored frozen over a desiccant such as calcium chloride, it should be allowed to warm to 20°C (for about 1 h) before use.)

Protocol 4. *Continued*

- GSH standards: prepare daily in the correct % SSA stock with chelator, for example for tissues homogenized in 5% SSA, use 4.31% SSA (prepare by diluting 5 ml of 5% SSA to 5.8 ml); for plasma prepare 3.33% SSA by diluting 1 ml of 10% to 3 ml. For tissues such as kidney, liver, brain, spleen, etc., standard curves from 0 to 4 nmol, in 1 nmol increments, are usually adequate with 5 µl of GSSG reductase solution. For tissues that are depleted of GSH and many cultured cells, a standard curve from 0 to 2 nmol in 0.5 nmol increments with 5 µl of enzyme. For plasma, cerebrospinal fluid or GSH-depleted cultured cells, a standard curve from 0 to 1 nmol, in 0.25 nmol increments, using 10–25 µl of GSSG reductase solution is recommended.
- DTNB solution: DTNB in stock buffer to a final concentration of 6 mM and stored at −20°C in aliquots
- GSSG reductase solution: 266 U/ml in stock buffer (Sigma, yeast Type III), store at 4°C, gently mix before using
- GSH stock: 100 mM in water, store at −20°C, prepare at least every two weeks; GSSG at 50 mM can also be used, but note that most commercial sources contain ethanol
- It is very important that the total volume (and concentration) of SSA be constant in the assay for both the samples and the standards because GSSG reductase is inhibited by SSA and small changes in its amount can dramatically alter the rates.
- The amount of NADPH may be increased by as much as 100% with age.

Method

1. Fill cuvettes with 700 µl of daily buffer, 100 µl of DTNB solution, and water (200 µl minus sample volume).

2. Place in cuvette holder (30°C) or in a water bath at 30°C, for at least 5 min.

3. Add sample and mix (Parafilm over the top using a thumb works well).

4. Add enzyme, mix rapidly, and place in a spectrophotometer.

5. After the first 0.5 *A* unit, which is not recorded, take measurements of about 0.5 *A* at 412 or 410 nm. The sampling interval depends on the spectrophotometer; for example, the Cary 219 has a continuous chart recorder and we measure 0.5 to 1.0 *A*.

6. Assay samples in duplicate. (The lines from the recorder are parallel, or the rate of increase of absorbance should be the same for duplicates.)

7. After all the samples have been run, generate a standard curve using appropriate standards; the sample rates should be within the rates of the standards. Note: it is important to run a 0 value because GSSG reductase contains a small amount of GSH and this blank indeed has a measurable rate; duplicates of this blank are tedious to obtain, but the use of clean cuvettes facilitates the process.

i Quantification

A graph of rate of change of absorbance versus the amount (nmol) of GSH is drawn, and the values for the samples are calculated using a linear regression program. The concentration of GSH + 0.5 GSSG is obtained by multiplying by the appropriate sample dilution factors, such as 5.8 ml/g; the 5.8 comes from the volume of SSA that the tissue was homogenized in plus the volume of the tissue, about 0.8 ml/g.

ii Plasma and other low level GSH samples

Plasma samples from rodents and humans, and rodent cerebrospinal fluid have low GSH levels, of about 25, 5, and 3 μM, respectively. The assay of samples with low GSH values necessitate care with the recycling assay. Usually larger sample volumes (25 μl) and more enzyme (10 or even 25 μl) are used. However, the rate of the blank increases with increasing enzyme, so greater care must be taken to obtain the blank value. Also each assay can take much longer to run (10 vs 2 min). This low level assay is not as linear as the higher level assay, so that the rates are usually calculated from a smaller segment of the run, about 0.5–0.75 absorbance units in a very uniform manner.

3.1.2 Plate reader assay for GSH + 0.5 GSSG

The plate reader assay, although somewhat time consuming to set up, is the fastest way to run a large number of samples. This method has been successfully applied to various samples, in particular for cultured cells. Samples, such as tissues, with high levels of GSH may have to be diluted 5- to 20-fold. There is slight variation with enzyme solutions, such as age, storage conditions, etc. and with instruments; thus, it is desirable that linearity and limits be established with a test run.

Protocol 5. Automated recycling assay

Reagents and equipment

- 96-well plate reader with kinetics, 405 filter
- Plate buffer: daily buffer (7 ml) mixed with DTNB solution (1 ml) (see *Protocol 4*)
- Enzyme solution: prepare 10 U/ml of stock buffer (as in *Protocol 4*)
- GSH standards in the appropriate concentration of SSA with chelator
- 96-well plates, flat bottom
- Adjustable multi-channel pipette

Method

1. Program the plate reader for 405 nm, 0 lag time, 30 sec read interval, 5 min run time, absorbance limit 2.0, temperature 30°C; mixing after initial reading is optional.
2. Enter the standard values into the program together with dilution factors.
3. Set up plates with standards and samples at least in triplicate. Note that a blank (SSA, no GSH, no enzyme) is used in the software program to subtract the background value of reagent automatically; an enzyme blank (0 standard value; enzyme, SSA, no GSH) is used as the zero standard.

219

Protocol 5. *Continued*

4. Add plate buffer (80 µl) to the wells using a multi-channel pipette.

5. Add sample (10 µl; other volumes are possible if the plate buffer is adjusted) carefully.

6. Add enzyme solution (10 µl) rapidly with a multi-pipetter, and rapidly put the plate into the reader and start the run.

7. When the correct dilution values are programmed, obtain direct values for GSH + 0.5 GSSG using a kinetic software program.

3.2 DTNB–GSSG reductase recycling assay for GSSG

There are several procedures for the determination of GSSG in biological samples (8–10, 19). GSSG reductase can be used to measure GSSG levels by following the change in absorbance of NADPH at 340 nm or fluorimetrically; the normally low levels of GSSG make this procedure difficult. A sensitive method for determining GSSG, masks GSH by derivatization and then uses the same DTNB–GSSG reductase recycling assay described in *Protocol 4* using plasma and other low level samples. The most common methods for masking GSH is to derivatize the thiol with either *N*-ethylmaliamide (NEM) or 2-vinylpyridine (2VP). NEM reacts faster with thiols, but since it inhibits GSSG reductase, it must be carefully removed, usually by column filtration, before analysis. NEM is not specific for thiols, but also reacts with the ε-amino groups of lysine and the imidazole group of histidine. When NEM reacts with GSH, two chiral products are formed. These adducts are unstable and hydrolyse to two products; also, rearrangement of the adduct is possible (22). 2VP reacts somewhat more slowly, but at the concentrations used, does not affect GSSG reductase, thus this procedure is described.

Protocol 6. DTNB–GSSG reductase recycling assay for GSSG

Reagents and equipment

- All reagents and equipment as described in *Protocol 4*
- Triethanolamine (TEA), undiluted
- Microcaps (Drummond) or a positive displacement pipetter
- Microcentrifuge tubes
- 2-vinyl pyridine (Aldrich Chemical Co.); use undiluted, store at −20°C; when it becomes brown or viscous with age, replace with a new bottle; 2VP has a low vapour pressure and may be irritating, thus a fume hood is used for derivatizations

Method

1. Place supernatant (100 µl) at the bottom of a microfuge tube (1.5 ml).

2. Add 2VP (2 µl) with mixing.

3. Add TEA to the side of the tube wall (above the sample) to prevent a high local pH, which can lead to oxidation.

4. Cap the tube and mix contents vigorously.

5. Adjust the pH of the SSA supernatant to about 6.8, but not higher than 7.2, with TEA. For 100 µl of 4.31% SSA sample supernatant, 6 µl of TEA is appropriate.[a]

6. Check the pH with short-range pH paper, if it is over 7.2, repeat the derivatization because of possible sample oxidation

7. Assay the derivatized samples as described in *Protocol 4*. Use GSSG as the standard at half the concentration (same number of GSH equivalents) of GSH.

[a] TEA is viscous; the exact amount used for any type of sample can only be determined empirically.

4. Column methods for the determination of GSH

There are numerous column methods available for the determination of GSH. A discussion of the relative merits of each is beyond the scope of this chapter, but has been discussed in detail elsewhere (9). One advantage of a column method for GSH assay is that the specific activity of radiolabelled metabolites can be determined. Also many column methods allow for the simultaneous determination of GSH and other thiols such as cysteine. Two convenient methods are described, one based on ion exchange with visible light detection, and the other is an HPLC method with fluorescent detection.

4.1 Ion-exchange chromatography of GSH and other thiols

GSH and other thiols are derivatized with 2-vinylpyridine (2VP) and chromatographed on an amino acid analyser using ion-exchange chromatography (11, 19). These derivatives are still amino acids and react with ninhydrin that is detected with a visible detector.

Protocol 7. Ion-exchange chromatography of GSH

Reagents and equipment
- Amino acid analyser (Durrum 500 or Beckman 7300)
- Derivatization reagents as in *Protocol 6*

A. *Procedure*

Derivatize samples with 2VP as described in *Protocol 6*.

B. *Analysis*

Inject samples of about 20–40 nmol. Several procedures are possible, for example, with a Durrum 500 analyser, the retention times for GSH,

---

Protocol 7. *Continued*

γ-glutamylcysteine, and cysteine are about 78, 72, and 165 min, respectively; glutamate elutes at about 30 min. These peaks are well resolved
from the 20 normal amino acids. The procedure can be shortened by
increasing temperature or changing the eluting buffers. For a Beckman
amino acid analyser using a hydrolysate procedure, the retention
times are less.

C. *Quantification*

1. Prepare and run standards of known concentration and a colour value
   (area/nmol) obtained.

2. Determine the concentration of GSH in samples by dividing the area of
   the sample by the colour value and multiplying by the dilution factor.

4.2 Ion-exchange chromatography for GSH and GSSG

Samples are reduced first, then derivatized, and run on an amino acid
analyser.

Protocol 8. Ion-exchange chromatography of GSH and GSSG

Reagents and equipment
- Dithiothreitol (DTT), 1 M, prepared fresh
- All other reagents as described in *Protocol 7*

A. *Procedure*

1. Adjust pH samples (100 μl) to 6.8–7 with TEA (4.31% SSA samples
   usually require 6 μl; see *Protocol 6*).

2. Add DTT (5 μl; 1 M) with mixing.

3. After 1 h, add 2VP (3 μl) in a hood.

4. After 1 h, run samples on the analyser. Sample are stable at −20°C for
   several weeks.

B. *Quantification*

1. Calculate the concentration of GSH plus 0.5 GSSG as described in
   Protocol 7.

2. Determine the amount of GSSG in the sample, by subtracting the
   amount of GSH obtained from the sample using *Protocol 7* from the
   valued obtained in *Protocol 8* (note: results are usually given either as
   GSH equivalents or divided by 2 to give GSSG levels).

4.3 Determination of GSH using monobromobimane derivatization and HPLC

GSH and many biological thiols, such as cysteine, γ-glutamylcysteine, etc., react with monobromobimane (mBBr; 3,7-dimethyl-4-bromomethyl-6-methyl-1,5-diazobicyclo[3.3.0]octa-3,6-diene-2,8-dione) (20) (*Figure 1*). Such mBBr derivatives are then separated by HPLC and quantified using fluorescence detection (9, 11, 21).

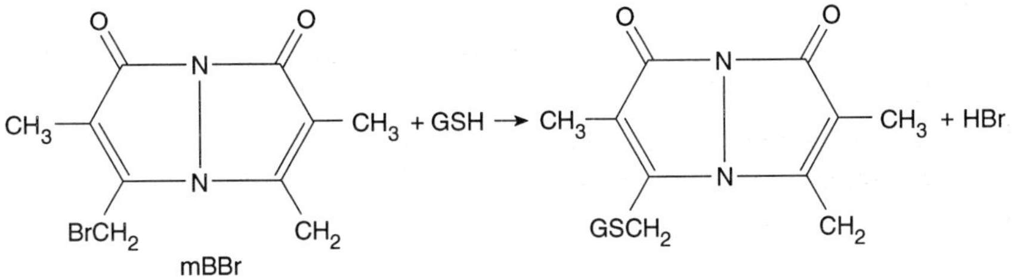

Figure 1. Reaction of mBBr with GSH to form a fluorescent adduct.

Protocol 9. Determination of GSH using HPLC

Reagents and equipment

- mBBr (Thiolyte MB from Calbiochem): prepare a 100 mM solution in acetonitrile and store in a brown bottle at −20°C; it is stable for several weeks
- N-Ethyl morpholine (NEMP) (Aldrich Chemical Co.), 7.8 M (undiluted); dilute to 1 M just prior to use; store in a brown bottle covered with foil at room temperature; it is stable for several months. To remove any impurities, we distil (using a water-vacuum rotory evaporator) NEMP taking the middle fraction (starting after about 20 ml and continuing for about a 50 ml collection; NEMP should not be distilled to dryness to prevent contaminating compounds, possibly containing peroxides).
- Acetic acid

- GSH and other thiol standards, prepared in the appropriate % SSA immediately before derivatization
- HPLC system, sample injector, pumps, timed solvent changer, fluorescence detector (o-phthalaldehyde filters); temperature jacketed column (25°C) is preferred to maintain steady retention times
- HPLC column: Waters C18 10 μm or Altex C18 5 μm (3.9 mm × 25 cm)
- Buffer A: 14.2% methanol and 0.25% acetic acid (pH 3.9 with sodium hydroxide)
- Buffer B: 90% methanol and 0.25% acetic acid (pH 3.9; little or no sodium hydroxide is needed) (Note: all percentages are v/v; the percentage of methanol varies from 12.4 to 14.2% with columns and temperature.)

A. Procedure

1. Carry out the derivatization in subdued light to minimize mBBr degradation.

2. Add water (340 μl) and sample (120 μl, 4.31% SSA with chelator) to a 1.5 ml tube.

3. Add NEMP (100 μl, 1 M), followed rapidly by mBBr (20 μl, 0.1 M) with mixing (the pH should be about 7.6).

Protocol 9. *Continued*

4. Allow mixture to stand in the dark for 20 min.

5. Add acetic acid (20 µl) with mixing.

6. Store samples at −20°C in the dark until analysis. Samples are stable for several weeks.

B. *Analysis*

1. Equilibrate the column for 12 min (1 ml/min) with buffer A.

2. Inject the derivatized sample (100 µl or appropriate volume depending on the GSH concentration and fluorometer).

3. Continue elution with buffer A for 30 min.

4. Flush the column with buffer B for 8 min.

5. Re-equllibrate the column with buffer A.

6. Elute SSA,[a] cysteine, cysteinylglycine, γ-glutamylcysteine, GSH and mBBr at 5, 10, 13, 15, 19, and 32 min, respectively.

C. *Quantification*

1. Prepare and run standards, including a blank, in duplicate in the same manner as the samples.

2. Use standard curves to obtain the GSH values of the samples. The fluorescence detector is a major determinate of overall sensitivity. Injection volumes of between 10 and 200 µl have been successfully used. Amounts of between 10 and 200 nmol per injection have been used; these are generally adequate ranges for tissue samples. (Note: when cysteine is to be determined, the areas for cysteine standards should be close to those for GSH, at least 90%; however, since cysteine levels are usually lower than GSH levels, lower cysteine standards are run.)

[a] SSA is fluorescent, and for ease of quantification, the integration is not turned on until after it elutes (about 8 min) in a broad peak. There is considerable variation in run times from laboratory to laboratory because of slight changes in pH, temperature, and buffer batches. A new C18 column is routinely washed with buffer A followed by buffer B, alternating every few hours, for a few days, to 'settle in'; a new column is needed when peak widths are too great or when the peaks begin to form doublets. Although samples are stable when frozen in the dark, when an autosampler is used, the autosampler window is blocked to keep light from the samples, and only about 12 h of samples are loaded at a time (if a refrigerated autosampler is available, this time can be increased).

4.3.1 Alternative derivatization procedure

The use of NEMP to neutralize samples is satisfactory. However, the need to redistil NEMP has also led to the use of Tris buffer to pH adjust samples. The following procedures have been determined empirically. This procedure has been applied to low level samples. The use of Tris buffer adds a peak in the chromatogram, but it is easily resolved from the peaks of interest. DPTA is an excellent chelator and its use seems to improve peak height. In this procedure the sample (360 μl, 3.33% SSA with chelator) is placed in a microfuge tube and DPTA (142 μl; 2 mM) is added. Tris–HCl buffer (80 μl, 2 M, pH 9.0 at 21 °C) is then added followed rapidly by the addition of mBBr (5 μl, 0.1 M) with mixing. After 20 min in the dark, acetic acid (13 μl) is added.

Samples with lower levels of GSH, such as plasma samples or cultured cells, may be derivatized by the procedure as described above. A standard curve using 3.33% SSA with chelator in the range 0–1.5 nmol per injection volume (200 μl is suggested) is prepared.

4.3.2. Comments

The mBBr derivatization method is useful for determining a large number of biologically important thiols. Its use for the determination of disulphides is limited by the need for sample reduction. DTT is an excellent reductant; however, since it reacts with mBBr it must be removed, by extraction or column methods, before derivatization; these procedures are time consuming and require great care. The usual excess of mBBr in the derivatization reaction is a 5- to 10- fold excess. Higher levels of mBBr may lead to the formation of degradation and other peaks. Other bimane derivatives have been prepared. The chlorobimane derivative (mBCl) has been directly applied to cultured cells, and has been used in a fluorescence-activated cell sorting (FACS) assay for GSH. This use of mBCl depends upon the GSH S-transferase catalysed formation of a GSH bimane adduct. The exposure time for cells to mBCl must be determined for each cell line; the use of this assay for FACS analysis of human samples, such as lymphocytes, is problematic because individuals may show a wide range of variation in the activity of the GSH S-transferase for the conjugation reaction.

5. Conclusions

There are numerous methods available to measure GSH (and GSSG). The choice of method depends upon the nature of the sample, equipment available, and whether GSH, with or without other thiols, or GSSG determinations are required. Several methods that are relatively easy and dependable have been described in this chapter.

Acknowledgment

This work was supported in part by a grant (AI 31804) from the National Institutes of Health, National Institute of Allergy and Infectious Diseases.

References

1. Meister, A. (1989). In *Coenzymes and cofactors: glutathione* (ed. D. Dolphin, R. Poulson, and O. Avramovic), pp. 367–474. Wiley, New York.
2. Meister, A. and Anderson, M. E. (1983). *Annu. Rev. Biochem.*, **52**, 711.
3. Meister, A. (1992). *Biochem. Pharmacol.*, **44**, 1905.
4. Meister, A. (1994). *J. Biol. Chem.*, **269**, 9397.
5. Meister, A. (1995). In *Methods in enzymology* (ed. L. Packer). Vol. 251, in press, Academic Press, New York.
6. Anderson, M. E. and Meister, A. (1983). *Proc. Natl Acad. Sci. USA*, **80**, 707.
7. Anderson, M. E., Underwood, M., Bridges, R. J., and Meister, A. (1989). *FASEB J.*, **3**, 2527.
8. Anderson, M. E. (1989). In *Coenzymes and cofactors: glutathione* (ed. D. Dolphin, R. Poulson, and O. Avramovic), pp. 339–366. Wiley, New York.
9. Fahey, R. C. (1989). In *Coenzymes and cofactors: glutathione* (ed. D. Dolphin, R. Poulson and O. Avramovic), pp. 304–337. Wiley, New York.
10. Anderson, M. E. (1985). In *Handbook for oxygen radical research* (ed. A. Greenwald), pp. 317–323. CRC Press, Boca Raton. FL.
11. Anderson, M. E. (1985). In *Methods in enzymology* (ed. A. Meister), Vol. 113, pp. 548–555. Academic Press, New York.
12. Anderson, M. E. and Meister, A. (1980). *J. Biol. Chem.*, **255**, 9530.
13. Abbott, W. and Meister, A. (1986). *Proc. Natl Acad. Sci. USA*, **83**, 1246.
14. Anderson, M. E., Powrie, F., Puri, R. N., and Meister, A. (1985). *Arch. Biochem. Biophys.*, **239**, 538.
15. Martinsson, J., Jain, A., Frayer, W., and Meister, A. (1989). *Proc. Natl Acad. Sci. USA*, **86**, 5296.
16. Ellman, G. L. (1959). *Arch. Biochem. Biophys.*, **82**, 70.
17. Owens, C. W. I. and Belcher, B. J. (1965). *Biochem. J.*, **94**, 705.
18. Tietze, F. (1969). *Anal. Biochem.*, **27**, 502.
19. Griffith, O. W. (1980). *Anal. Biochem.*, **106**, 207.
20. Kosower, E. M., Pazhenchevsky, B., and Hershkowitz, E. (1978). *J. Am. Chem. Soc.*, **100**, 6516.
21. Newton, G. L., Dorian, R., and Fahey, R. C. (1981). *Anal. Biochem.*, **114**, 383.
22. Fahey, R. C. (1989). In *Coenzymes and cofactors: glutathione* (ed. D. Dolphin, R. Poulson and O. Avramovic), pp. 303–337. Wiley, New York.

16

Glutathione peroxidase: activity and steady-state level of mRNA

DARET K. ST CLAIR and CHING K. CHOW

1. Introduction

Glutathione peroxidase (glutathione: H_2O_2 oxidoreductase, EC 1.11.1.9) was first reported by Mills in 1957 in erythrocytes (1) and later in the liver of various species (2). The enzyme catalyses the reduction of hydroperoxides by reduced glutathione (GSH) to their corresponding alcohols. The ability of GSH peroxidase to reduce various hydroperoxides has led to the proposal of the role of the enzyme in protecting tissues against oxidative damage to critical biomembranes and macromolecules (3, 4).

Based on their substrate specificity, two types of GSH peroxidase, selenium (Se)-dependent and Se-independent enzymes, are generally recognized. Se-dependent GSH peroxidase, which is capable of utilizing hydrogen peroxide (H_2O_2) and a variety of organic hydroperoxides as substrates, was isolated in crystalline form and characterized in mechanistic and kinetic studies (5, 6) before it was recognized as a selenoprotein (6, 7). The Se-independent activity is attributed to isoenzymes of GSH S-transferases, particularly GSH S-transferase B (8), acting on organic hydroperoxides but not on hydrogen peroxide. GSH is the electron donor of physiological importance for both types of GSH peroxidases. Although the activity of Se-GSH peroxidase is generally positively correlated with the status of dietary selenium, Se-independent GSH peroxidase activity is found to increase in the tissues of Se-deficient animals (9, 10).

The classic Se-dependent GSH peroxidase is found mainly in the cytosol of mammalian and avian tissues. This cytosolic selenoenzyme (C-GSH peroxidase) is composed of four 19–23 kDa subunits, each of which contains Se in the form of a single selenocysteine residue (11, 12). C-GSH peroxidase utilizes hydrogen peroxide, cumene peroxide, *tert*-butyl hydroperoxide and fatty acid hydroperoxides as substrates. Another selenoenzyme, phospholipid hydroperoxide (PH) GSH peroxidase, also a soluble enzyme, has been purified from cell sap of different rat and pig organs (13, 14). It is a monomeric enzyme, of approximately 20 kDa, and contains one selenium atom at the active site. In addition to the relatively polar substrates utilized by C-GSH peroxidase, PH-GSH peroxidase exhibits a unique peroxidase activity on hydroperoxides of phospholipids and cholesterol (13, 14). However, it is possible that C-GSH peroxidase, in conjunction with the action of phospholipases, may also play a role in reducing peroxidized membrane lipids.

In addition to C-GSH peroxidase and PH-GSH peroxidase, an enzymatically, structurally and antigenically distinct Se-dependent GSH peroxidase is found in human plasma (15, 16). The plasma GSH peroxidase (P-GSH peroxidase) is glycosylated and appears to be synthesized by and secreted from hepatic cells. P-GSH peroxidase is also a tetramer of 24–25 kDa and can metabolize H_2O_2 and fatty acid hydroperoxides effectively but metabolizes hydroperoxides of phospholipid and cholesterol poorly (17, 18). Recently, another immunologically distinct cytosolic Se-GSH peroxidase has been found in the gastrointestinal (GI) tract of human and rodents and human liver (19). This selenoenzyme, GI-GSH peroxidase, is a tetrameric protein composed of 22 kDa monomers and does not cross-react with antisera against human C-GSH peroxidase or P-GSH peroxidase. Similar to C-GSH peroxidase and P-GSH peroxidase, GI-GSH peroxidase catalyses the reduction of H_2O_2, organic hydroperoxides, and fatty acid hydroperoxides with GSH (19).

2. Measurement of GSH peroxidase activity

As with most enzymes, GSH peroxidase activity can be assayed by more than one method. The methods available for assaying GSH peroxidase activity are almost exclusively based on the measurement of the rate of GSH consumption, either directly or indirectly (20–25). The most commonly used assay procedure for GSH peroxidase activity is the spectrophotometric method using GSH reductase and NADPH originally described by Paglio and Valentine (20). Many modifications of this indirect measurement of GSH oxidation has been reported. The assay is based on the reduction of oxidized glutathione (GSSG), generated from the action of GSH peroxidase to catalyse the reduction of hydroperoxides (ROOH) by GSH, back to GSH by GSH reductase utilizing the reducing equivalent of NADPH. This method measures the rate of NADPH oxidation at 340 nm as the indication of GSH peroxidase activity.

$$2 \text{ GSH} + \text{ROOH} \xrightarrow{\text{GSH peroxidase}} \text{ROH} + \text{GSSG}$$

$$\text{GSSG} + \text{NADPH} \xrightarrow{\text{GSH reductase}} 2 \text{ GSH} + \text{NADP}$$

Most of the assay procedures applying this principle are conducted in 50–100 mM Tris or phosphate buffer, pH 7.0–7.6 (1–10, 22, 23). The procedure described in *Protocol 1* can be used to assay the activity of C-, P- and

16: Glutathione peroxidase

GI-GSH peroxidase. The substrate requirements and assay conditions for measuring the activity of PH-GSH peroxidase are described by Ursini *et al.* (13–16, 21).

Protocol 1. Spectrophotometric assay of GSH peroxidase activity

Equipment and reagents

- Spectrophotometer with recorder or data processor
- Tissue homogenizer
- Cell sonicator
- Refrigerated centrifuge
- Tris–HCl buffer, 50 mM, pH 7.6
- Phosphate buffer, 50 mM, pH 6.6
- Sodium chloride, 0.89%
- 10 mM (N-[2-hydroxyethyl]piperazine-N'-[ethanesulphonic acid] (Hepes) in 0.25 M sucrose, pH 7.4

- 1.15% KCl in 50 mM Tris buffer, pH 7.4
- Drabkin's reagent (1.6 mM KCN, 1.2 mM $K_3Fe(CN)_6$ and 23.8 mM $NaHCO_3$)
- Ethylenediaminetetra-acetic acid (EDTA), disodium salt
- Sodium azide (NaN_3)
- Reduced glutathione (GSH)
- NADPH, tetrasodium salt
- GSH reductase, type III
- Hydrogen peroxide (H_2O_2), 30%
- Cumene hydroperoxide, 80%

A. *Sample preparation*

(a) *Plasma and red blood cells*

1. Collect 0.5–2 ml of blood into a heparinized tube, mix gently, and centrifuge the blood immediately (1000 *g*, 10 min, 4°C).

2. Aliquot plasma and store at −70°C until analysis.

3. Remove the buffy coat and wash the red cells twice with approximately 5 vol. of ice-cold saline.

4. After the second washing, lyse the red cells with approximately 20 vol. of 50 mM phosphate buffer, pH 6.6, and store at −70°C until analysis. Prior to analysis, mix the haemolysate with an equal volume of double-strength Drabkin's reagent to convert all haemoglobin into the stable cyanmethaemoglobin form.

(b) *Tissues*

1. Prepare 10% homogenate with ice-cold 0.25 M sucrose buffered with 10 mM Hepes, pH 7.4, or 1.15% KCl in 0.05 M phosphate buffer, pH 7.6, and centrifuge at 4°C to obtain subcellular fractions (including cytosolic fraction) as needed; store at −70°C until analysis.

2. Sonicate or freeze–thaw subcellular organelles two or three times prior to use. A further dilution of enzyme preparation is usually needed for tissues with high GSH peroxidase activity, such as liver and kidney.

(c) *Cultured cells*

1. Lyse the cell in distilled water or hypotonic buffer (e.g. 50 mM Tris–HCl buffer, pH 7.6).

2. Sonicate or freeze–thaw the cell lysate prior to use.

Protocol 1. *Continued*

B. *Assay procedure*

1. Prepare coupling reagent by dissolving 33.6 mg disodium EDTA (2 mM), 6.5 mg NaN_3 (1 mM), 30.7 mg GSH (1 mM), 16.7 mg NADPH (0.2 mM), and 100 units of GSH reductase in 100 ml 50 mM Tris–HCl buffer, pH 7.6.[a]

2. Prepare hydroperoxide substrate by diluting 113.3 μl of 30% hydrogen peroxide (1 mM) or 912 μl of 80% cumene hydroperoxide (4.8 mM) in 1 L 0.05 M Tris–HCl buffer, pH 7.6.[b]

3. Incubate 10–100 μl of the sample preparation with 965–875 μl of the coupling mixture at a desirable assay temperature[c] for 2–3 min.

4. Add 25 μl of a hydroperoxide substrate to start the reaction[d] (total volume 1 ml[e]). The final concentrations of hydrogen peroxide and cumene peroxide are 0.25 mM and 1.2 mM, respectively.

5. Monitor the decrease of absorbency at 340 nm for the rate of disappearance of NADPH in a thermostated spectrophotometer with a recorder or data processor for 1–2 min.[f]

[a] The amount of the coupling reagent needed is approximately 1 ml/assay. Freshly prepared reagent is good for 6–8 h at room temperature or 24 h at 4°C. Sodium azide is added to inhibit catalase activity.
[b] Hydrogen peroxide substrate needs to be prepared fresh and cumene hydroperoxide substrate can be stored at −20°C for over one month.
[c] Reaction temperature at 37°C or room temperature (25°C) is often used.
[d] Prior to the measurement of enzyme activity, a linear reaction with respect to the amount of sample (enzyme) preparation used and reaction time has to be established. Also, blank values for the reaction mixture (without substrate) and hydroperoxide substrate (without sample) are to be obtained.
[e] The reaction volume can be increased or decreased as needed.
[f] A net decrease of 0.01–0.05 A units/min is a desirable range. If the enzyme activity is too high (the rate of enzymic reaction decreases rapidly and becomes non-linear), the sample preparations need to be diluted appropriately. If the enzyme activity is too low, a larger amount of enzyme preparation may be used or the reaction time increased.

2.1 Calculation of GSH peroxidase activity

The activity of GSH peroxidase is calculated based on the molar extinction coefficient of NADPH which at 340 nm is equal to 6220. One unit of an enzyme activity is normally defined as μmol substrate utilized or product produced/minute at 25 or 37°C. The specific activity of an enzyme can be expressed as units/ml (e.g. plasma), units/mg haemoglobin (e.g. haemolysate) or units/mg protein (e.g. liver cytoplasm). The activity of GSH peroxidase is usually expressed as nmol/min (or munits) per ml or mg protein. Based on the decrease of A at 340 nm/min resulting from the amount of sample preparation (enzyme source) employed in the assay, the activity of GSH peroxidase is calculated as follows:

GSH peroxidase activity (nmol/min/ml or mg protein)

$$= A/\text{min} \div 6220 \times 10^6/(\text{volume of sample (in ml) or amount of protein in}$$
$$\text{the sample used (in mg))} \times \text{reaction volume (in ml)}$$
$$= A/\text{min} \times 161/(\text{ml sample or mg protein in the sample used}) \times \text{ml}$$
$$\text{reaction volume.}$$

When 1 ml reaction volume is used the enzyme activity is equal to $A \times 161$/(ml sample or mg protein in the sample preparation used). For example, if 50 μl of a plasma sample is used and the rate of decrease at 340 nm is 0.031/min is measured, the enzyme activity in that plasma sample is $0.031 \times 161/0.05 = 10.0$ nmol/min/ml.

As stated above, at least four different Se-dependent GSH peroxidases and one Se-independent GSH peroxidase are known to exist in mammalian tissues. Since the cytosolic, plasma and gastrointestinal Se-GSH peroxidases have similar substrate specificity, they can be assayed using either organic hydroperoxides, fatty acid hydroperoxides or H_2O_2 as substrate. The activity due to PH-GSH peroxidase can be specifically measured using hydroperoxides of phospholipids or cholesterol as substrate (13–16, 21). As Se-independent GSH peroxidase does not reduce H_2O_2, its activity can be calculated based on the differences between the total activity (using organic hydroperoxide as substrate) and that of Se-GSH peroxidase (using H_2O_2 as substrate). Although tissue localization and substrate specificity provide useful information to differentiate various types of GSH peroxidase, specific information concerning the presence and localization of individual enzyme protein can only be obtained immunologically.

The activity of GSH peroxidase also has been assayed by measuring the rate of GSH oxidation spectrophotometrically using 5,5′-dithio-bis(2-nitrobenzoic acid) as the reagent (24). However, the method for measuring the rate of GSH utilized directly is relatively more cumbersome and subject to interference. More recently, Pascual *et al.* (25) have developed a more specific method for the determination of GSH peroxidase activity. The assay is based on the separation and quantitation of GSH and GSSG by high-performance capillary electrophoresis. Although the capillary electrophoretic method is highly specific and sensitive, it is more time-consuming and requires more sophisticated instrumentation.

3. Detection of GSH peroxidase expression

All four Se-containing enzymes with GSH peroxidase activity characterized to date are encoded by nuclear DNA and have been mapped to: human chromosomes 21, 3, and X for C-GSH peroxidase; chromosome 19 for PH-GSH peroxidase; chromosome 5 for P-GSH peroxidase; and chromosome 14 for GI-GSH peroxidase (26). Although each of these enzymes is expressed in more than one tissue, the C-GSH peroxidase and PH-GSH peroxidase are found in most of the tissues that have been examined, whereas the P-GSH peroxidase and GI-GSH peroxidase enzymes appear to have limited tissue expression. The C-GSH peroxidase is expressed at high levels in erythrocytes, liver, and kidney (27) and the PH-GSH peroxidase is present in testis at a high level. The P-GSH peroxidase is detected in plasma, milk, and lung (28). In humans, P-GSH peroxidase mRNA is expressed in liver, kidney, lung, heart, placenta, and breast tissue, but it is not expressed in rat liver (29). GI-GSH peroxidase is found in human gastrointestinal tract and liver (19). In rodent tissue, GI-GSH peroxidase mRNA is detected only in the GI tract but not in other tissues including liver (19). Except for the mouse GI-GSH peroxidase, the cDNA sequence of all four GSH peroxidase isoenzymes contains a UGA codon coding for selenocysteine (19, 27, 30, 31).

The expression of GSH peroxidase can be detected at multiple levels, e.g. primary transcription, steady-state level of mRNA, and translatable level of mRNA. The expression of GSH peroxidase primary transcripts can be determined by nuclear run-off transcription assays; the steady-state level of GSH peroxidase mRNA can be determined by Northern analysis, dot blot, RNase protection, etc., and the level of translatable GSH peroxidase mRNA can be determined by *in vitro* translation coupled to immunoprecipitation with specific antibodies against each GSH peroxidase. Among these, the expression at baseline or following induction of GSH peroxidase is most commonly determined by measuring the steady-state level of mRNA.

Various methods have been described and used successfully for preparation of intact RNA. The selection of the method is based on the amount of the tissue or cells; the type of the samples (culture cells versus solid tissues); the number of samples to be prepared; and the availability of equipment. The following procedures for isolation of total RNA from mammalian cells work well with a wide range of samples.

The procedure described in *Protocol 2A* for isolation of total RNA is modified from that of Chingwin *et al.* (32). The use of guanidine thiocyanate, which is a denaturing reagent, and β-mercaptoethanol, a reducing agent, will inactivate RNases. RNA is separated from DNA and protein by centrifugation in CsCl. This method has been used successfully in isolation of intact RNA from various tissues, including tissues that are rich in RNase, such as pancreas. The procedure described in *Protocol 2B* for preparing total RNA is modified from that of Chomczynski and Sacchi (33) which uses the guanidine thiocyanate and β-mercaptoethanol to inhibit RNase and phenol/chloroform extraction to separate RNA from contaminants. This procedure does not require ultracentrifugation. Thus a large number of samples can be processed simultaneously. However, DNA contamination could be a problem if the sample concentration is high. The following procedure is designed for isolating RNA from 1 g of tissue or 10^8 cells.

Protocol 2. Preparation of total RNA

Equipment and reagents

- Ultracentrifuge or high speed centrifuge
- Microcentrifuge
- Tissue grinder or polytron homogenizer
- Guanidine thiocyanate, 4 M solution (100 ml) prepared by mixing of 50 g guanidine thiocyanate (4 M); 0.5 g of sodium-*N*-lauryl sarcosine (0.5%), and 5.0 ml of 0.5 M EDTA, pH 8.0 (25 mM) in 65 ml sterile double-distilled water. Heat to ~45°C, adjust to pH 7.0, filter through a 0.45 μm filter, and adjust volume to 99.9 ml. Add 0.1 ml of β-mercaptoethanol (0.1%) just before use. The solution without β-mercaptoethanol can be kept in a brown bottle at room temperature for several months.
- CsCl, 2.4 M (100 ml): dissolve 40 g CsCl (2.4 M), 0.83 ml 3M NaAc, pH 5.2 (25 mM), and 2.0 ml 0.5 M EDTA, pH 8.0 (10 mM), in DEPC-treated water as in 5.7 M CsCl

- CsCl, 5.7 M (100 ml): dissolve 95.9 g CsCl (5.7 M), 0.83 ml 3M sodium acetate (NaAC), and pH 5.2 (25 mM) 2.0 ml 0.5 M EDTA, pH 8.0 (10 mM) in 80 ml diethyl pyrocarbonate (DEPC)-treated water. Autoclave for 20 min at 15 lb/in^2, cool, and adjust volume to 100 ml with sterile DEPC-treated water.
- 0.1% DEPC-treated water: dissolve 0.1% DEPC in double-distilled water, allow to stand at 22–37°C for at least 12 h, and autoclave for 20 min at 15 lb/in^2 on liquid cycle
- Denaturing solution (100 ml): mix 50 g guanidine thiocyanate (4 M), 0.74 g sodium citrate (25 mM) and 0.5 g sodium *N*-lauryl sarcosine (0.5%) in 65 ml of sterile double-stilled water. Heat to ~45°C, adjust to pH 7.0, filter through a 0.45 μm filter, and adjust volume to 99.9 ml. Add 0.1 ml β-mercaptoethanol (0.1%) just before use.

A. *Isolation of total RNA by CsCl centrifugation*

1. Remove tissue and mince well with scissors.

2. To the minced tissue, add 6 7 volumes of 4 M guanidine thiocyanate solution and homogenize with a tissue grinder or a Polytron homogenizer for 30 sec–2 min.

3. Prepare ultraclear or polyallomer centrifuge tube according to the rotor to be used. For the Beckman SW41 rotor, add 3 ml of 5.7 M CsCl solution to the bottom of the tube, then gently layer 1 ml of 2.4 M CsCl solution on top of the 5.7 M CsCl.

4. Layer 8 ml of tissue homogenate on top of the 2.4 M CsCl cushion.

5. Equalize all the tubes with guanidine thiocyanate homogenization buffer, and centrifuge for 18–22 h at 34 000 r.p.m. (100 000 *g*) at 22°C.

6. After centrifugation, the RNA should appear as pellet at the bottom of the tube. Use a Pasteur pipette to aspirate out the supernatant and the viscous DNA layer. Sometimes, it is difficult to see the RNA pellet so take great care to remove the CsCl without disturbing the pellet.

7. Dissolve the RNA pellet in sterile DEPC-treated 1 mM EDTA in 10 mM Tris–HCl (TE), pH 7.5, and transfer to a sterile microcentrifuge tube; rinse the bottom of the centrifuge tube twice with a small volume of DEPC-treated TE and pool in the microcentrifuge tube. The volume of TE needed to dissolve the pellet depends upon the size of the pellet: start with 300 μl for the 8 ml homogenate and wash with 100 μl each time. If the pellet is difficult to dissolve, add more TE and heat it to 45°C to dissolve.

Protocol 2. *Continued*

8. Extract the RNA solution with an equal volume of chloroform: *n*-butanol (4:1, v/v) and transfer the aqueous phase to a clean microcentrifuge tube. Add a small volume of TE to the organic phase and back-extract the RNA. Pool the aqueous phases from the two extractions.

9. Extract the aqueous phase with an equal volume of chloroform: isoamyl alcohol (24:1, v/v).

10. Precipitate the RNA with 0.1 vol. of 3 M (NaAc), pH. 5.2, and 2 vol. of 100% ethanol. Store at −20°C overnight.

11. Collect the RNA by centrifugation at 12 000 *g* in a microcentrifuge for 10–15 min at 4°C. Wash the pellet briefly with 70% ethanol and re-centrifuge for 5 min. Remove the ethanol and allow the pellet to dry.

12. Redissolve the RNA pellet with 0.1% DEPC-treated water. Determine the RNA concentration by taking a small volume of RNA solution and measuring the absorption at 260 and 280 nm.

13. Aliquot the RNA into microcentrifuge tubes, add 3 vol. of ethanol, and store the RNA at −70°C until needed. To recover the RNA, add 0.1 vol. of 3 M NaAc, mix well, and centrifuge at 12 000 *g* at 4°C.

B. *Isolation of total RNA by acid phenol/chloroform extraction*

1. Place tissue or cell pellet (free of cultured medium) into 12 ml of prechilled denaturing solution and homogenize in a Dounce homogenizer or Polytron homogenizer for 30 sec–2 min.

2. Add 1.2 ml of 2 M NaAc, pH. 4.0, and mix thoroughly.

3. Extract with an equal volume of phenol/chloroform:isoamyl alcohol (25:24:1) by: mixing, chilling on ice for 15 min, then centrifuging for 20 min at 10 000 *g* at 4°C in a high speed centrifuge.

4. Remove the aqueous phase to a new test tube and re-extract with an equal volume of chloroform:isoamyl alcohol.

5. Precipitate the RNA by adding an equal volume of isopropanol and incubate the mixture at −20°C for at least 1 h.

6. Collect the RNA by centrifugation at 12 000 *g* for 10–15 min at 4°C. Wash the pellet briefly with 70% ethanol, then re-centrifuge for 5 min. Remove the ethanol and allow the pellet to dry.

7. Redissolve the RNA pellet in 5 ml of denaturing solution and repeat steps 2–6.

8. Redissolve the RNA pellet with 0.5–1 ml DEPC-treated water. Determine the RNA concentration by taking a small volume of RNA solution and measuring the absorption at 260 and 280 nm.

> **9.** Aliquot the RNA into microcentrifuge tubes, add 3 vol. of ethanol, and store the RNA at $-70\,°C$ until needed. To recover the RNA, add 0.1 vol. of 3 M NaAc, mix well, and centrifuge at $12\,000\,g$ at $4\,°C$.

The quality and quantity of the mRNA prepared by either of these two methods are generally good and sufficient for multiple analyses of GSH peroxidase expression by Northern or dot and slot hybridization analyses. However, for tissues which the expression of a particular GSH peroxidase is relatively low, enrichment of the mRNA in the total RNA preparation may be needed. This can be accomplished by selection of poly(A)$^+$ RNA from the total RNA preparation using oligo(dT)–cellulose column as described by Aviv and Leder (34). In this method, columns of oligonucleotide chains of deoxythymidine (oligo(dT) bound to cellulose are used that bind, and thus retain, the polyadenine nucleotide tails that occur on all eukaryote mRNA (poly(A)$^+$ RNA). Oligo(dT)–cellulose can be obtained commercially. Although other methods of isolation, such as the use of magnetic beads, also exist and are available commercially in our experience, the oligo(dT)–cellulose method works well and is simple to perform.

4. Analysis of steady-state mRNA by Northern hybridization

Analysis of GSH peroxidase mRNA can be performed using nitrocellulose or positively charged nylon membranes after the RNA has been denatured by glyoxal and dimethyl sulphoxide (DMSO) or formaldehyde. In our experience, nitrocellulose usually yields a slightly cleaner background than the nylon membrane. However, nitrocellulose is not durable enough to withstand many rounds of washing and rehybridization, whereas nylon membrane usually lasts for many rounds of hybridization and washing. We, therefore, prefer to use the nylon membrane when the RNA will be hybridized with many probes sequentially. The following procedure for Northern analysis works well for both nitrocellulose and many types of nylon membrane. However, for a particular type of membrane, these conditions may not be optimum and the reader should refer to the instructions of the manufacturer for that particular membrane.

To perform Northern analysis, the RNA is first denatured by glyoxal and DMSO or formaldehyde and separated by electrophoresis through agarose gels. Because glyoxal reacts with ethidium bromide, the gels have to be run in the absence of the dye, and the glyoxal has to be removed before staining the gels. Therefore, the use of formaldehyde is preferred for denaturation and separation.

Protocol 3. Analysis of steady-state mRNA by Northern hybridization

Equipment and reagents

- Horizontal gel electrophoresis tank and power supply
- Capillary transfer system
- Variable temperature water bath
- Hybridization oven (optional)
- Nitrocellulose or nylon membrane
- Whatman 3MM paper
- UV illuminator/photographic system
- UV cross-linker or vacuum oven
- 10× electrophoresis running buffer (1000 ml): dissolve 41.6 g 3-(N-morpholino) propanesulphonic acid (0.2 M) in 800 ml DEPC-treated 100 mM sodium acetate (8.2 g), and adjust to pH 7.0 with 2 M NaOH. Then mix with 20 ml 0.5 M DEPC-treated EDTA (10 mM), adjust to 1 litre, filter through a 0.45 μm filter and store at room temperature protected from light
- Tracking dye (10 ml): 0.04 g bromophenol blue and 0.04 g xylene cyanol in 5 ml of 6× electrophoresis running buffer and 5 ml glycerol
- Formaldehyde–agarose gel (100 ml) prepared by boiling 1.1 g agarose in 74 ml water. After cooling to 60°C, mix the gel with 10 ml of 10× running buffer and 16 ml of 37% formaldehyde, cast in a chemical hood, and allow to solidify for 30–60 min.
- Prehybridization solution (100 ml): 50 ml 50% deionized formamide mixed with 25 ml 5× SSPE (from 20 × stock), 10 ml 5× Denhardt's solution (from 50× stock), 1 ml 0.1% sodium dodecyl sulphate (from 10% stock), 1 ml 10 μg/ml sonicated salmon sperm DNA (from 10 mg/ml stock) and 13 ml DEPC-treated water

- 20× standard saline–citrate (SSC) (1 litre) 175.3 g NaCl (3 M) and 88.2 g sodium citrate (0.3 M) in 800 ml water. Adjust the volume to 1 litre at pH 7.0
- 20× standard saline phosphate (SSPE) (1 litre): 175.3 g NaCl (3 M), 24.0 g NaH_2PO_4 (0.2 M), and 7.4 g Na_2 EDTA (0.02 M) in 800 ml water; adjust the pH to pH 7.4 with 2 M NaOH, and the volume to 1 litre
- 50× Denhardt's solution (100 ml): 1 g Ficoll (Type 400, Pharmacia) 1 g bovine serum albumin (Fraction V), and 1 g polyvinylpyrrolidone in 100 ml water, filter through a 0.45 μm filter
- Denatured salmon sperm DNA is prepared by dissolving salmon sperm or herring sperm DNA at a concentration of 10 mg/ml in water by stirring. Shear the DNA by passing through an 18 gauge hypodermic needle 10–20 times or by sonicating for 2 min. Then boil the DNA solution for 10 min, aliquot, and store at −20°C. Just before use, heat the DNA for 5 min in a boiling water bath, followed by a quick chill in ice water.
- Wash solution A (100 ml): 100 ml 20× SSC, 870 ml DEPC-treated water, 10 ml 10% SDS, and 20 ml 1 M sodium pyrophosphate. Water and 20× SSC need to be mixed first before adding SDS to prevent precipitation
- Wash solution B (1000 ml): 5 ml 20× SSC, 965 ml DEPC-treated water, 10 ml 10 SDS, and 20 ml 1 M sodium pyrophosphate, prewarmed to the desired temperature.
- Molecular weight markers (RNA Ladder or 28S and 18S rRNA)

Method

1. Prepare samples by adding 4.5 μl RNA (10–20 μg total RNA or 1–5 μg poly(A)$^+$(RNA), 2 μl 10× electrophoresis running buffer, 3.5 μl 37% formaldehyde and 10 μl formamide (ultrapure, deionized) in a centrifuge tube. Then incubate the samples at 65°C for 15 min, immediately cool on ice, and add 4 μl of tracking dye.

2. Submerge 1.1% formaldehyde–agarose gels in 1× running buffer. Load the molecular weight markers or samples to the well under the buffer, and run the gel at 3–4 V/cm.

3. When the bromophenol blue dye has migrated through approximately three-quarters the length of the gel, stop electrophoresis, cut

out the lane containing the molecular weight marker, and stain with 0.5 mg/ml ethidium bromide for 30–60 min. Photograph the gel alongside a ruler under ultraviolet illumination. Make sure that the ruler indicates the position of the wells. If the integrity of the RNA needs to be confirmed, the entire gel can be stained briefly with ethidium bromide and photographed. However, prolonged staining of the RNA will lead to inefficient transfer of RNA to the filter.

4. Transfer the RNA from the gel to nitrocellulose or positively charged nylon membrane using capillary transfer, vacuum transfer, or electro-blotting. Vacuum and electroblotting transfer should be performed according to the instruction of the manufacturer. Capillary transfer can be performed as follows:

   (a) Rinse the gel with DEPC-treated water several times to remove formaldehyde.

   (b) Place the gel, sample side up, on the wick of a transfer set-up, consisting of a tray filled with 10× SSC, a solid support, and a wick made of 3MM paper. Cut the lower corner of the gel to mark the orientation of the samples. Remove any air bubbles that may be trapped beneath the gel and the filter paper.

   (c) Place a piece of nitrocellulose or nylon filter which has been cut to the same size with the gel and immerse in 10× SSC for 10 min. Do not use the filter unless it is uniformly wet after a few minutes in the SSC. Make sure that there are no air bubbles between the gel and the filter. Surround the gel with plastic wrap or Para-film to prevent liquid short-circuiting without passing through the gel.

   (d) Place three pieces of 3MM paper which have been wetted with 2× SSC on top of the filter.

   (e) Place a stack of paper towels just smaller than the 3MM papers on top of the 3MM papers.

   (f) Place a small weight on top of the paper towels. Allow the RNA to transfer for 10–18 h. As the paper towels become wet, they should be replaced.

   (g) Remove the paper towels, turn over the gel and filter, and mark the position of the well on the filter with a nitrocellulose marking pen or soft lead pencil.

   (h) Briefly rinse the filter in 6× SSC and allow the filter to dry at room temperature.

   (i) Place the dried filter between 3MM papers and bake the filter at 60–80 °C for 2 h in a vacuum oven. UV crosslinking can be used in place of baking.

Protocol 3. *Continued*

5. Place filter in a sealable bag or plastic pan and add prehybridization solution to cover the entire filter. Prehybridize at 42°C for 2–6 h.

6. Label GSH peroxidase cDNA with [α-^{32}P]dCTP using either nick translation or the random hexamers methods. Types of probe other than cDNA can also be used to detect RNA transferred to nitrocellulose or nylon filter; these include oligonucleotides or single-stranded RNA prepared from primer extension, etc.

7. Perform hybridization by adding radiolabelled cDNA probe (2–5 × 10^6 c.p.m./ml) that has been denatured by boiling for 5 min and immediately cooled in ice/water or dry ice/alcohol. Hybridize at 42°C for 24–48 h.

8. Wash the filter for 2 × 15 min with wash solution A at room temperature.

9. Wash the filter for 2 × 30 min with wash solution B at 60–68°C (depending upon the stringency needed).

10. Blot the filter dry and cover with plastic wrap, then expose it for 1–3 days to X-ray film with intensifying screen at −70°C (Kodak XAR-5 or equivalent).

11. Develop the X-ray film, mark the position of the molecular weight markers on the film. The size of (C)GSH peroxidase should appear at 0.9–1.0 kb. The relative quantity of each RNA species on the film can be determined from the relative density of each RNA band on the autoradiograph.

Northern analysis is a convenient method for determining the steady-state level of a particular mRNA because it provides information on the size and quantity of each mRNA species simultaneously. However, if multiple dilutions of mRNA are desired to determine the amount of mRNA, a slot blot or dot blot analysis is also often used although the information obtained from these alternatives may not be as specific or as informative. In addition, if the quantity of the samples is a problem, the expression of GSH peroxidase can be quantified by the polymerase chain reaction after the mRNA has been converted into cDNA. Since increases in the steady-state level of an mRNA could be the result of increased transcription of the gene or increased stability of the mRNA, if a significant increase in the steady-state mRNA is found, nuclear run-on transcription should be performed to determine the change in GSH peroxidase expression.

The success of the expression study is critically dependent on the integrity of the RNA. Thus, care must be made to minimize the activity of RNases liberated during cell fragmentation or contaminated during the RNA preparation and analysis. All solutions used for the preparation of RNA must be treated with inhibitor of RNase and all glassware must be baked at 180°C for several hours. The investigator's hands are also a potential source of RNase, thus disposable gloves must be worn at all times when working with RNA.

If adherent cultured cells are used to prepare intact RNA, the guanidine thiocyanate solution should be added directly to the monolayer cells after removal of the culture medium and washing once with phosphate-buffered saline. The monolayer cells can then be scraped off the plate in the homogenization buffer and homogenization can be accomplished with a Dounce homogenizer.

RNA measurement based on absorption at 260 nm does not always accurately reflect the amount of intact RNA; variations in the amount of RNA used for each sample is normally adjusted by reprobing the same nitrocellulose blot for an RNA species that does not change with the treatment condition under study.

References

1. Mills, G. C. (1957). *J. Biol. Chem.*, **229**, 189.
2. Mills, G. C. (1960). *Arch. Biochem. Biophys.*, **86**, 1.
3. Wendel, A. (1980). In *Enzymatic basis of detoxification* (ed. B. Jakoby), Vol. 1, pp. 333–353. Academic Press, New York.
4. Chow, C. K. (1979). *Am. J. Clin. Nutr.*, **32**, 1066.
5. Flohe, L., Gunzler, W. A., and Schock, H. H. (1973). *FEBS Lett.*, **32**, 132.
6. Flohe, L., Loschen, G., Gunzler, W. A., and Eichele, E. (1972). *Hoppe-Seylers Z. Physiol. Chem.*, **353**, 989.
7. Rotruck, J. T., Pope, A. L., Ganther, H. E., Swanson, A. B., Hafeman, D. G., and Hoekstra, W. G. (1973). *Science*, **179**, 588.
8. Stadtman, T. C. (1979). *Adv. Enzymol.*, **48**, 1.
9. Hafeman, D. G., Sunde, R. A., and Hoekstra, W. G. (1979). *J. Nutr.*, **104**, 580.
10. Smith, P. J., Tappel, A. L., and Chow, C. K. (1973). *Nature*, **247**, 392.
11. Forstrom, J. W., Zakowski, J. J., and Tappel, A. L. (1978). *Biochemistry*, **17**, 2639.
12. Wendel, A., Kerner, B., and Graupe, K. (1978). *Hoppe-Seyler's Z. Physiol. Chem.*, **359**, 1035.
13. Ursini, F., Maiorino, M., Valente, M., Ferri, L., and Gregolin, C. (1982). *Biochim. Biophys. Acta*, **710**, 197.
14. Maiorino, M., Thomas, J. P., Girotti, A. W., and Ursini, F. (1991). *Free Radic. Res. Commun.*, **12–13**, 131.
15. Maddipati, K. R., Gasparski, C., and Marnett, L. J. (1987). *Arch. Biochem. Biophys.*, **254**, 9.
16. Takahashi, K., Avissar, N., Whitin, J., and Cohen, H. (1987). *Arch. Biochem. Biophys.*, **256**, 677.
17. Avissar, N., Whitin, J. C., Allen, P. Z., Wagner, D. D., Liegey, P., and Cohen, H. J. (1989). *J. Biol. Chem.*, **264**, 15850.
18. Thomas, J. P., Maiorino, M., Ursini, F., and Girotti, A. W. (1990). *J. Biol. Chem.*, **265**, 454.
19. Chu, F.-F., Doroshow, J. H., and Esworthy, R. S. (1993). *J. Biol. Chem.*, **268**, 2571.

20. Paglio, D. E. and Valentine, W. N. (1967). *J. Lab. Clin. Med.*, **70**, 158.
21. Lawrence, R. A. and Burk, R. F. (1976). *Biochem. Biophys. Res. Commun.*, **71**, 952.
22. Reddy, C. C., Tu., C.-P., Burgess, J. R., Ho., C.-Y., Scholz, R. W., and Massaro, E. J. (1981). *Biochem. Biophys. Res. Commun.*, **101**, 970.
23. Maiorino, M., Gregolin, C., and Ursini, F. (1990). In *Methods in enzymology* (ed. L. Packer and A. N. Glazer), Vol. 186, pp. 448–457. Academic Press, New York.
24. Chow, C. K. and Tappel, A. L. (1972). *Lipids*, **7**, 518.
25. Pascual, P., Martinez-Lara, E., Barcena, J. A., Lopez-Barea, J., and Toribio, F. (1992). *J. Chromatogr.*, **581**, 49.
26. Chu, F.-F. (1994). *Cytogenet. Cell Genet.*, **66**, 96.
27. Chambere, I., Frampton, J., Goldfarb, P., Affara, N., McBain, W., and Harrison, P. R. (1986). *EMBO J.*, **5**, 1221.
28. Roveri, A., Cassasco, A., Maiorino, M., Dalan, P., Calligaro, A., and Ursini, F. (1992). *J. Biol. Chem.*, **267**, 6142.
29. Chu, F.-F., Esworthy, R. S., Doroshow, J. H., Doan, K., and Liu, X.-F., (1992) *Blood*, **79**, 3233.
30. Schuckelt, R., Brigelius-Flohe, R., Maiorino, M., Roveri, A., Reumkens, J., Strassburger, W., Ursini, F., Wolf. B., and Flohe, L. (1993). *Free Radic. Res. Commun.*, **14**, 343.
31. Takahashi, K., Akasaka, M., Yamamoto, Y., Kobayashi, C., Mizoguchi, J., and Koyama, J. (1990). *J. Biochem. (Tokyo)*, **108**, 145.
32. Chirgwin, J. M., Przybyla, A. E., MacDonald, R. J., and Rutter, W. J. (1979). *Biochemistry*, **18**, 5294.
33. Chomczynski, P. and Sacchi, N. (1987). *Anal. Biochem.*, **162**, 156.
34. Aviv, H. and Leder, P. (1972). *Proc. Natl Acad. Sci. USA*, **69**, 1408.

Superoxide dismutase

SARA GOLDSTEIN and GIDON CZAPSKI

1. Introduction

The superoxide anion radical ($O_2^{\bullet-}$) is considered to be a highly toxic entity in many biological systems (1–4). It is formed in normal metabolism as well as through the action of many drugs, poisons, and radiation (1–4). It is also involved in radiation damage, DNA damage, phagocytosis, ageing, cancer, etc. Superoxide dismutases (SODs) are metalloenzymes that catalyse very efficiently the dismutation of superoxide ions into oxygen and hydrogen peroxide. It has been widely recognized that these enzymes, which are present in almost all living organisms, provide a defence system that is essential for their survival under aerobic conditions. The deleterious role of superoxide during ischaemia (5), which occurs in organ transplantation and many surgical interventions and during heart and brain events, suggests that SOD may have a potential clinical use. This led many investigators to search for SOD mimics, a search which arose because human SODs have short metabolic half-lifes (<10 min), and do not penetrate into the cells.

The alternative to human SOD may include modified human SOD (6), which has a longer metabolic half-life, and SOD mimics that may involve metal compounds (7–12) or non-metal compounds (13), which may have a longer metabolic half-life and penetration ability into the cells. Furthermore, there are additional essential requirements for efficient SOD mimics, working *in vivo*, that will be discussed below. Unless these requirements are fulfilled, the study with these mimics should not be carried on.

2. Mechanism of the catalysis of superoxide dismutation by SOD and SOD mimics

The mechanism of the catalysis of superoxide dismutation by SOD as well as by many other metal compounds and non-metal compounds has been suggested to proceed via the 'ping-pong mechanism', in which the metal or the non-metal compound oscillates between two oxidation states (7–13). This mechanism is illustrated here with copper ions. In this case, Cu(II) may oscillate between Cu(II) and Cu(I) or between Cu(II) and Cu(III):

$$Cu(II) + O_2^{\cdot} \rightleftharpoons Cu(I) + O_2 \tag{1}$$

$$Cu(I) + O_2^{\cdot} + 2H^+ \longrightarrow Cu(II) + H_2O_2 \tag{2}$$

or

$$Cu(II) + O_2^{\cdot} + 2H^+ \longrightarrow Cu(III) + H_2O_2 \tag{3}$$

$$Cu(III) + O_2^{\cdot} \longrightarrow Cu(II) + O_2 \tag{4}$$

The overall reaction according to both of these mechanisms is given by reaction 5:

$$O_2^{\cdot} + O_2^{\cdot} + 2H^+ \longrightarrow O_2 + H_2O_2 \tag{5}$$

These two mechanisms cannot be distinguished kinetically, provided back reaction 1 is negligible. Assuming the steady-state approximation for Cu(I) or Cu(III), rate Equation 6 is derived:

$$-d[O_2^{\cdot}]/dt = k_{cat}[Cu(II)]_o[O_2^{\cdot}] \tag{6}$$

where

$$k_{cat} = 2k_1 k_2/(k_1 + k_2) \quad \text{or} \quad k_{cat} = 2k_3 k_4/(k_3 + k_4) \tag{7}$$

The higher the values of k_1 (k_3) and k_2 (k_4), the more efficient the catalyst. If k_1 (k_3) and k_2 (k_4) differ substantially, k_{cat} approaches the value of the lower rate constant. If reaction 1 cannot be neglected, k_r replaces k_{cat} in eqn 6 and k_r is given by eqn 8:

$$k_r = 2k_1 k_2/(k_1 + k_2 + k_{-1}[O_2]/[O_2^{\cdot}]) \tag{8}$$

In the case of bovine SOD, $k_1 = k_2 = (2-3) \times 10^9 \text{ M}^{-1} \text{ sec}^{-1}$, $k_{-1} = 0.44 \text{ M}^{-1} \text{ sec}^{-1}$ (which can be neglected), and hence it catalyses $O_2^{\cdot}$ dismutation very efficiently. If for SOD mimics, $k_{-1}[O_2]/[O_2^{\cdot}] > (k_1 + k_2)$, then $k_r < k_{cat}$, and therefore such a SOD mimic is less efficient (*Table 1*).

Table 1. Relative SOD activity

$[O_2^{\cdot}]$ (M)	k_{-1} (M^{-1} sec^{-1})					
	1	10	10^2	10^3	10^4	10^5
10^{-6}	1	1	1	1	1	0.994
10^{-9}	1	1	0.994	0.94	0.63	0.14
10^{-10}	1	0.994	0.94	0.63	0.14	0.016
10^{-11}	0.994	0.94	0.63	0.14	0.016	0.0016

The calculation of k_r/k_{cat} was based on eqns 7 and 8, assuming $[O_2] = 0.24$ mM and $k_1 = k_2 = 2 \times 10^9$ M^{-1} sec^{-1}.

3. Methods of determining SOD activity

The difficulty in assaying SOD activity arises from the nature of its substrate ($O_2^{\bullet-}$), which is a free radical with a short half-life in neutral aqueous solutions. More than 20 direct and indirect methods for assaying SOD activity are described in the literature. We shall discuss the difference between the main and most used methods and point out their limitations, as well as their advantages and disadvantages.

3.1 Direct methods

When using a direct method for determining the SOD-like activity, $O_2^{\bullet-}$ is generated at initially relatively high concentrations (>1 μM). The decay of $O_2^{\bullet-}$ absorbance is followed spectrophotometrically in the UV region ($\epsilon_{245} = 2350$ M^{-1} cm^{-1} (14)) in the absence and in the presence of a testing compound. This method affords the most reliable way of ascertaining SOD activity and its reaction mechanism. With this method, it is easy to discriminate between a catalytic and a non-catalytic compound, as under catalytic conditions the initial concentration of $O_2^{\bullet-}$ should always exceed that of the tested compounds.

In the absence of a tested compound, $O_2^{\bullet-}$ decays in a second order process (*Figure 1*; plot a), and its rate is highly pH dependent (*Figure 2*, ref. 14).

$$-d[O_2^{\bullet-}]/dt = 2k_{dis}[O_2^{\bullet-}]^2 \tag{9}$$

In the presence of a testing compound, the decay of O_2^- is given by Equation 10.

$$-d[O_2^{\bullet-}]/dt = 2k_{dis}[O_2^{\bullet-}]^2 + k_{cat}[cat]_o[O_2^{\bullet-}] \tag{10}$$

Under catalytic conditions where $[cat]_o < [O_2^{\bullet-}]_o$, $O_2^{\bullet-}$ will decay in a first order process only if $k_{cat}[cat]_o \gg 2k_{dis}[O_2^{\bullet-}]_o$ (*Figure 1*; plot b). In this case, the observed first order rate constant will depend linearly on $[cat]_o$, and k_{cat} can be determined by plotting the observed first order rate constants as a function of $[cat]_o$. Under these conditions, the direct method is suitable only for very efficient catalysts for which k_{cat} exceeds by two orders of magnitude the self-dismutation rate of $O_2^{\bullet-}$ at the same pH. However, if in the presence of a tested compound the second order decay of $O_2^{\bullet-}$ is accelerated, but its spontaneous dismutation cannot be neglected, as $k_{cat}[cat]_o \approx 2k_{dis}[O_2^{\bullet-}]_o$, the calculation of k_{cat} will be more complicated. If the tested compound is not a catalyst but only a very efficient scavenger of $O_2^{\bullet-}$, there will initially be a very fast decrease in the absorbance of $O_2^{\bullet-}$ until most of the compound is diminished, and then $O_2^{\bullet-}$ will continue its spontaneous slower decay (*Figure 1*; plot c).

3.1.1 Pulse radiolysis

The pulse radiolysis technique is one of the most powerful tools for studying reactions of free radicals. The power of this technique is based on the

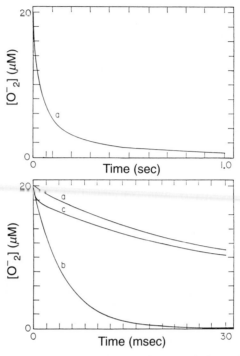

Figure 1. Plot a: the spontaneous decay of 20 μM O_2^- which is formed when air-saturated solutions containing 0.05 M $NaHCO_2$ and 2 mM phosphate buffer (pH 6.8) are pulse irradiated. The pulse duration was 1.5 μsec with 200 mA current of 5 MeV electrons. The decay of O_2^- was followed at 250 nm; Plot b: as in plot a, but in the presence of 2 μM catalyst, for which $k_1 = k_2 = 1 \times 10^8$ M^{-1} sec^{-1}; plot c: as in plot b, but $k_2 = 0$.

capability of producing a large variety of free radicals within less than 1 μsec in physically observable concentrations.

When air- or oxygen-saturated solutions containing formate ions are irradiated, all the primary free radicals formed by the radiation are converted into O_2^- according to the following sequence of reactions (15):

$$H_2O \longrightarrow \text{'OH} (2.75), e_{aq}^- (2.75), H^\bullet (0.60), H_2O_2 (0.75) \qquad (11)$$

(The numbers in parentheses are G-values, which represent the number of molecules formed per 100 eV absorbed by the solution (15).)

$$e_{aq}^- + O_2 \rightarrow O_2^- \qquad\qquad k_{12} = 2 \times 10^{10} \text{ M}^{-1} \text{ sec}^{-1} \qquad (12)$$

$$H^\bullet + O_2 \rightarrow HO_2^\bullet \qquad\qquad k_{13} = 2 \times 10^{10} \text{ M}^{-1} \text{ sec}^{-1} \qquad (13)$$

$$HO_2^\bullet \rightleftharpoons O_2^- + H^+ \qquad\qquad pK_a = 4.75 \qquad (14)$$

$$\text{'OH} + HCO_2^- \rightarrow H_2O + CO_2^- \qquad k_{15} = 3 \times 10^9 \text{ M}^{-1} \text{ sec}^{-1} \qquad (15)$$

$$CO_2^- + O_2 \rightarrow CO_2 + O_2^- \qquad\qquad k_{16} = 3.5 \times 10^9 \text{ M}^{-1} \text{ sec}^{-1} \qquad (16)$$

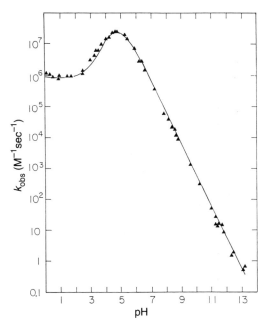

Figure 2. Observed second order rate constant for the decay of O_2^-/HO_2 plotted as a function of pH. Data taken from reference 8.

Protocol 1. Determination of SOD activity using the pulse radiolysis technique

Equipment and reagents
- Pulse radiolysis set-up
- Sodium formate, 1 M
- Sodium acetate buffer, 0.1 M (pH 4–6)
- Sodium phosphate buffer, 0.1 M (pH 6–8)
- Borax buffer, 0.1 M (pH 8–9)
- EDTA, 0.1 mM

Method

1. Use distilled water, which is further purified by a Milli-Q reagent grade water system (Millipore) involving reversed osmosis, ion exchange, and filtration, and yielding a final resistance of water >10 MΩ /cm.

2. The final concentration of formate ions should be higher than 10 mM. Be aware that this concentration controls the ionic strength of the solution, and that k_{cat} decreases with the increase in ionic strength.

3. Adjust the pH of the solutions with NaOH or $HClO_4$ and for the pH range 4.0–9.5 use buffers. The concentration of the buffer should not exceed 1 mM.

4. The initial concentration of O_2^- should exceed the concentration of the tested compound.

245

Protocol 1. *Continued*

5. Irradiate the solutions with a pulse from the accelerator. The O_2^- yield depends on the pulse duration, and the measured absorbance depends on the optical path length of the irradiated cell. These two parameters are specific for each pulse radiolysis set-up.

6. Follow the absorbance of O_2^- at 245–250 nm.

7. If the tested compound catalyses O_2^- dismutation efficiently, so that the self-dismutation can be neglected, plot the observed first order rate constant of O_2^- decay as a function of the initial concentration of the tested compound. The slope of such a plot yields k_{cat}. If the decay of O_2^- is accelerated in the presence of the tested compound, but the spontaneous dismutation cannot be neglected, use an appropriate computer program in order to determine k_{cat}.

8. If the solutions are contaminated with relatively high concentrations of metal impurities, the decay of O_2^- in the absence of a tested compound might appear more like a first order than a second order one. This can be avoided by the addition of 1–2 μM EDTA. Do not add EDTA to systems where EDTA may sequester the metal out of the tested complex. If you cannot add EDTA, you will be able to determine SOD activity of only very efficient catalysts, for which $k_{cat} > 5 \times 10^7$ M^{-1} sec^{-1} at physiological pH.

3.1.2 Stopped-flow kinetics

Unfortunately, the pulse radiolysis technique is of limited use due to the complicated and very expensive equipment necessary for this method. The use of the stopped-flow technique is much cheaper and more available to researchers. The principle of this technique is that KO_2 dissolved in dry and deaerated dimethyl sulphoxide (DMSO) solution is mixed with a buffer solution, and the decay of O_2^- is followed at 245–250 nm in the absence and in the presence of a tested compound (16, 17). Since the time of mixing of the two solutions is within 2 msec, the decay of O_2^- can be studied only at pH > 7, where the half-life of about 100 μM O_2^- is longer than 10 msec (*Figure 2*).

Protocol 2. Determination of SOD activity using the stopped-flow technique

Equipment and reagents

- Stopped-flow apparatus
- Potassium superoxide (Sigma)
- HPLC-grade DMSO (Aldrich)

- Sodium phosphate buffer, 0.1 M (pH 6–8)
- Hepes buffer, 0.1 M (Sigma)
- EDTA, 0.1 mM

Method

1. Use distilled water, which is further purified as in *Protocol 1*.

2. Prepare dry DMSO solutions of KO_2 before each experiment under a dry, inert atmosphere of argon in a dry glovebox using dried glassware. Using a mortar and pestle, grind about 100 mg of the yellow solid of KO_2 with a few drops of DMSO and transfer the slurry to a flask containing 25 ml DMSO. Stir for 30 min and filter. This procedure gives ~2 mM O_2^- in DMSO. Load one of the stopped-flow syringes with this solution.

3. Load the second stopped-flow syringe with a buffer. Use Hepes buffer when free metal ions are tested and phosphate buffer where metal complexes are tested. It is recommended that the buffers are filtered through a Millipore 0.22 μm (pore-size) filter prior to use. The final concentration of the buffer should not exceed 100 mM. Add 1–2 μM EDTA to the buffer solutions when possible.

4. The injection volumes of the buffer solutions should exceed that of the KO_2/DMSO solutions. Use actual ratio of 17/1 to get about 100 μM initial concentration of O_2^-.

5. Follow steps 5–8 exactly as in *Protocol 1*. If the solutions contain traces of metal impurities at relatively high concentrations, the initial concentration of O_2^- will be much lower than 100 μM, and the decay will obey first order kinetics. This will enable you to determine k_{cat} of only very efficient catalysts.

In a modification of the above method, O_2^- is formed from UV irradiation of oxygenated water in one arm of a flow system, and then mixed with the tested compound. This equipment is scarcely used, but is a unique way to study SOD activity with a flow system (18).

3.1.3 The direct spectrophotometry assay with KO2

A conventional spectrophotometry, with KO_2 as source of O_2^-, offers a reliable alternative assay to pulse radiolysis and stopped-flow techniques, which are not commonly available in most laboratories, and is preferred to any of the indirect methods for determining the kinetics and the mechanism of the catalysis. However, due to the relatively fast decay of O_2^- in neutral solutions, this assay is carried out at pH 9.5 (19).

Protocol 3. Determining k_{cat} by direct spectrophotometry

Equipment and reagents

- UV–visible spectrophotometer
- Potassium superoxide (Sigma)
- 2-Amino-2-methyl-1 propanol (AMProp)–HCl buffer, 50 mM at pH 9.5

- Standard 1 cm quartz cuvettes
- NaOH, 50 mM
- Diethylenetriaminepenta-acetic acid (DTPA), 10 mM

Method

1. Use distilled water, which is further purified as in *Protocol 1*.

2. Add 3 ml of the AMProp–HCl buffer to the cuvettes. When assaying for a non-metal compound, add also 0.2 mM DTPA.

3. Dissolve a piece of KO_2 (50–100 mg) in 12.5 ml of ice-cold 50 mM NaOH with or without 0.5 mM DTPA. When the intense bubbling begins to lessen (after about 30 sec), transfer about 7.5–15 μl to the reaction medium in the cuvette, mix rapidly, and monitor the decay of $O_2^{\bullet-}$ absorbance at 245–250 nm.

4. Follow steps 5–8 as in *Protocol 1*.

3.2 The indirect method

When using an indirect method, $O_2^{\bullet-}$ is generated either chemically or enzymatically with a constant flux, and is allowed to react with a detector molecule, D, which scavenges the radical.

$$O_2^{\bullet-} + D \xrightarrow{k_D} O_2 + D^- \tag{17}$$

The yield of D^- or the initial rate of its formation can be followed by its absorbance, luminescence, electron paramagnetic resonance, etc.

Assuming the steady-state approximation for $O_2^{\bullet-}$, the rate of D^- formation in the absence of a tested compound is given by eqn 18, whereas in its presence eqn 19 replaces eqn 18.

$$d[D^-]/dt = V_o = \text{flux} \tag{18}$$

$$d[D^-]/dt = V_c = V_o k_D[D]_o/(k_D[D]_o + k_{cat}[cat]_o) \tag{19}$$

Thus, in the presence of a tested compound, which competes with D for $O_2^{\bullet-}$, the yield of D^- or the initial rate of its formation will decrease. Rearrangement of eqn 19 yields eqn 20, and a plot of V_o/V_c or $[D^-]_o/[D^-]_c$ as a function of $[cat]_o$ should yield a straight line. From the slope of the line, k_{cat} can be calculated.

$$V_o/V_c = [D^-]_o/[D^-]_c = 1 + k_{cat}[cat]_o/k_D[D]_o \tag{20}$$

However, using any of the indirect methods assumes the simple 'ping-pong' mechanism, and neglects any side reactions. The use of indirect methods does not give any insight into the reaction mechanism, and hence does not guarantee the absence of side reactions. If side reactions occur and are neglected, the method will give erroneous results. In these methods $O_2^{\cdot}$ is formed with a constant flux for a time period, Δt. Thus, the total concentration of $O_2^{\cdot}$ formed is $\Delta t \times$ flux.

Assuming that the 'ping-pong' mechanism takes place and no side reactions interfere, the limitation of the indirect assay arises from the following requirements:

(a) Reaction 17 should efficiently compete and prevail the self-dismutation of $O_2^{\cdot}$.

(b) $[D]_o > \Delta t \times$ flux $= [O_2^{\cdot}]_{Total}$

(c) $\Delta t \times$ flux $> [cat]_o$.

(d) $k_D[D]_o \sim k_{cat}[cat]_o$.

These trivial requirements indicate that with indirect methods only values of k_{cat} higher by at least an order of magnitude than k_D can be determined. Therefore, in order to be able to determine relatively less-efficient SOD mimics, one has to choose a detector molecule with low k_D. However, in order to compete efficiently with the spontaneous dismutation of $O_2^{\cdot}$, $k_D[D]_o$ should remain constant. This will require the use of very high concentrations of D when k_D is low, and therefore side reactions may take place and may lead to misinterpretation of the results.

Many indirect methods are described in the literature with various detector molecules. These include cytochrome c (20), nitroblue tetrazolium (NBT) (21), luminol (22), lucigen (23), pyrogallol (24), nitrite (25), adrenalin (20), and tetranitromethane (20). For more information about several other detector molecules see reference 26. However, there are several serious pitfalls and problems arising with each of these methods, which may lead to misinterpretation of the results. Therefore, only the cytochrome c and the NBT assays are described here as these seem at the present time to be the most reliable and easily performed methods for determining the SOD activity of a compound in a test tube. The indirect assays of SOD activities in tissues and biological fluids are not discussed here (for more information see refs 26 and 27).

3.2.1 Sources of $O_2^{\cdot}$

$O_2^{\cdot}$ can be generated with a constant flux either enzymatically or chemically (26). The best-known enzymatic system and the most frequently used is the xanthine/xanthine oxidase system. γ-Radiolysis, UV photolysis, and electrochemical reduction of oxygen are considered as purely chemical sources of $O_2^{\cdot}$. $O_2^{\cdot}$ can also be chemically generated by photoreduction of flavins, autoxidation of quinones and hydroquinones, and by NADH oxidation by phenazine methosulphate (26). We shall describe the xanthine/xanthine oxidase and the γ-radiolysis systems as the best and most convenient sources for $O_2^{\bullet-}$.

i Xanthine/xanthine oxidase

Xanthine oxidase produces $O_2^{\bullet-}$ when it oxidizes aerobically xanthine to urate. The amount of $O_2^{\bullet-}$ production depends on the pH, pO_2 and substrate concentration (26).

Protocol 4. Preparation of xanthine/xanthine oxidase

Equipment and reagents

- UV–visible spectrophotometer
- Standard 1 cm quartz cuvettes
- Xanthine oxidase grade I from buttermilk (Sigma)
- Xanthine (Sigma)
- EDTA, 10 mM
- Phosphate buffer, 0.1 M

Method

1. Use distilled water, further purified as in *Protocol 1*.

2. Dilute the xanthine oxidase to give the desired rate of $O_2^{\bullet-}$ production close to the time of assay.

3. Add EDTA at a final concentration of 0.1 mM in order to complex traces of metal impurities that may inactivate the enzyme. Avoid the use of EDTA when a metal compound is being tested.

4. Reaction mixtures should contain no more than 50 μM xanthine to avoid excess substrate inhibition.

5. Measure the activity of the xanthine oxidase by monitoring the conversion of xanthine into urate at 295 nm. Dilute it to 10^{-2} units/ml in 50 mM phosphate buffer at pH 7.8 and 0.1 mM EDTA. The change in ϵ is approximately 11000 M^{-1} cm^{-1} under these conditions. 1 unit of xanthine oxidase will convert 1 μmol of xanthine into uric acid per min.

---

ii γ-Radiolysis

When air- or oxygen-saturated solutions containing formate ions are γ-irradiated, all the primary free radicals generated by the radiation are converted into $O_2^{\bullet-}$ according to the sequence of reactions 11–16.

3.2.2 The cytochrome *c* assay

The rate constant of the reaction of $O_2^{\bullet-}$ with ferricytochrome *c* (cyt(III)) depends on the pH and the ionic strength of the solution. In the presence of 0.1 mM EDTA and 50 mM phosphate buffer (pH 7.8), $k = (2.6 \pm 0.1) \times 10^5$

M^{-1} sec^{-1} and in the presence of 1 mM phosphate buffer and 10 mM formate it increases to 1.1×10^6 M^{-1} sec^{-1} (28). The formation of ferrocytochrome c (cyt(II)) is followed spectrophotometrically at 550 nm (ϵ = 21 000 M^{-1} cm^{-1} (29)).

Protocol 5. Determination of k_{cat} using the cytochrome c assay

Equipment and reagents

- As in *Protocol 4*
- Continuous radiation source
- Small glass bulbs (5–10 ml)
- 50 μM Ferrocytochrome c

- Ferricytochrome c type VI from horse heart (Sigma)
- Dithionite
- 1 M sodium formate

Method

1. Use distilled water, which is further purified as in *Protocol 1*.

2. Commercial cyt(III) is often contaminated with cyt(II). In order to determine the concentration of cyt(III) in the stock solution, measure the absorbance at 550 nm before and after reduction with dithionite (ϵ = 8900 and 21 000 M^{-1} cm^{-1} at 550 nm for cyt(III) and cyt(II), respectively).

3. Check before you start the assay whether the tested compound or its reduced form reacts with either cyt(II) or cyt(III). This can be done by following the change in the absorbance at 550 nm when 1–5 μM of the tested compound or its reduced form is added to 20 μM cyt(III) at pH 7.8 (50 mM phosphate buffer). A similar procedure can be carried out in the presence of 20 μM cyt(II). Solutions of cyt(II) are prepared by reducing cyt(III) with equimolar concentrations of dithionite. If either the tested compound or its reduced form react with cyt(III) or cyt(II), do not proceed.

4. Use reaction mixtures containing 50 μM xanthine, 0.1 mM EDTA, 10 μM cyt(III) and 50 mM phosphate buffer at pH 7.8 and at 25°C in a volume of 3 ml. Avoid the use of EDTA when a metal compound is tested.

5. Add an amount of xanthine oxidase (~6 nM), which is sufficient to give a ΔA_{550} of 0.025/min. This corresponds to a flux of 1.2 μM O_2^-/min.

6. Check whether the tested compound inhibits the xanthine oxidase by examining its effect on the rate of the conversion of xanthine into urate at 295 nm in the absence of cyt(III).

7. The concentration of the tested compound should not exceed 1 μM.

8. Follow the formation of cyt(II) at 550 nm in the presence and in the absence of a tested compound, and measure the initial rate of cyt(II) formation during the first 1–3 min.

Protocol 5. *Continued*

9. Plot V_o/V_c as a function of $[cat]_o$. A straight line should be obtained with unity intercept. The slope of this line equals 2.6 k_{cat}.

10. When γ-radiation is used as the source of O_2^-, use reaction mixtures containing 10 mM formate ions, 20 μM cyt(III), 0.1 mM EDTA, and 1 mM phosphate buffer at pH 7.8 prepared in small glass bulbs. (Avoid the use of EDTA when you test a metal complex.)

11. Irradiate the solutions so that the total concentration of O_2^- should not exceed 20 μM, and measure the absorbance at 550 nm immediately after the irradiation. The yield of cyt(II) should be linearly dependent on the dose.

12. The concentration of the tested compound should not exceed the concentration of the total O_2^- formed by the radiation (< 20 μM).

13. At a constant dose, measure A_{550} in the absence and in the presence of a tested compound.

14. Plot $[cyt(II)]_o/[cyt(II)]_c$ as a function of $[cat]_o$. A straight line should be obtained with an intercept 1. The slope of this line equals $22 \times k_{cat}$.

3.2.3 The NBT assay

In this method NBT is used as the detector molecule, and the reduction of NBT^{2+} to blue formazan (MF^+) by O_2^- is followed spectrophotometrically at 550 nm. The rate constant of the reaction of O_2^- with NBT has been determined to be $(5.88 \pm 0.12) \times 10^4 \, M^{-1} \, sec^{-1}$, independent of pH over the range 5.7–10.5 (30). The γ-radiolysis cannot be used as a source for O_2^- in this assay due to the very fast reaction of NBT^{2+} with CO_2^-, which is formed via reaction 15.

Protocol 6. Determination of k_{cat} using the NBT assay

Equipment and reagents
- As in *Protocol 4*
- NBT (Sigma)

Method

1. Use distilled water, further purified as in *Protocol 1*.

2. Check before you start the assay whether the tested compound or its reduced form reacts with NBT. This can be done by following the change in the absorbance at 550 nm when 1–5 μM of the tested compound or its reduced form is added to 100 μM NBT at pH 7.8 (50 mM phosphate buffer). If either the tested compound or its reduced form reacts with NBT, do not proceed.

3. Reaction mixtures contain 50 μM xanthine, 0.1 mM EDTA, 100 μM NBT, and 50 mM phosphate buffer at pH 7.8 and at 25°C in a volume of 3 ml. Avoid the use of EDTA when a metal compound is tested.

4. Add an amount of xanthine oxidase, which is sufficient to give an ΔA_{550} of 0.025/min. This corresponds to a flux of about 1 μM $O_2^{\bullet}$/min.

5. Check whether the tested compound inhibits the xanthine oxidase by examining whether it effects the rate of the conversion of xanthine into urate at 295 nm in the absence of NBT.

6. The concentration of the tested compound should not exceed 1 μM.

7. Measure the initial rate of the formation of MF^+ during the first 1–3 min.

8. Plot V_o/V_c as a function of $[cat]_o$. A straight line should be obtained with unity intercept. The slope of this line equals $5.88 \times k_{cat}$.

4. Properties required for SOD mimics operating *in vitro* to work also *in vivo*

The SOD activity of a compound can be determined either directly or indirectly. However, the ability of a compound to catalyse $O_2^{\bullet}$ dismutation *in vitro*, does not necessarily indicate that this compound will be an efficient SOD mimic *in vivo*. There are some additional requirements for an efficient SOD mimic *in vitro* to act also efficiently *in vivo*, which include that these mimics should be non-toxic, non-immunogenic and preferentially not expensive. These compounds should also have a high metabolic half-life and should be able to penetrate into the cells. It is also required from a metal compound to have a high stability constant, otherwise, the metal will be sequestered out of the complex by the cell components, which are present at relatively very high concentrations. In addition, both metal and non-metal compounds should not form ternary complexes with the cell components, as these ternary complexes may be less efficient catalysts, if at all (31).

From the thermodynamic point of view, it is required that $\Delta E_1 (\Delta E_3) > 0$ and $\Delta E_2 (\Delta E_4) > 0$. Therefore, under standard conditions and at pH 7, where $E^0_{(O_2^{\bullet}, H^+/H_2O_2)} = 0.94$ V and $E^0_{(O_2/O_2^{\bullet})} = -0.16$ V (32),

$$-0.16 \text{ V} < E^0_{Cu(II)/Cu(I)} (E^0_{Cu(III)/Cu(II)}) < 0.94 \text{ V}$$

However, *in vivo* concentrations of $O_2^{\bullet}$, O_2 and H_2O_2 differ from standard conditions (1 M for each of the components). Typical concentrations of O_2, $O_2^{\bullet}$, and H_2O_2 under *in vivo* conditions are given in *Table 2*. Under these conditions, the reduction potential of an efficient SOD mimic should be in a range 0.28–0.64V, and for less-efficient catalysts 0.16–0.76V (*Table 2*).

Table 2. $E^0_{Cu(II)/Cu(I)}{}^a$ required for an efficient SOD mimic *in vivo*

	$k_1/k_2 = 0.01$	$k_1/k_2 = 1$	$k_1/k_2 = 100$
$[O_2] = 2.4 \times 10^{-4}$ M	> 0.16 V	> 0.28 V	> 0.39 V
$[O_2^{\bar{\cdot}}] = 10^{-11}$ M			
$[O_2^{\bar{\cdot}}] = 10^{-11}$ M	< 0.53 V	< 0.64 V	< 0.76 V
$[H_2O_2] = 10^{-6}$ M			

^{a}In the case of $E^0_{Cu(III)/Cu(II)}$ k_3/k_4 should be used instead of k_1/k_2.

References

1. Oberly, L. W. (ed.) (1982). *Superoxide dismutase*, Vols I and II. CRC Press, Boca Raton, FL.
2. Rotelio, G. (ed.) (1986). *Superoxide and superoxide dismutase in chemistry, biology and medicine*. Elsevier Science Amsterdam.
3. Harman, D. (1993). In *Free radicals: from basic science to medicine* (ed. G. Poli, E. Albani, and M. O. Dizani), pp. 124–143. Birkhauser Verlag, Basel.
4. Von Sonntag, C. (ed.) (1987). *The chemical basis of radiation biology*. Taylor & Francis, London.
5. McCord, J. M. (1988). *J. Free Radic. Biol. Med.*, **4**, 9.
6. Somack, R., Saifer, M. G. P., and Williams, L. D. (1991). *Free Radic. Res. Commun.*, **12–13**, 553.
7. Brigelius, R., Spottl, R., Bors, W., Lengfelder, E., Saran, M., and Weser, U. (1974). *FEBS Lett.*, **47**, 72.
8. Klug-Roth, D. and Rabani, J. (1976). *J. Phys. Chem.*, **80**, 587.
9. Einstein, J. and Bielski, B. H. J. (1980). *J. Am. Chem. Soc.*, **102**, 4916.
10. Goldstein, S. and Czapski, G. (1983). *J. Am. Chem. Soc.* **105**, 7276.
11. Goldstein, S. and Czapski, G. (1985). *Inorg. Chem.*, **24**, 1087.
12. Goldstein, S. Czapski, G., and Meyerstein, D. (1990). *J. Am. Chem. Soc.*, **112**, 6489.
13. Samuni, A., Murali-Krishna, C., Riesz, P., Finkelstein, E., and Russo, A. (1988). *J. Biol. Chem.*, **263**, 17921.
14. Bielski, B. H. J. and Allen, A. O. (1977). *J. Phys. Chem.*, **81**, 1048.
15. Buxton, J. V., Greenstock, C. L., Helman, W. P., and Ross, A. B. (1988). *J. Phys. Ref. Data*, **17**, 513.
16. Bull, C., McClune, G. J., and Fee, J. A. (1983). *J. Am. Chem. Soc.*, **105**, 5290.
17. Riely, P. D., Rivers, W. J., and Weiss, R. H. (1991). *Anal. Biochem.*, **196**, 344.
18. Bielski, B. H. J. (1984). In *Methods in enzymology* (ed. L. Packer), Vol. 105, pp. 81–83. Academic Press, Orlando, FL.
19. Bolann, B. J., Henriksen, H., and Ulvik, R. J. (1992). *Biochim. Biophys. Acta*, **1156**, 27.
20. McCord. J. M. and Fridovich, I. (1969). *J. Biol. Chem.*, **244**, 6049.
21. Beauchamp, C. and Fridovich, I. (1971). *Anal. Biochem.*, **44**, 276.
22. Hodgson, E. K. and Fridovich, I. (1973). *Photochem. Photobiol.*, **18**, 451.
23. Storch, J. and Ferber, E. (1988). *Anal. Biochem.*, **169**, 262.
24. Karklund, S. and Marklund, G. (1974). *Eur. J. Biochem.*, **47**, 469.

25. Sun, Y., Oberley, L. W., and Li, Y. A. (1988). *Clin. Chem.*, **34**, 497.
26. Greenwald, R. A. (ed.) (1985). *CRC handbook of methods for oxygen radical research*. CRC Press, Boca Raton, FL.
27. DiSilversto, R. A., David, C., and David, E. A. (1990). *J. Free Radic. Biol. Med.*, **10**, 507.
28 Butler, J., Koppenol, W. H., and Margoliash, E. (1982). *J. Biol. Chem.*, **257**, 10747.
29. VanGelder, B. F. and Slater, E. (1962). *Biochim. Biophys. Acta*, **58**, 593.
30. Bielski, B. H. J., Shiue, G. G., and Bajuk, S. (1980). *J. Phys. Chem.*, **84**, 830.
31. Czapski, G. and Goldstein, S. (1991). *Free Radic. Res. Commun.*, **12–13**, 167.
32. Koppenol, W. H. (1989). In *CRC critical reviews in membrane and lipid peroxidation* (ed. C. Vigo-Pelfrety), Vol. I, pp. 1–13. CRC Press, Boca Raton, FL.

18

Deuterated vitamin E:
measurement in tissues
and body fluids

GRAHAM W. BURTON and MALGORZATA DAROSZEWSKA

1. Introduction

Vitamin E, being the major (≥90%) lipid-soluble, chain-breaking antioxidant in mammalian tissues and body fluids (1–3), is important for the *in vivo* protection of lipid membrane and lipoprotein structures against free radical lipid peroxidation. The most abundant forms of vitamin E in foodstuffs are α- and γ tocopherols (α T and γ T; *Figure 1*).

α-T, the most biologically active form of vitamin E, is not only the most efficient radical-trapping scavenger (i.e. chain-breaking antioxidant) among the tocopherols but also one of the best in the entire class of simple phenolic antioxidants (4, 5).

In vitro studies have shown that the antioxidant activity of α-T is controlled by the chroman head group (4–7), whereas transport between and retention within membranes and lipoproteins is affected largely by the phytyl group (8–10) (see *Figure 1*).

The potential importance of vitamin E in controlling lipid peroxidation *in vivo* makes it essential to know more about its absorption, transport, and distribution behaviour in tissues. Previous studies of the intestinal absorption, transport and uptake into tissues of vitamin E *in vivo* have relied upon the use of either radiolabelled tocopherol (11, 12) or large doses of the unlabelled vitamin (13–15). The limitations imposed by these two methods, namely, the hazards associated with the use of radiolabelled substances or the disadvantages of using pharmacological scale doses, have severely hindered progress in obtaining useful information concerning the behaviour of vitamin E *in vivo*. These problems have now been circumvented with the development of a method using deuterium-labelled vitamin E in conjunction with gas chromatography–mass spectrometric (GC–MS) measurement.

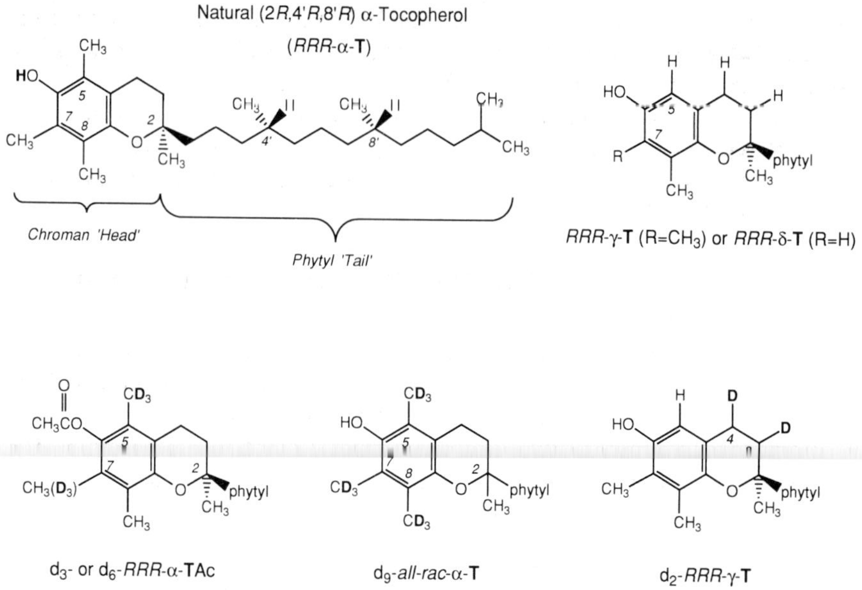

Figure 1. Structures of α- and γ-tocopherols indicating positions of deuterium labelling.

2. Synthesis and administration of deuterium-labelled tocopherols

2.1 Syntheses of deuterated tocopherols

Deuterium-labelled tocopherols are not available commercially. However, multigram amounts of natural 2R,4'R,8'R-α-T (RRR-α-T) containing three and six atoms of deuterium per molecule can be prepared by deuteriomethylation of γ- and δ-T, respectively (16–18). The latter are available in abundant quantities from natural source materials, such as soya bean or sunflower oils. The deuteriomethylation reaction is a simple adaptation of existing methylation techniques for converting γ- and δ-T into α-T. The deuterium atoms are introduced specifically into one or both of two, non-labile, metabolically stable, aromatic methyl positions (5-CD$_3$-α-T (d$_3$-α-T); 5,7-(CD$_3$)$_2$-α-T (d$_6$-α-T; *Figure 1*). These particular forms of deuterated α-T may be ingested safely because deuterium is a stable (i.e. non-radioactive) isotope. An internal standard, 2-*ambo*- or all-rac-5,7,8-(CD$_3$)$_3$-α-T (d$_9$-α-T), also has been prepared in a similar way by deuteriomethylation of the corresponding tocols obtained by condensation of hydroquinone with phytol or isophytol, respectively (17).

The purity of each deuterium-substituted α-T at its nominally stated level of deuteration (i.e., tri- or hexadeuterated), typically 83–92%, is limited by the deuterium content of the starting materials (perdeuterated paraformaldehyde and aqueous deuterium chloride) and the degree to which adventitious amounts of water and other solvents with exchangeable hydrogen can be excluded during the synthesis (17, 18).

The deuterium labelling method has been extended to include the use of γ-T labelled with two deuterium atoms (d_2-γ-T; *Figure 4*) (18) in conjunction with a highly labelled (d_{17}) γ-T internal standard derived by deuteration of γ-tocotrienol (19).

2.2 Administration of deuterated tocopherols to humans and animals

Deuterated vitamin E is usually given to humans as a weighed amount enclosed in a gelatin hard shell, slip-joint capsule and consumed with or shortly after a meal. Administration to laboratory animals (e.g. rats) is either by direct incorporation of deuterated tocopheryl acetate (α-TAc) into their food as the sole source of vitamin E, or by gavage of a solution in a suitable solvent, e.g. tocopherol-stripped corn oil.

3. Extraction of vitamin E from tissues, cell suspensions, and fluids

Tocopherols are extracted from plasma, red cells, and animal tissues and analysed essentially as has been described previously (10, 16, 20–22). Tissues and fluids are stored at $-80\,^{\circ}$C prior to extraction by *Protocol 1* and purification by high performance liquid chromatography (HPLC).

Protocol 1. Basic method for extraction of vitamin E from biological fluids

Equipment and reagents

- Ethanol (absolute)
- *n*-Heptane (HPLC grade)
- *n*-Decane (Aldrich Gold Label, >99%)
- Disposable glass test tubes (13 × 100 mm)
- Positive displacement pipettes (Gilson Pipetman 1 ml, Microman M25 (25 μl) and M50 (50 μl)), pipette tips
- Pasteur pipettes
- HPLC vials (2 ml)
- Vortex-stirrer
- Clinical bench-top centrifuge
- d_9-α-T[a] internal standard in *n*-decane (0.2 mM and 0.02 mM)
- d_{17}-γ-T[a] internal standard in *n*-decane (0.6 mM), if analysing for γ-T

A. *Plasma, lymph and serum samples*

1. Pipette 1 ml of fluid sample (e.g. plasma), thawed on ice, into a disposable test tube.[b]

2. Add vitamin E internal standard (50 μl of 0.2 mM d_9-α-T (10 nmol) and 25 μl of 0.6 mM d_{17}-γ-T (15 nmol)) if γ-T analysis required.

3. Add 1 ml of absolute ethanol, vortex-stir for 10 sec.

Protocol 1. *Continued*

4. Add 1 ml of heptane, vortex-stir for 1 min.

5. Centrifuge mixture for 3 min at 1700 g.

6. Collect top layer with a Pasteur pipette, place into an HPLC vial and store the capped vial at $-20\,^{\circ}C$ until the HPLC purification step (*Protocol 4*).

B. *Synovial fluid*

1. Pipette 1 ml of synovial fluid thawed on ice into a disposable test tube.

2. Add 50 μl of 0.02 mM d_9-α-T internal standard (1 nmol).

3. Proceed as for steps 3–6 in *Protocol 1A*.

[a]Custom synthesized (see section 2.1)
[b]Smaller volumes (e.g. 200 μl) of plasma may be analysed. The volumes of ethanol and *n*-heptane are correspondingly reduced to maintain a 1:1:1 ratio of aqueous fraction:ethanol:*n*-heptane. The amount of internal standards added is reduced accordingly.

The sodium dodecyl sulphate (SDS) method (*Protocol 2*), an elaboration of the extraction method used for plasma, is used for all tissues except adipose tissue and skin.

Protocol 2. The SDS method for extraction of vitamin E from cells and tissues

Equipment and reagents

- d_9-α-T (and d_{17}-γ-T if analysing for γ-T) internal standard in *n*-decane (as described in *Protocol 1*)
- Sodium ascorbate (Sigma)
- Glass-distilled water
- SDS, 0.5 M aqueous solution
- Ethanol (absolute)
- *n*-Heptane (HPLC grade)
- Positive displacement pipettes (Gilson Pipetman 1 ml, Microman M25 (25 μl) and M50 (50 μl))
- Disposable glass test tubes (13 $\times$ 100 mm)
- Centrifuge tubes (50 ml)
- Pipette tips
- Pasteur pipettes
- HPLC vials (2 ml)
- Analytical balance
- Homogenizer (Brinkmann PT10/35) equipped with Brinkmann Polytron[R] PTA 10TS probe
- Vortex-stirrer
- Clinical bench-top centrifuge

A. *Red blood cells*

1. Pipette 1 ml of a thawed red blood cell suspension of known haematocrit value into a centrifuge tube.

2. Add 50–100 mg sodium ascorbate, vitamin E internal standard (25 μl of 0.2 mM d_9-α-T (5 nmol), and, if required, 25 μl of 0.6 mM d_{17}-γ-T (15 nmol)), and vortex-stir for 10 sec.[a]

3. Add 1 ml of 0.5 M SDS solution and vortex-stir for 10 sec.

4. Add 2 ml of absolute ethanol; vortex-stir for 10 sec.

5. Add 3 ml of heptane; vortex-stir for 1 min.

6. Proceed as for steps 5 and 6 in *Protocol 1A.*

B. *Platelets*

1. Pipette 0.8 ml of a thawed saline platelet suspension[b] into a disposable test tube.

2. Add 50–100 mg sodium ascorbate, vitamin E internal standard (25 μl of 0.02 mM d_9-α-T (0.5 nmol), and, if required, 25 μl of 0.6 mM d_{17}-γ-T (15 nmol)); vortex-stir for 10 sec.

3. Add 0.5 ml of 0.5 M SDS solution; vortex-stir for 10 sec.

4. Add 2 ml of absolute ethanol; vortex-stir for 10 sec.

5. Add 1 ml of heptane; vortex-stir for 1 min.

6. Proceed as for steps 5 and 6 in *Protocol 1A.*

C. *Buccal cells*

1. Measure 1 ml of buccal cell suspension in saline solution[c] into a disposable test tube.

2. Add 50–100 mg sodium ascorbate and internal standard (25 μl of 0.02 mM d_9-α-T (0.5 nmol) and, if required, 25 μl of 0.6 mM d_{17}-γ-tocopherol (15 nmol)); vortex-stir for 10 sec.

3. Add 1 ml of 0.5 M SDS solution; vortex-stir for 10 sec.

4. Proceed as for steps 4–6 in *Protocol 2B.*

D. *Tissues (except adipose tissue and skin)*

1. Weigh approximately 0.5 g or less of tissue into a centrifuge tube, mince, and add 2 ml of distilled water.

2. Add vitamin E internal standard (25 μl of 0.2 mM d_9-α-T (5 nmol), and, if required, 25 μl of 0.6 mM d_{17}-γ-T (15 nmol)).

3. Homogenize in centrifuge tube.

4. Add 1 ml of 0.5 M SDS solution; vortex-stir for 10 sec.

5. Add 3 ml of absolute ethanol; vortex-stir for 10 sec.

6. Add 3 ml heptane; vortex-stir for 1 min.

7. Proceed as for steps 5 and 6 in *Protocol 1A.*

E. *Special application: bile*

1. Place 1–1.5 ml of thawed bile in a disposable test-tube containing 50 mg sodium ascorbate and vitamin E internal standard (40 μl of

Protocol 2. *Continued*

    0.2 mM d_9-α-T (8 nmol), and, if required, 40 μl of 0.6 mM $d_{17}\gamma$-T (24 nmol)).

2. Add 2 ml of 0.5 M SDS solution; vortex-stir for 10 sec.

3. Add 2 ml of ethanol; vortex-stir for 10 sec.

4. Add 2 ml heptane; vortex-stir for 1 min.

5. Proceed as for steps 5 and 6 in *Protocol 1A*.

---

[a] Sodium ascorbate is added to reduce any Fe(III) to Fe(II) and thus prevent direct oxidative loss of α-T.

[b] The washed platelets are stored frozen in saline (0.3–1.2 $\times$ 10^9 cells/ml saline) at $-80\,°C$.

[c] Human buccal cells from the oral mucosa of the cheek are readily obtained by gently brushing with a new, soft bristle tooth brush and collecting saline rinsings of the tooth brush and the inside of the mouth, centrifuging at 1400 r.p.m. (*c.* 380 *g*) for 15 min at 5°C in a bench top centrifuge, removing the supernatant, resuspending in saline, and storing frozen at $-80\,°C$.

The potassium hydroxide (KOH) method (*Protocol 3*) is used for adipose tissue because of the large amount of interfering lipid and for skin because the strongly alkaline conditions of the method facilitate the physical disruption of the tissue.

Protocol 3. The KOH method for extraction of vitamin E from tissues

Equipment and reagents

- d_9-α-T and d_{17}-γ-T (if required) internal standards in *n*-decane (as described in *Protocol 1*)
- Glass-distilled water
- Saturated aqueous solution of KOH
- Ethanol (absolute) containing 1% of sodium ascorbate
- Positive displacement pipettes (Gilson Pipetman 1 ml, Microman M25 (25 μl) and M50 (50 μl)) pipette tips

- *n*-Heptane (HPLC grade)
- Screw-topped centrifuge tubes (20 ml)
- Pasteur pipettes
- HPLC vials
- Analytical balance
- Vortex-mixer
- Electrically heated evaporation block (Thermolyne dri-bath)
- Ice bath
- Clinical, bench-top centrifuge

A. *General procedure*

1. Weigh *c.* 0.1 g of tissue into a screw-topped centrifuge tube, mince, and add 1 ml of distilled water.

2. Add internal standard (25 μl of 0.2 mM d_9-α-T (5 nmol) and, if required, 25 μl of 0.6 mM d_{17}-γ-T (15 nmol)) and 2 ml of ethanol containing 1% sodium ascorbate; vortex-stir for 10 sec.

3. Add 0.3 ml saturated KOH, vortex-stir for 10 sec, cap tube firmly and heat at 70°C in an electrically heated block for 30 min.[a]

4. Cool tube in ice.

5. Add 1 ml of water, 3 ml heptane and vortex-stir for 1 min.

6. Proceed as for steps 5 and 6 in *Protocol 1A*.

B. *Special application for small tissue samples: human myocardial biopsies (23)*

1. Thaw the myocardial tissue sample (approximately 2–10 mg),[b] blot dry, weigh, place into a screw-top centrifuge tube and add 0.5 ml water.

2. Add internal standard (15 μl of 0.2 mM of d$_9$-α-T (3 nmol) and, if required, 15 μl of 0.6 mM of d$_{17}$-γ-T (9 nmol)) and 2 ml of absolute ethanol containing 1% sodium ascorbate; vortex-stir for 10 sec.

3. Proceed as for steps 3–6 in *Protocol 3A*.

[a] For small samples (*c.* 25 mg) the sample is not minced, the amount of internal standard is reduced to *c.* 1 nmol and the heating time is reduced to 10 min.

[b] Transmural left ventricular biopsy specimens are obtained with a Trucut needle while the patients are on cardiopulmonary bypass, with the onset of reperfusion immediately after cross-clamp release and after 20 min of reperfusion. The biopsied tissue samples, immersed immediately in ice-cold saline, are frozen in liquid nitrogen and stored at −80°C until needed for analysis.

4. Purification and derivatization

4.1 Purification

In order to obtain successful GC–MS results the lipid extract must first be partially purified by HPLC to obtain a tocopherol-enriched fraction that is largely free of other, potentially interfering, lipids. *Protocol 4* illustrates how this is done in the authors' laboratory. (Mention of particular equipment brand names does not imply that the procedure cannot be performed equally well with comparable products from other manufacturers and suppliers.)

Protocol 4. HPLC purification of lipid extracts

Equipment and reagents

- High pressure liquid chromatograph Varian model 5000
- 5 μm LiChrosorb Si 60 column (250 mm × 4 mm; Merck)
- Varian Fluorichrom detector equipped with a deuterium lamp supply, a 220 nm interference excitation filter, and a 2 mm thick Schott VG-1 glass band emission filter
- Varian autosampler model 9090
- Fraction collector (FOXY™ series 2130–00, ISCO Inc.)
- Varian DS 654 control station and data handling system
- 13 × 100 mm glass collection tubes
- *n*-Heptane, methyl *tert*-butyl ether (HPLC grade)

Protocol 4. *Continued*

HPLC operating conditions

1. Mobile phase: 90% heptane/10% methyl *tert*-butyl ether.
2. Flow rate: 2 ml/min with a sample run time of 6 min.
3. Temperature: 20°C.
4. Injection volume: 700 μl of lipid extract.
5. Sample collection: tocopherol fraction eluting at 1.8–3.0 min (refrigerate at *c.* −20°C if GC–MS analysis is not to be performed immediately).

4.2 Derivatization

Tocopherols are usually converted into their trimethylsilyl ethers before analysis by GC–MS (*Protocol 5*). The tocopheryl silyl ether elutes with a more symmetrical peak shape that improves the accuracy of peak area measurement. Also, the greater volatility of the silylated lipid samples results in less sample contamination of the mass spectrometer, thus reducing mass spectrometer down-time for source cleaning.

Protocol 5. Derivatization of purified lipid extracts

Equipment and reagents

- Pyridine (silylation grade), *N,O-bis* (trimethylsilyl)-trifluoroacetamide (BSTFA) with 1% trimethylsilane (Pierce)
- Vials with tin foil-lined caps
- Electrically heated evaporation block (Thermolyne dri-bath)

Method

1. Place vial containing collected HPLC fraction under a stream of N_2 and evaporate solution to dryness.
2. Add 100 μl pyridine and 50 μl BSTFA to remaining residue.
3. Cap vials and heat mixture at 65°C in electrically heated block for at least 15 min.

Adipose tissue samples and other samples containing a lot of unlabelled α-T (i.e. d_0-α-T) compared to d_3-α-T are analysed directly by GC–MS as the free tocopherols. The absence of the silyl group eliminates the contribution that silicon isotopes, present in the d_0-α-T silyl ether, make to the d_3 peak area, reducing the natural abundance isotope correction from *c.* 2.4% to *c.* 0.6% (see Section 5.1).

5. GC–MS analysis of tocopherols

The relative amounts of deuterated and non-deuterated α-tocopherols (d_0-α-T) present in lipid extracts of biological tissues and fluids are determined by

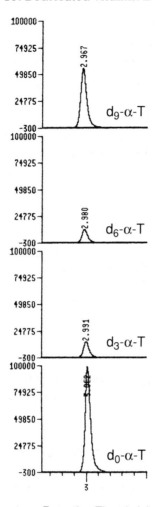

Retention Time (min)

Figure 2. Example of graphical output from GC–MS analysis of d_0-, d_3-, d_6-, and d_9-α-Ts (internal standard) in the lipid extract of a plasma sample obtained from a subject 6 h after taking an oral dose of 100 mg of a 1:1 mixture of d_3- and d_6-*RRR*-α-T succinate esters. The α-Ts were analysed as their silyl ethers using the single ion monitoring mode for each parent ion, i.e. 502.4 (d_0), 505.4 (d_3), 508.4 (d_6), and 511.4 (d_9) mass units. The vertical scale is given in arbitrary units. The numbers above each peak denote the retention time. The calculated peak areas, expressed relative to d_9-α-T, are given in *Table 1*.

injecting the (silylated) HPLC-purified tocopherol fractions into a commercially available, bench-top type, gas chromatograph–mass spectrometer (GC–MS) (see *Protocol 6*) (16). As the α-tocopheryl silyl ether emerges from the gas chromatograph and enters the mass spectrometer, it is resolved simultaneously into its various component parent ions (*Figure 2*). The absolute concentration of each tocopherol is readily determined by relating the peak area of its parent ion to that of the internal standard, d_9-α-T (*Figure 1*), added in known amount to the sample just prior to extraction.

Protocol 6. GC–MS analysis

Equipment

- Hewlett Packard (HP) 5890 gas chromatograph
- HP Ultra 1 fused silica capillary column (12 m × 0.2 mm ID, cross-linked methyl silicone bonded phase)
- HP 7673A autosampler
- HP 5970A Series Mass Selective Detector
- HP 59970 MS Chem Station
- HP polypropylene vials with Teflon-lined septum caps

GC–MS operating conditions

1. Injection port temperature: 300°C.
2. Oven temperature: 280°C.
3. Split ratio: 30:1.
4. Column pressure: 10 psi.
5. Volume of sample injected: typically 1 µl.
6. Mass selective detector is used in the single ion monitoring mode (SIM).
7. Ions monitored: 502.4 (d_0), 505.4 (d_3), 508.4 (d_6), and 511.4 (d_9) mass units for the silyl ethers or 430.4, 433.4, 436.4, and 439.4 mass units, respectively, for free tocopherols.

5.1 Calculation of tocopherol concentrations

GC–MS batch reports are transferred to a personal computer (Microsoft DOS operating system) as ASCII files for processing by wordprocessor (WordPerfect, DOS version 5.1 (Novell)) into a form amenable to spreadsheet and database analysis (Lotus 123 (Lotus Development Corp.); Paradox, DOS version 2 (Borland)). Concentrations of d_0-α-T, d_3-α-T and d_6-α-T are obtained from the peak areas of the corresponding parent ions in the mass spectrum expressed relative to the peak area of the d_9-α-T internal standard (see *Figure 2*). Often, natural abundance isotopes (mostly ^{13}C, ^{29}Si, ^{30}Si) present in a tocopherol or its derivative contribute to the peak area of the parent ion of an isotopic tocopherol with a higher degree of deuterium substitution. The contribution of this isotopic contamination must be subtracted from the measured peak area of the deuterated parent ion in order to obtain the true peak area. In the present situation in which d_0-, d_3-, d_6-, and d_9-α-T are used, the necessary corrections can be made in a sequential manner by estimating the contribution of d_0 to d_3, d_3 to d_6, and d_6 to d_9.

The relative peak areas obtained from the GC–MS batch report, reported as percentages ($\%d_0$, $\%d_3$, $\%d_6$) relative to the d_9-α-T internal standard ($\%d_9 = 100\%$), are corrected, where appropriate, using eqns 1–3.

$$\%d_3^{corr} = \%d_3 - c \times \%d_0 \tag{1}$$

$$\%d_6^{corr} = \%d_6 - c \times \%d_3^{corr} \tag{2}$$

$$\%d_9^{corr} = \%d_9 - c \times \%d_6^{corr} \tag{3}$$

The correction factor, c, takes the value of 0.024 or 0.006, depending on whether or not the tocopherol is silylated. The concentrations of d_0-, d_3-, and d_6-α-T ($[d_0]$, $[d_3]$, and $[d_6]$, respectively) in the sample are calculated using eqns 4–6 in which d_9^{std} is the amount of d_9-α-T internal standard used, d_3^{pur}, d_6^{pur}, and d_9^{pur} are the isotopic purities of each deuterated tocopherol (typically 0.8–0.9; i.e. the fractional extent to which each tocopherol contains its nominal amount of deuterium), and volume (or weight) is the quantity of fluid or tissue analysed.

$$[d_0] = \%d_0 \times (d_9^{std} \times d_9^{pur} / \%d_9^{corr}) / \text{volume (or weight)} \tag{4}$$

$$[d_3] = (\%d_3^{corr} / d_3^{pur}) \times (d_9^{std} \times d_9^{pur} / \%d_9^{corr}) / \text{volume (or weight)} \tag{5}$$

$$[d_6] = (\%d_6^{corr} / d_6^{pur}) \times (d_9^{std} \times d_9^{pur} / \%d_9^{corr}) / \text{volume (or weight)} \tag{6}$$

The results of a sample calculation applied to the data corresponding to the plots displayed in *Figure 2* are given in *Table 1*

6. Sensitivity and limitations

In the authors' experience, the size of a sample necessary to obtain reliable data requires that there be at least 100 ng (c. 0.2 nmol) of α-T present in the final, concentrated 50 μl volume from which 1 μl is drawn for GC–MS analysis. However, it is possible to detect as little as 40 pg of tocopherol and we have, for example, successfully measured uptake of deuterated α-T into human heart tissue using biopsy samples as small as 1 mg (23).

The reliability of measurements and, therefore, the applicability of the

---

Table 1. Concentrations (μM) of d_0-, d_3- and d_6-α-T ($[d_0]$, $[d_3]$ and $[d_6]$) in plasma of a subject 6 h after an oral dose of 100 mg of a 1:1 mixture of d_3- and d_6-*RRR*-α-T succinate esters. Values are calculated from peak area data ($\%d_0$–$\%d_9$), corresponding to the plots displayed in *Figure 2*, using eqns 1–6

$\%d_0$	$\%d_3$	$\%d_6$	$\%d_9$	$\%d_3^{corr}$	$\%d_6^{corr}$	$\%d_9^{corr}$	$[d_0]$	$[d_3]$	$[d_6]$
161.5	26.5	21.0	100.0	22.6	20.4	99.5	17.2	2.84	2.66

Correction, $c = 0.024$ (used in Equations 1–3); $d_9^{std} = 11.1$ nmol; $d_3^{pur} = 0.85$; $d_6^{pur} = 0.82$; $d_9^{pur} = 0.86$; plasma volume = 0.9 ml.

deuterated tocopherol technique is limited by the amount that naturally occurring isotopes contribute to the peak area of the labelled tocopherol. For example, when unlabelled d_0-α-T and d_3-α T arc analysed as their silyl ethers, the contribution from the natural abundance isotopes (^{13}C, ^{29}Si, ^{30}Si) present in unlabelled (d_0) α-TSi(CH$_3$)$_3$ to the d_3-α-TSi(CH$_3$)$_3$ parent ion peak is 2.37% of the peak area of the unlabelled parent ion. Limiting situations arise when the amount of d_3-α-TSi(CH$_3$)$_3$ is low ($< 5\%$) relative to the amount of unlabelled (d_0) α-TSi(CH$_3$)$_3$. Under these circumstances, the contribution of the d_0 silyl ether to the d_3 peak area is comparable to the true peak area of the d_3 silyl ether. In this situation it is better, if possible, to use d_6-α-T, because the d_6 parent ion peak is sufficiently different in mass that it is not affected by isotopic contributions from the d_0 ion.

Another option is to analyse d_3-α-T directly as the free phenol, which obviates the substantial contributions of silicon isotopes (the d_3-α-T peak area correction is reduced from 2.37% to 0.63% of the unlabelled α-T peak area). The drawbacks to this option are that underivatized tocopherols tend to 'tail' on the GC column, thus lowering the accuracy of the peak area integration, and, in the authors' experience, repeated use of underivatized samples leads to a more rapid decline in GC–MS performance.

7. Applications

A big advantage inherent in the use of deuterated tocopherols is that, unlike radiolabelled vitamin E, the compounds may be ingested without risk. Furthermore, as the deuterium appears not to undergo any measurable, metabolically mediated exchange, deuterated tocopherols can be used conveniently

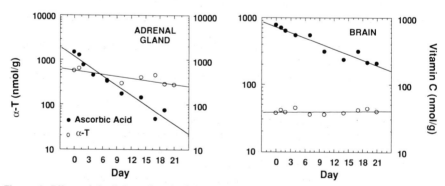

Figure 3. Effect of declining vitamin C (ascorbic acid) levels on concentrations of α-T in adrenal gland and brain tissue from guinea pigs at selected times during a 3-week period during which the animals were fed a diet containing a scorbutic level of vitamin C (10 mg/kg diet) and an adequate level of vitamin E (35 mg α-TAc/kg diet). The animals were maintained beforehand for 2 weeks on a diet containing the same level of vitamin E and an adequate level of vitamin C (250 mg/kg diet). Each point corresponds to a single animal.

and readily in human and animal studies. Thus, these compounds have been used in both short-term and long-term feeding studies in laboratory animals (16, 21, 24) and in single or multiple dose studies in humans (19, 21–23, 25–27), monkeys (28), dogs (29), weanling piglets and calves, and some large zoo animals.

A further advantage of using the GC–MS method with a deuterated internal standard is the enhanced precision of vitamin E measurements. We have used this to examine the influence of dietary vitamin C on levels of vitamin E (α-T) in guinea-pig tissues (E. Pietrzak, unpublished results). *Figure 3* shows that as vitamin C levels decline vitamin E levels also decline in adrenal gland tissue but remain steady in brain tissue.

The GC–MS SIM method is easily modified to allow simultaneous analysis of d_0-γ-T in tissue and fluid samples using d_{17}-γ-T as the internal standard (28).

References

1. Burton, G. W., Joyce, A., and Ingold, K. U. (1983). *Arch. Biochem. Biophys.*, **221**, 281.
2. Cheeseman, K. H., Burton, G. W., Ingold, K. U., and Slater, T. F. (1984). *Toxicol. Pathol.*, **12**, 235.
3. Cheeseman, K. H., Emery, S., Maddix, S. P., Slater, T. F., Burton, G. W., and Ingold, K. U. (1988). *Biochem. J.*, **250**, 247.
4. Burton, G. W. and Ingold, K. U. (1981). *J. Am. Chem. Soc.*, **103**, 6472.
5. Burton, G. W. and Ingold, K. U. (1986). *Accounts Chem. Res.*, **19**, 194.
6. Burton, G. W., Doba, T., Gabe, E. J., Hughes, L., Lee, F. L., Prasad, L., and Ingold, K. U. (1985). *J. Am. Chem. Soc.*, **107**, 7053.
7. Burton, G. W. and Ingold, K. U. (1989). *Ann. NY Acad. Sci.*, **570**, 7.
8. Niki, E., Kawakami, A., Saito, M., Yamamoto, Y., Tsuchiya, Y., and Kamiya, Y. (1985). *J. Biol. Chem.*, **260**, 2191.
9. Niki, E., Komuro, E., Takahashi, M., Urano, S., Ito, E., and Terao, K. (1988). *J. Biol. Chem.*, **263**, 19809.
10. Cheng, S. C., Burton, G. W., Ingold, K. U., and Foster, D. O. (1987). *Lipids*, **22**, 469.
11. Gallo-Torres, H. E. (1980). In *Vitamin E: a comprehensive treatise* (ed. L. J. Machlin), Vol. pp. 193. Marcel Dekker, New York.
12. Gallo-Torres, H. E. (1980). In *Vitamin E: a comprehensive treatise* (ed. L. J. Machlin), Vol. pp. 170. Marcel Dekker, New York.
13. Bieri, J. G. (1972). *Ann. NY Acad. Sci.*, **203**, 181.
14. Machlin, L. J. and Gabriel, E. (1982). *Ann. NY Acad. Sci.*, **393**, 48.
15. Vatassery, G. T., Brin, M. F., Fahn, S., Kayden, H. J., and Traber, M. G. (1988). *J. Neurochem.*, **51**, 621.
16. Ingold, K. U., Burton, G. W., Foster, D. O., Hughes, L., Lindsay, D. A., and Webb, A. (1987). *Lipids*, **22**, 163.
17. Ingold, K. U., Hughes, L., Slaby, M., and Burton, G. W. (1987). *J. Labelled Comp. Radiopharm.*, **24**, 817.

18. Hughes, L., Slaby, M., Burton, G. W., and Ingold, K. U. (1990). *J. Labelled Comp. Radiopharm.*, **28**, 1049.
19. Traber, M. G., Burton, G. W., Hughes, L., Ingold, K. U., Hidaka, H., Malloy, M., Kane, J., Hyams, J., and Kayden, H. J. (1992). *J. Lipid Res.*, **33**, 1171.
20. Burton, G. W., Webb, A., and Ingold, K. U. (1985). *Lipids*, **20**, 29.
21. Burton, G. W., Ingold, K. U., Foster, D. O., Cheng, S. C., Webb, A., Hughes, L., and Lusztyk, E. (1988). *Lipids*, **23**, 834.
22. Traber, M. G., Ingold, K. U., Burton, G. W., and Kayden, H. J. (1988). *Lipids*, **23**, 791.
23. Weisel, R. D., Mickle, D. A. G., Finkle, C. D., Tumiati, L. C., Madonik, M. M., Ivanov, J., Burton, G. W., and Ingold, K. U. (1989). *Circulation*, **80**, Suppl. III, 14.
24. Burton, G. W., Wronska, U., Stone, L., Foster, D. O., and Ingold, K. U. (1990). *Lipids*, **25**, 199.
25. Traber, M. G., Sokol, R. J., Burton, G. W., Ingold, K. U., Papas, A. M., Huffaker, J. E., and Kayden, H. J. (1990). *J. Clin. Invest.*, **85**, 397.
26. Traber, M. G., Burton, G. W., Ingold, K. U., and Kayden, H. J. (1990). *J. Lipid Res.*, **31**, 675.
27. Traber, M. G., Sokol, R. J., Kohlschutter, A., Yokota, T., Muller, D. P., Dufour, R., and Kayden, H. J. (1993). *J Lipid Res*, **34**, 201.
28. Traber, M. G., Rudel, L. L., Burton, G. W., Hughes, L., Ingold, K. U., and Kayden, H. J. (1990). *J. Lipid Res.*, **31**, 687.
29. Traber, M. G., Pillai, S. R., Kayden, H. J., and Steiss, J. E. (1993). *Lipids*, **28**, 1107.

Further reading

1. Burton, G. W. and Ingold, K. U. (1993). In *Vitamin E in health and disease* (ed. L. Packer and J. Fuchs), pp. 329–344. Marcel Dekker, New York.
2. Burton, G. W., Ingold, K. U., Zahalka, H., Dutton, P., Hodgkinson, B., Hughes, L., Foster, D. O., and Behrens, W. A. (1993). In *Vitamin E—its usefulness in health and in curing diseases* (ed. M. Mino, H. Nakamura, A. T. Diplock, and H. J. Kayden), pp. 51–61. Japan Science Society Press, Tokyo.
3. Burton, G. W. and Traber, M. G. (1990). *Annu. Rev. Nutr.*, **10**, 357.

19

Selective and sensitive measurement of vitamin C, ubiquinol-10, and other low-molecular-weight antioxidants

DETLEF MOHR and ROLAND STOCKER

1. Introduction

Inadvertent modification of biological macromolecules by free radical-mediated oxidation is implicated in the pathogenesis of a number of human diseases, such as cancer and atherosclerosis. The latter in particular has received much attention recently. Since oxidation of lipids carried in the circulation by lipoproteins can precede that of the apoprotein, considerable interest has focused on the oxidation of lipoprotein lipids and its inhibition by endogenous antioxidants. In particular, the antioxidative action of ascorbate, α-tocopherol, ubiquinol-10, carotenoids, and bile pigments have been investigated in more or less detail. The selective and sensitive detection of endogeneous anti-oxidants in human plasma and lipoproteins has, therefore, become increasingly important. Although numerous, simple and mostly colorimetric assays are often employed for the measurement of many of the antioxidants, their suit-ability for the selective detection of antioxidants in complex biological mix-tures such as human plasma, is limited. Also, many of the extremely labile (i.e. autoxidizable) antioxidants are present in small concentrations which renders colorimetric assays particularly problematic. Lengthy derivatization procedures required to increase sensitivity can lead to artificial underestima-tion or even complete loss of the compounds of interest. In this situation, high performance liquid chromatographic (HPLC) methods are useful alter-natives and can guarantee a high degree of selectivity and sensitivity. In par-ticular, with the availability of simpler and more reliable electrochemical cells, HPLC with electrochemical detection (HPLC-EC) has become a frequently used detection method. This detection method provides high sensitivity and additional selectivity, due to the different and characteristic redox activity of different antioxidants.

This chapter describes separation and detection methods for the most commonly found antioxidants in human plasma. These methods are used routinely in our laboratory.

Protocol 1. Preparation of human plasma extracts

Equipment and reagents

- Two Dispensettes, 10 ml (Brinkmann)
- Hexane, HPLC grade (Mallinckrodt)
- H₂O, nanopure, Type I HPLC grade, double deionized (Continental Water Systems)
- Methanol, HPLC grade (E. Merck)
- Acetic acid, 96% HPLC grade (Merck)
- Methanol containing 0.02 vol% acetic acid
- Propan 2 ol, HPLC grade (Mallinckrodt)
- Extraction tubes (culture tubes, screw-capped with Teflon lining, 16 × 150 mm, Kimble)

- Argon, high purity (CIG)
- Human blood plasma (obtained from fresh heparinized human blood)
- Rotary evaporator (Büchi)
- 25 ml pear-shaped evaporator flasks
- Bench-top centrifuge, GS-6R (Beckman)
- 0.2 μm nylon filter, HPLC certified (Gelman)
- Na₂EDTA (Merck), saturated in nanopure water
- DL-Homocysteine (Sigma), prepared as a 1% (w/v) solution in H₂O and stored at 4°C

A. *Preparation of plasma (or low density lipoprotein (LDL)) extract for HPLC-EC systems*

1. Saturate hexane with nanopure water by mixing approximately 1/3 vol. H₂O with 2/3 vol. hexane.

2. Allow the two phases to separate and pipette 10 ml of the top layer (hexane) into an extraction tube.

3. Add 200 μl acetic acid (HPLC grade) (0.02%) to 1000 ml of methanol (HPLC grade).

4. Add 2 ml of this acidified methanol to the extraction tube containing 10 ml hexane.

5. Gently pass a stream of argon through the biphasic extraction mixture for 30 sec, close, and keep the tube on ice and away from (room) light if stored for extended periods.

6. Add 200 μl of plasma or 500 μl of LDL (0.1–2 mg protein/ml) to the extraction tube and shake the mixture vigorously for 15 sec.

7. Place the extraction tube into a bench-top centrifuge and centrifuge (5 min, 600 *g*, 4°C), to separate the phases.

8. Pipette 9 ml of the hexane (top) layer into an evaporation flask and evaporate under vacuum using the rotory evaporator.

9. Release vacuum under argon (for ubiquinol-10).

10. Add 180 μl of argon-flushed propan-2-ol to the lipid-residue and ensure that all residues are washed from the flask wall. This process should be done as fast as possible, as antioxidants (especially ubiquinol-10) present in a thin lipid film are particularly prone to autoxidation.

11. Use this extract for the determination of vitamin E (see *Protocol 4*), ubiquinol-10, ubiquinone-10, lycopene, and β-carotene (see *Protocol 5*).

12. Preferentially, transfer extracts into septum-sealed autosampler vials and pass a gentle stream of argon through the vial for 30 sec. This treatment is absolutely required for the reductive determination of ubiquinone-10. If autosampler vials are not available, Teflon-lined, screw-capped vials can be used as an alternative.

13. Immediately after the extraction and phase separation (i.e. when the hexane phase is being dried), remove 1 ml of the aqueous methanol (bottom) layer and pass it through 0.2 μm nylon filter.

14. Use this filtered, aqueous methanol extract for the determination of ascorbate and urate (see *Protocol 2*) and bilirubin (see *Protocol 3*).

15. For determination of vitamin C redox-status, add to a 180 μl aliquot of the filtered aqueous methanol phase 20 μl of homocysteine (1%) and incubate for 30 min at 25°C.

B. *Preparation of plasma extract (1) for HPLC-systems equipped with UV–visible or fluorescence detector*

1. Add 100 μl blood plasma to a mixture of 180 μl methanol and 20 μl EDTA-saturated water, vortex and keep on ice for 10 min.

2. Centrifuge tubes (5000 r.p.m. in GS-6R rotor for 2 min) to pellet precipitate.

3. Use the supernatant for the determination of ascorbate and urate (2–4).

2. Properties of selected natural antioxidants

2.1 Water-soluble, non-proteinaceous antioxidants

2.1.1 Vitamin C

Vitamin C has long been known to be essential for the protection of humans against scurvy. The ascorbic activity of vitamin C lies in the role of ascorbic acid (the reduced form of vitamin C) as an essential cofactor in hydroxylation reactions involved in the biosynthesis of stable crosslinked collagen (5). This and other metabolic functions of ascorbate depend on its strong reducing potential. The same property makes this vitamin an excellent antioxidant, capable of scavenging a wide variety of different oxidants. For example, ascorbate has been shown to effectively scavenge $O_2^{\cdot-}$, H_2O_2, HOCl, aqueous peroxyl radicals, and singlet oxygen (6–14). During its antioxidation action, ascorbate undergoes a two-electron oxidation to dehydroascorbic acid (the oxidized form of vitamin C) with intermediate formation of the relatively unreactive ascorbyl radical (15). Although dehydroascorbic acid is relatively unstable and hydrolyses readily to L-2,3-diketogulonic acid, it can be reduced back to ascorbate by a variety of cells or thiols such as homocysteine. There-fore, both ascorbate and dehydroascorbic acid are biologically active forms of vitamin C. Ascorbate is able to interact synergistically with membrane-bound (16) and lipoprotein confined α-tocopherol (17), i.e. it readily reduces α-tocopheroxyl radical back to α-tocopherol (18). Ascorbate and α-tocopherol may be classified as phase-transfer active antioxidants.

Following extraction (see *Protocol 1A*), ascorbate in human blood plasma is detected best by HPLC with electrochemical detection (see *Protocol 2*) as shown in *Figure 1* (19, 20; for review see ref. 21). Unlike with vitamin E, it is impossible to measure both reduced and two-electron oxidized forms of vitamin C simultaneously; dehydroascorbic acid is electrochemically inactive (22, 23). Reduced vitamin C is analysed directly by HPLC-EC whereas dehydroascorbic acid must first be reduced with homocysteine to be detected by HPLC-EC, the result giving the total amount of vitamin C (reduced plus oxidized forms). In a more concentrated extract (see *Protocol 1B*) plasma ascorbate can also be quantified by HPLC equipped with UV_{268nm} (2) or fluorescence detection (3, 4).

2.1.2 Uric acid

Uric acid is the metabolic degradation product of xanthine and efficiently scavenges certain oxygen radicals (24, 25). It has also been shown to stabilize ascorbic acid in human plasma at physiological concentrations (26). Unlike ascorbate, uric acid is not able to reduce α-tocopheroxyl radical (17, 27). Uric acid in human blood plasma is detected with high sensitivity by HPLC-EC (see *Protocol 2*) (*Figure 1*), but can also be quantitated by HPLC equipped with UV_{265nm} detection.

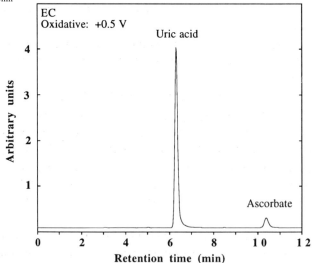

Figure 1. Typical separation and electrochemical detection of uric acid and ascorbate present in the aqueous methanol extract of a human plasma sample. Human plasma extract was prepared according to and HPLC conditions were as described in *Protocol 2*. The injection volume was 20 μl.

Protocol 2. Determination of ascorbate and uric acid by HPLC-EC (*Figure 1*)

Equipment and reagents

- LC-18, 5 μm particle size HPLC column, 25 × 0.46 cm, with guard column (Supelco)
- HPLC pump, PM-60 (BAS)
- Amperometric detector, LC-4B (BAS)
- Dual glassy carbon electrode (BAS)
- Electrochemical cell, CC-4 (BAS)
- Integrator, CR-4A (Shimadzu)
- pH meter (Orion)
- L-Ascorbic acid (Aldrich)
- Uric acid, sodium salt (Sigma)

- Sodium acetate, suprapur (Merck)
- Acetic acid, 96% HPLC grade (Merck)
- Na_2EDTA (Merck)
- H_2O, nanopure, Type I HPLC grade, double deionized (Continental Water Systems)
- Methanol, HPLC grade (Mallinckrodt)
- Dodecyl-triethylammonium-phosphate (Q12), (Regis)
- Helium, ultra high purity (CIG)

Method

1. Prepare 40 mM sodium acetate solution (13.1 g sodium acetate in 3688 ml H_2O), add 300 ml methanol, 800 mg EDTA (0.45 mM), 10 ml Q12 (1.5 mM), and adjust the pH of this mobile phase to 4.75 using HPLC grade acetic acid.

2. Degas mobile phase thoroughly by passing a stream of helium through it; equilibrate analytical column overnight at a flow rate of 0.9 ml/min (recycling of the mobile phase). The column requires this length of time for complete coating with the ion-pairing agent Q12.

3. Assemble the electrochemical cell making sure that no air is trapped inside the cell.

4. Set the oxidative potential at the amperometric detector to + 500 mV, switch the cell on and allow the system to equilibrate for approximately 1 h.

5. Set the amperometric detector sensitivity to 100 nA.

6. Inject 20 μl of the extract, prepared according to *Protocol 1A*, and monitor for 15 min. Under these conditions the retention times for urate and ascorbate are 6.3 min and 10.4 min, respectively (*Figure 1*).

7. Inject 20 μl of the homocysteine-treated sample (see *Protocol 1A*). The difference in ascorbate area units between steps 6 and 7 is used to calculate the amount of dehydroascorbic acid present in the sample.

8. Prepare standard curves by injecting equal volumes of different freshly prepared dilutions of the standard. Sodium urate can be dissolved in H_2O only by sonicating the salt first in a minimum volume of water and then slowly making up to the final volume, e.g. 1 mg/ml, by adding small amounts of water to the milky suspension.

2.1.3 Bilirubin

Human albumin-bound bilirubin contributes to the non-enzymatic anti-oxidant defences in human plasma as judged by *in vitro* studies (28–30). It efficiently protects albumin-bound linoleic acid from peroxyl radical-induced oxidation *in vitro* (31). A water-soluble form of the pigment has been shown to synergize with α-tocopherol in liposomal lipid antioxidation (32). More recently, free- and albumin-bound bilirubin have been shown to inhibit lipid oxidation in human blood plasma and isolated LDL exposed to lipophilic peroxyl radicals and this antioxidant activity was dependent on the presence of vitamin E and most likely involved interaction of bilirubin with α-tocopheroxyl radical, trapped within lipoproteins (30). Bilirubin in human plasma can be detected in the aqueous methanol extract (see *Protocol 1A*) by the HPLC method reported in *Protocol 3*.

Protocol 3. Determination of (plasma) free bilirubin and biliverdin by HPLC equipped with UV/Vis detector (33)

Equipment and reagents

- LC-C18, 5 μm particle size HPLC column, 25 × 0.46 cm, with guard column (Supelco)
- HPLC pump, 2150 (LKB, Pharmacia)
- UV/Visible detector, automated wavelength change UVIS (Linear)
- Integrator, CR-4A (Shimadzu)
- pH Meter (Orion)
- Biliverdin (Porphyrin Products)

- Bilirubin (Sigma), recrystallized as described in (33)
- Di-*n*-octylamine (Aldrich)
- Methanol, HPLC-grade (Mallinckrodt)
- Acetic acid, HPLC-grade (Merck)
- H_2O, nanopure, Type I HPLC grade, double deionized (Continental Water Systems)

Method

1. A methanolic 0.1 M di-*n*-octylamine/H_2O (95:5 (v/v)) solution is prepared and adjusted to pH 7.7 by adding the appropriate amount of acetic acid.

2. Equilibrate the analytical column at a flow rate of 1 ml/min overnight.

3. Program UV/Vis detector to 650 nm from 0 min and 460 nm from 7 min run time.

4. Inject free bilirubin (prepared by dissolving recrystallized bilirubin in 50 mM NaOH) and biliverdin standard. Molar extinction coefficients at λ_{max} for bilirubin and biliverdin are $\varepsilon_{453} = 60\,700$ (chloroform) and $\varepsilon_{377} = 66\,600$ (methanol), respectively (34).

5. Under these conditions the retention times for biliverdin and bilirubin are 5.4 and 12.5 min, respectively.

2.2 Lipid-soluble, non-proteinaceous antioxidants

2.2.1 α-Tocopherol and other vitamin E compounds

α-Tocopherol is quantitatively the major lipid-soluble antioxidant in human plasma (35–37) and LDL (17, 38, 39). Other isomers of vitamin E, such as β-, γ- and δ-tocopherol, are either present in very low concentrations or not detectable at all. Judged by their rate of reaction with peroxyl radicals (40), the antioxidant activity decreases in the order α > γ > β > δ, in analogy with the biological potencies of these different forms of vitamin E (35). α-Tocopheroxyl radical, the one electron oxidation product of α-tocopherol, is readily reduced by ascorbate, albumin-bound bilirubin (see above) and ubiquinol-10 (see below). The required radical phase-transfer process may be essential for the integrity of colloidal biological systems, such as lipoproteins in plasma. Recent investigations (17, 39, 41) have shown that impairment of radical-export from lipoproteins is the basis for a pro- rather than antioxidant activity of α-tocopherol alone. The various isomers of vitamin E in human blood plasma and isolated lipoproteins are detected best by HPLC-EC (see *Protocol 4*). A typical chromatogram is shown in *Figure 2*.

Protocol 4. Determination of α- and γ-tocopherol and the two-electron oxidation product α-tocopherylquinone (*Figure 2*)

Equipment and reagents

- The equipment is essentially the same as described in *Protocol 5*
- d-α-Tocopherol (Henkel or Kodak)
- d-γ-Tocopherol (Henkel or Kodak)
- α-Tocopherylquinone (Kodak)

- Sodium perchlorate 1 × H_2O (Merck)
- Methanol, HPLC grade (Mallinckrodt)
- Helium, ultra high purity (CIG)
- H_2O, nanopure, Type I HPLC grade, double deionized (Continental Water Systems)

Method

1. Prepare 4% (v/v) 50 mM $NaClO_4$ (7.0 g in 1000 ml H_2O) in methanol.

2. Degas mobile phase thoroughly by passing a gentle stream of helium through it; equilibrate column overnight at 1.2 ml/min (recycling mobile phase).

3. Assemble the electrochemical cells at the same time making sure no air is trapped inside the cell.

4. Set the potential of the up-stream electrode to −700 mV (reductive mode) and that of the down-stream electrode to +600 mV (oxidative mode).

5. Switch the cells on and allow the system to equilibrate for approximately 3 h.

6. Set the amperometric detector sensitivity to 50 nA for each detector.

7. Prepare α- and γ-tocopherol and α-tocopherylquinone standard

---

Protocol 4. *Continued*

curves by injecting equal amounts of diluted standards. The concentration of the standards are determined by UV-spectroscopy using molar extinction coefficients at λ_{max} of $\varepsilon_{294} = 3058$ (ethanol), $\varepsilon_{298} = 3870$ (ethanol) and $\varepsilon_{268} = 18\,204$ (ethanol) for α- and γ-tocopherol and α-tocopherylquinone, respectively (42).

8. Inject 10 μl of human plasma hexane extract as prepared in *Protocol 1*.

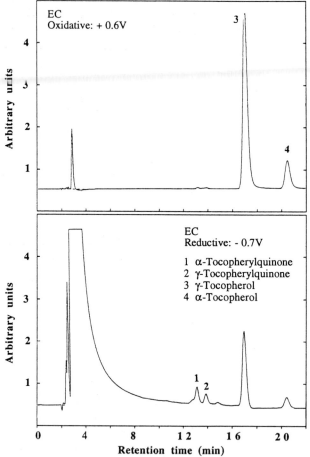

Figure 2. Separation and electrochemical detection of α- and γ-tocopherols and their corresponding quinones. The chromatogram was obtained by injection of an α- and γ-tocopherol standard mixture previously oxidized with the peroxyl radical initiator 2,2′-azo*bis*(2,4-dimethylvaleronitrile). The presence of α-tocopherylquinone was confirmed by co-elution with an authentic standard. Peak 2 was assigned to γ-tocopherylquinone on the basis of its electrochemical property, its expected retention time, and the fact that it is formed during oxidation of γ-tocopherol. Chromatographic conditions were as described in *Protocol 4*.

2.2.2 Ubiquinol-10 (reduced coenzyme Q10)

Ubiquinol-10 is also an effective, lipid-soluble antioxidant (43, 44) present in lipoproteins of human blood plasma. Its antioxidant properties have been reviewed (41, 45, 46). In the presence of both ubiquinol-10 and α-tocopherol (lipoprotein) lipids are better protected from peroxyl radical mediated oxidation than in the presence of the vitamin alone (43, 44, 47, 48), and this appears to be due to the ability of ubiquinol-10 to eliminate α-tocopheroxyl radical by reducing it back to regenerate α-tocopherol (41, 49, 50). In plasma from healthy people about 80% of coenzyme Q10 is present in the reduced, i.e. antioxidant active, form (44, 48, 51, 52). Dietary supplementation with

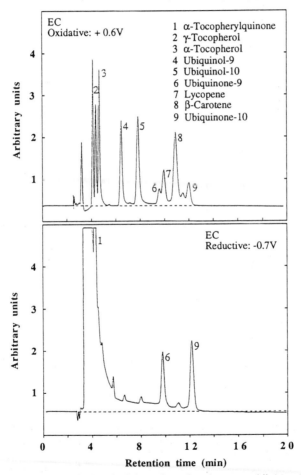

Figure 3. Separation and electrochemical detection of major natural lipid-soluble antioxidants in their reduced and oxidized forms. The chromatogram was obtained from injection of a standard mixture of the indicated compounds. Chromatographic conditions were as described in *Protocol 5*.

ubiquinone-10, the antioxidant inactive form of coenzyme Q10 results in an approximately fourfold increase in ubiquinol-10 (reduced) levels in plasma and low density lipoprotein. The redox-status of coenzyme Q, however, remains constant at about 80% (48). Ubiquinol-10 is most sensitive to autoxidation and great care must be taken to determine its concentration accurately. Anaerobic HPLC-EC is the best available method to determine the redox-status of coenzyme Q10 and other coenzymes (Q9). The method described in *Protocol 5* has been specifically designed to give maximal selectivity and highest sensitivity for the determination of ubiquinol- and ubiquinone-10 in human plasma or lipoproteins.

2.2.3 Carotenoids (lycopene and β-carotene)

The peroxyl radical trapping activity of β-carotene and other carotenoids, is dependent on the partial pressure of oxygen applied (53). β-Carotene is less efficient under air, but becomes moderately active at lower partial pressure of oxygen (pO_2) (28, 53) that prevails in biological tissues. At the very low pO_2 of 4 torr, β-carotene inhibited adriamycin-enhanced microsomal lipid peroxidation even more efficiently than α-tocopherol. In contrast to β-carotene, the antioxidant action observed with retinol did not show a dependency on oxygen concentration (54). While lycopene and β-carotene can be detected in the hexane extract (see *Protocol 1A*) using the HPLC-EC method described in *Protocol 5*, more specific HPLC methods are available, particularly if the analyst is interested in the ratio of *cis*- to *trans*-isomers of carotenoids (55, 56). For a typical chromatogram derived using *Protocol 5* see *Figure 3*.

Protocol 5. Determination of ubiquinols-9 and -10, ubiquinones-9 and -10, α-tocopherol, γ-tocopherol, lycopene, and β-carotene by HPLC-EC (*Figure 3*)

Equipment and reagents

- LC-C18, 5 μm particle size HPLC column, 25 × 0.46 cm, with guard column (Supelco)
- HPLC pump, PM-60 (BAS)
- Two amperometric detectors, LC-4B (BAS)
- Electrochemical cell, CC-4 (BAS)
- Dual glassy carbon electrode (BAS)
- Integrator, CR-4A (Shimadzu)
- Water bath, set at 50°C (Ratek)
- Ubiquinone-10 (Mitsubishi Gas Chemicals or Fluka)
- Ubiquinone-9 (Fluka)
- d-α-Tocopherol (Henkel or Kodak)
- Lycopene (Sigma)
- β-Carotene (Sigma)
- Lithium perchlorate (LiClO$_4$•3 H$_2$O) (Merck)
- Ethanol, analytical grade (BDH)
- Methanol, HPLC grade (E. Merck)
- Propan-2-ol, HPLC grade (Mallinckrodt)
- Helium, ultra high purity (CIG)

Method

1. Prepare mobile phase by mixing ethanol/methanol/propan-2-ol (74:22:4, by vol.) i.e. 2950 ml ethanol, 900 ml methanol, and 150 ml

propan-2-ol. Add 12.8 g LiClO$_4$ (40 mM) and mix until salt is dissolved completely.

2. Degas mobile phase thoroughly by passing a gentle stream of helium through it; equilibrate column overnight at 1.0 ml/min (recycling of mobile phase).

3. Assemble the electrochemical cell according to manufacturer's instruction making sure no air is trapped inside the cell.

4. Increase the flow rate to 1 ml/min and set the potential of the upstream electrode to -700 mV (reductive mode) and that of the downstream electrode to $+600$ mV (oxidative mode).

5. Switch the cells on and allow the system to equilibrate for approximately 3 h.

6. Set the sensitivity to 20 nA for each amperometric detector.

7. Prepare standards of ubiquinols-9 and -10 immediately before use according to *Protocol 6* (approximate time required: 1 h).

8. Inject 10 µl of ubiquinols-9 or -10 standard (10–30 µM) and check for signs of oxidative degeneration of the standard on the column (*Figure 4*, if observed exchange column, if oxidative degeneration persists, refer to *Protocol 7*).

9. Prepare standards of ubiquinones-9 and -10 (molar extinction coefficients at λ_{max} for ubiquinones-9 and -10 are $\varepsilon_{275} = 14\,200$ (ethanol) and $\varepsilon_{275} = 14\,700$ (ethanol), respectively (57)) and ubiquinols-9 and-10 (see *Protocol 6*). Obtain standard curves by injecting equal amounts of variably diluted standards. Proceed equally for tocopherols (molar extinction coefficients are given in *Protocol 5*) and carotenoids (molar extinction coefficients at λ_{max} for α-carotene, β-carotene, and lycopene are $\varepsilon_{445} = 150\,360$ (petroleum ether), $\varepsilon_{451} = 134\,250$ (hexane), and $\varepsilon_{472} = 184\,575$ (hexane), respectively (58)).

10. Inject 20 µl of human plasma hexane extracts as prepared in *Protocol 1A* from sealed and argon-flushed autosampler vials.

11. After completion of analysis, reinject ubiquinols-9 and -10 standards to account for possible changes of system condition (e.g. oxidation of ubiquinols on column, *Figure 4*).

Protocol 6. Preparation of standards of ubiquinols-9 and -10

Equipment and reagents

- Bench-top centrifuge, GS-6R (Beckman)
- Extraction tube (culture tube, screw-capped, 16 × 150 mm, Kimble)
- UV/Vis-spectrophotometer, U-3210 (Hitachi)
- Two quartz cells, 1 cm (Lovibond)
- Ubiquinone-9 (Fluka)
- Ubiquinone-10 (Mitsubishi Gas Chemicals or Fluka)
- Sodium dithionite (Merck)
- Ethanol, analytical grade (BDH)

Protocol 6. *Continued*

- H₂O, nanopure, Type I HPLC-grade, double deionized (Continental Water Systems)
- Propan-2-ol, HPLC grade (Mallinckrodt)
- Argon, high purity (CIG)

Method

1. Dissolve ubiquinones-9 or -10 (0.3 mg) in ethanol (1 ml). Add H₂O (3 ml) and sodium dithionite (100 mg). Wrap tube in aluminium foil and shake well. Incubate for 30 min at 25°C.

2. Add hexane (2 ml) and pass a gentle stream of argon through the biphasic mixture. Shake well for 15 sec.

3. Place 5 ml of argon-flushed H₂O into a second tube, transfer hexane (top) layer of first tube and pass gentle stream of argon through the resulting biphasic mixture. Separate phases by centrifugation.

4. Transfer hexane (top) layer of second tube into 4 ml amber glass vial with screw cap and evaporate hexane by a gentle stream of argon in the fume hood. Add 3 ml argon-flushed propan-2-ol and record UV spectrum between 350 and 250 nm against propan-2-ol reference. Determine concentration of standards using the molar extinction coefficient for ubiquinol-10 and ubiquinol-9 at λ_{max}, $\varepsilon_{290} = 4003$ (ethanol) and $\varepsilon_{290} = 4108$ (ethanol), respectively (57).

5. Dilute standards to concentrations between 1 and 10 μM.

6. Transfer standards into septum-sealed auto-sampler vials and flush with a gentle stream of argon for 30 sec.

Protocol 7. Passivation of HPLC-EC system to prevent high background noise and oxidative degradation of autoxidation-sensitive antioxidants, particularly ubiquinols (to be done routinely approximately every 6 months)

Equipment and reagents

- H₂O, nanopure, Type I HPLC-grade, double deionized (Continental Water Systems)
- Nitric acid, 70%, Rhone Poulenc
- pH indicator paper, pH 0–14, Merck
- Old glassy carbon electrode
- Old reference electrode

Method

1. Disconnect the analytical column from the HPLC-EC system and replace working and reference electrode with old electrodes, which are no longer functional.

2. Flush the entire HPLC-EC system (including injection valve) thoroughly with nanopure H₂O making sure that no organic matter remains inside.

3. Prepare a half-concentrated nitric acid solution by adding the concentrated acid to an equal volume of water.

4. Flush the entire HPLC-EC system with half-concentrated nitric acid at a rate of 1 ml/min for 30 min collecting the waste in a separate container for safety reasons.

5. Replace nitric acid with nanopure water (several changes of water in the reservoir) and wash HPLC-EC system until the pH of the effluent matches the pH of the nanopure water used.

6. Equilibrate with fresh mobile phase and reconnect column and electrodes.

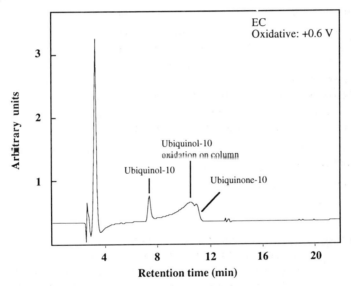

Figure 4. Example of unsatisfactory separation and analysis of a fresh ubiquinol-10 standard on an 'oxidizing' HPLC column. The column was used previously for the analysis of antioxidants in haemolysed rat plasma. The use of an LC-18 column dedicated for the analysis of ubiquinols in human plasma and LDL is recommended.

Acknowledgement

This work received support from the National Health & Medical Research Council of Australia grant 940915 to R.S.

References

1. Dhariwal, K. R., Hartzell, W. O., and Levine, M. (1991). *Am. J. Clin. Nutr.*, **54**, 712.

2. Margolis, S. A. and Davis, T. P. (1988). *Clin. Chem.*, **34**, 2217.
3. Speek, A. J., Schrijver, J., and Schreurs, W. H. P. (1984). *J. Chromatogr.*, **305**, 53.
4. Iwata, T., Yamaguchi, M., Hara, S., and Makamura, M. (1985). *J. Chromatogr.*, **344**, 351.
5. Jaffe, G. M. (1984). In *Handbook of vitamins: nutritional, biochemical, and clinical aspects* (ed. L. J. Machlin), pp. 199–244. Marcel Dekker, New York.
6. Nishikimi, M. (1975). *Biochem. Biophys. Res. Commun.*, **63**, 463.
7. Bodannes, R. S. and Chan, P. C. (1979). *FEBS Lett.*, **105**, 195.
8. Cabelli, D. E. and Bielski, B. H. J. (1983). *J. Phys. Chem.*, **87**, 1809.
9. Bendich, A., Machlin, L. J., Scandurra, O., Burton, G. W., and Wayner, D. D. M. (1986). *Adv. Free Radic. Biol. Med.*, **2**, 419.
10. Rose, R. R. (1987). *Ann. NY Acad. Sci.*, **498**, 506.
11. Halliwell, B., Wasil, M., and Grootveld, M. (1987). *FEBS Lett.*, **213**, 15.
12. Frei, B., England, L., and Ames, B. N. (1989). *Proc. Natl Acad. Sci. USA*, **86**, 6377.
13. Frei, B., Stocker, R., England, L., and Ames, B. N. (1990). *Adv. Exp. Med. Biol.*, **264**, 155.
14. Kwon, B.-M. and Foote, C. S. (1988). *J. Am. Chem. Soc.*, **110**, 6582.
15. Bielski, B. H. and Richter, H. W. (1975). *Ann. NY Acad. Sci.*, **258**, 231.
16. Niki, E. (1987). *Chem. Phys. Lipids*, **44**, 227.
17. Bowry, V. W. and Stocker, R. (1993). *J. Am. Chem. Soc.*, **115**, 6029.
18. Packer, J. E., Slater, T. F., and Willson, R. L. (1979). *Nature*, **278**, 737.
19. Behrens, W. A. and Madere, R. (1987). *Anal. Biochem.*, **165**, 102.
20. Kutnink, M. A., Hawkes, W. C., Schaus, E. E., and Omaye, S. T. (1987). *Anal. Biochem.*, **166**, 424.
21. Washko, P. W., Welch, R. W., Dhariwal, K. R., Wang, Y., and Levine, M. (1992). *Anal. Biochem.*, **204**, 1.
22. Ono, S., Takagi, M., and Wasa, T. (1953). *J. Am. Chem. Soc.*, **75**, 4369.
23. Huang, T. and Kissinger, P. T. (1989). *Current Separations*, **9**, 20.
24. Matsushita, S., Ibuki, F., and Aoki, A. (1963). *Arch. Biochem. Biophys.*, **102**, 446.
25. Ames, B. N., Cathcart, R., Schwiers, E., and Hochstein, P. (1981). *Proc. Natl Acad. Sci. USA*, **78**, 6858.
26. Sevanian, A., Davies, K. A., and Hochstein, P. (1985). *J. Free Radic. Biol. Med.*, **1**, 117.
27. Niki, E., Saito, M., Yoshikawa, Y., Yamamoto, Y., and Kamiya, Y. (1986). *Bull. Chem. Soc. Jpn.*, **59**, 471.
28. Stocker, R., Yamamoto, Y., McDonagh, A. F., Glazer, A. N., and Ames, B. N. (1987). *Science*, **235**, 1043.
29. Frei, B., Stocker, R., and Ames, B. N. (1988). *Proc. Natl Acad. Sci. USA*, **85**, 9748.
30. Neuzil, J. and Stocker, R. (1994). *J. Biol. Chem.*, **269**, 16712.
31. Stocker, R., Glazer, A. N., and Ames, B. N. (1987). *Proc. Natl Acad. Sci. USA*, **84**, 5918.
32. Stocker, R. and Peterhans, E. (1989). *Biochim. Biophys. Acta*, **1002**, 238.
33. Stocker, R., McDonagh, A. F., Glazer, A. N., and Ames, B. N. (1990). In *Methods in enzymology* (ed. L. Packer and A. N. Glazer), Vol. 186, pp. 301–309. Academic Press, New York.
34. McDonagh, A. F. (1979). In *The porphyrins* (ed. D. Dolphin), pp. 293–491. Academic Press, New York.

35. Burton, G. W. and Ingold, K. U. (1981). *J. Am. Chem. Soc.*, **103**, 6472.
36. Burton, G. W., Joyce, A., and Ingold, K. U. (1982). *Lancet*, **ii**, 327.
37. Ingold, K. U., Webb, A. C., Witter, D., Burton, G. W., Metcalfe, T. A., and Muller, D. P. R. (1987). *Arch. Biochem. Biophys.*, **259**, 224.
38. Esterbauer, H., Dieber-Rotheneder, M., Striegl, G., and Waeg, G. (1991). *Am. J. Clin. Nutr.*, **53**, 314.
39. Bowry, V. W., Ingold, K. U., and Stocker, R. (1992). *Biochem. J.*, **288**, 341.
40. Burton, G. W. and Ingold, K. U. (1986). *Acc. Chem. Res.*, **19**, 194.
41. Ingold, K. U., Bowry, V. W., Stocker, R., and Walling, C. (1993). *Proc. Natl Acad. Sci. USA*, **90**, 45.
42. Mayer, H. and Isler, O. (1971). In *Methods in enzymology* (ed. D. B. McCormick and L. D. Wright), pp. 241–348. Academic Press, New York.
43. Mellors, A. and Tappel, A. L. (1966). *J. Biol. Chem.*, **241**, 4353.
44. Frei, B., Kim, M., and Ames, B. N. (1990). *Proc. Natl Acad. Sci. USA*, **87**, 4879.
45. Beyer, R. E., Nordenbrand, K., and Ernster, L. (1987). *Chem. Scripta*, **27**, 145.
46. Beyer, R. E. and Ernster, L. (1990). In *Highlights in ubiquinone research* (ed. G. Lenaz, O. Barnabei, A. Rabbi, and M. Battani), pp. 191–213. Taylor & Francis, London.
47. Stocker, R., Bowry, V. W., and Frei, B. (1991). *Proc. Natl Acad. Sci. USA*, **88**, 1646.
48. Mohr, D., Bowry, V. W., and Stocker, R. (1992). *Biochim. Biophys. Acta*, **1126**, 247.
49. Maguire, J. J., Wilson, D. S., and Packer, L. (1989). *J. Biol. Chem.*, **264**, 21462.
50. Kagan, V., Serbinova, E., and Packer, L. (1990). *Biochem. Biophys. Res. Commun.*, **169**, 851.
51. Okamoto, T., Fukunaga, Y., Ida, Y., and Kishi, T. (1988). *J. Chromatogr.*, **430**, 11.
52. Yamamoto, Y. and Niki, E. (1989). *Biochem. Biophys. Res. Commun.*, **165**, 988.
53. Burton, G. W. and Ingold, K. U. (1984). *Science*, **224**, 569.
54. Vile, G. F. and Winterbourn, C. C. (1988). *FEBS Lett.*, **238**, 353.
55. Stahl, W., Schwarz, W., Sundquist, A. R., and Sies, H. (1992). *Arch. Biochem. Biophys.*, **294**, 173.
56. Stahl, W., Schwarz, W., and Sies, H. (1993). *J. Nutr.*, **123**, 847.
57. Hatefi, Y. (1963). *Adv. Enzymol.*, **25**, 275.
58. Cavina, G., Gallinella, B., Porra, R., Pecora, P., and Suraci, C. (1988). *J. Pharm. Biomed. Anal.*, **6**, 259.

20

Measurement of antioxidant gene expression

COLIN D. BINGLE

1. Introduction

As the reader will have already seen in the rest of this volume, the measurement of levels of antioxidant proteins using a variety of functional assays can yield valuable information to the researcher. To gain a better understanding of the processes regulating levels of antioxidant proteins within a particular experimental system it is becoming clear that studies aimed at investigating regulation at the genetic level are increasingly important. The present chapter details a number of basic protocols which should allow the researcher new to molecular biology to begin to study the steady-state levels of antioxidant gene expression. Such protocols have been used for example to study developmental regulation of antioxidant proteins (1) and their induction by hyperoxia (2), ozone (3), and inflammatory mediators (4). These protocols are by no means exhaustive and in depth discussion of the methods and theory of RNA analysis are given in references 5 and 6. Measurement of steady-state RNA levels also opens the way to further studies of the transcriptional status of antioxidant genes using techniques which allow direct analysis of the genes of interest.

2. Isolation of RNA

2.1 Precautions to reduce potential RNase contamination

One of the major problems encountered during the isolation and manipulation of RNA is that of contamination with RNases from both internal and external sources. Strict precautions must be adhered to in order to minimize the risk of accidental contamination. The following points will help to reduce the possibility of RNase-related problems arising in the laboratory.

(a) If possible dedicate a whole set of reagents and equipment for use whilst working with RNA. Clearly identify it, and work with it and store it in an area specifically designated for RNA work only.

(b) Sterile, single-use plasticware can be considered to be essentially RNase-free. Glassware can be soaked in a solution of 0.1% diethyl pyrocarbonate (DEPC) for at least 2 h, after which time it should be baked for at least 2 h at >180°C. Electrophoresis equipment, including combs, should be soaked in a solution of 0.1% DEPC for at least 10 min and then rinsed using DEPC-treated water.

(c) Solutions and water should be prepared with 0.1% DEPC and treated by autoclaving at 15 lb/in^2 for 20 min. Solutions containing Tris cannot be treated in this manner as DEPC reacts with amines and therefore extra care must be taken to ensure that solutions containing Tris are made with dedicated chemicals. In practice I have found that the use of sterile water prepared for irrigation in clinical settings can be used without any further treatment.

?.? Isolation of total RNA

A number of protocols have been developed for the isolation of total RNA from tissues and cells. In general it is recommended that a method be chosen in which the sample from which the RNA is to be extracted is treated rapidly with a potent denaturing agent such as guanidine-HCl or guanidine isothiocyanate. These methods further reduce the potential for RNase contamination and were developed to isolate RNA from samples such as pancreas which are rich in RNases (7). *Protocol 1* will lead to the successful isolation of total RNA from all tissue and cell sources.

Protocol 1. Isolation of total RNA from tissue and cells using guanidine isothiocyanate and caesium chloride

Equipment and reagents[a]

- 5.7 M Caesium chloride, 0.1 M EDTA
- Beckman TL 55 polyallomer tubes
- 3 M sodium acetate (pH 5.2)
- Beckman TL 100 centrifuge
- TL 55 swingout rotor
- Microcentrifuge
- TE (10 mM Tris–HCl, pH 7.4, 5 mM EDTA)
- Ethanol
- 10 ml syringes

- 19, 21, and 23 gauge needles
- 15 ml sterile plastic tubes
- RNase-free pestles and mortars
- Liquid nitrogen
- PBS (phosphate-buffered saline)
- GIT solution (4 M guanidine isothiocyanate, 5 mM sodium citrate (pH 7.0), 0.1 M β-mercaptoethanol, 0.5% Sarkosyl)

A. *Frozen tissue samples*

1. Using an RNase-free pestle and mortar homogenize the tissue sample (0.2–1 g) in liquid nitrogen until it forms a fine powder, making sure that it remains frozen at all times.

2. Rapidly transfer ground tissue to a 15 ml tube and resuspend in 2–10 ml of the guanidine isothiocyanate solution using a 10 ml syringe and 19 gauge needle.

---

3. Pass the homogenate through the needle 10 times and repeat with the 21 and 23 gauge needles until all the tissue is in solution.

B. *Cells from tissue culture*

1. Wash cells with sterile PBS and lyse by resuspending in guanidine isothiocyanate solution using a 10 ml syringe and 19 gauge needle. Lyse cell layers directly on plastic or pellet cell suspensions by centrifugation and lyse in Falcon tubes.

2. Pass homogenate through needles as for tissue above.

C. *Extraction of RNA*

1. Layer the homogenate on to 0.8 ml of 5.7 M CsCl solution in a Beckman TL55 polyallomer tube.[b]

2. Centrifuge at 100 000 *g* for 3 h at 20°C in a Beckman TL 100 centrifuge using a TL55 swinging-bucket rotor.

3. Discard the supernatant carefully trying not to contaminate the RNA pellet with protein/DNA from the CsCl gradient and wash the pellet with 200 μl of 70% ethanol.

4. Dissolve pellet in 100 μl TE, add 0.1 vol. of 3 M sodium acetate (pH 5.2) and 2.2 vol. of ethanol, mix, and freeze at −70°C for > 10 min.

5. Recover RNA by centrifugation for 30 min at full speed in a microcentrifuge at 4°C, wash pellet with 100 μl 70% ethanol, and resuspend in 100 μl TE. Store RNA at −70°C.

[a] Use Analar grade reagents for all RNA work.
[b] Other larger size rotors and centrifuges can be used but these use more sample and caesium chloride and also require longer centrifugation times (5).

Where levels of a particular transcript are found to be low in preparations of total RNA following Northern hybridization it may be necessary to isolate the mRNA (poly(A)$^+$ RNA) fraction (corresponding to that RNA that is destined for translation into protein), away from the ribosomal RNA. This can be achieved by the use of an oligo(dT)* affinity resin, consisting of chains of deoxythymidine linked to a supporting resin, which allows the preferential selection of the poly(A)$^+$-containing mRNA fraction (7). In practice it is probably best if the inexperienced researcher uses a commercially available mRNA isolation kit; these are produced by most molecular biology manufacturers. The kits provide all the reagents required and have the advantage that all reagents are known to be RNase free. Additionally the protocols have

*All eukaryotic mRNA has a tail of repeating adenine nucleotides (poly(A)$^+$ RNA) at the 3′ end that will bind by complementary base pairing to oligonucleotide chains of deoxythymidine (oligo (dT)) bound to a solid matrix.

generally been developed to minimize the time and complexity of the re-actions and in many cases poly(A)$^+$ RNA can be purified directly from tissue and cells without the need for prior isolation of total RNA.

2.3 Quantification of RNA

The concentration of RNA in a sample can be determined by measuring its absorbance at 260 nm. An RNA sample with an A_{260} of 1 has a concentration of 40 μg/ml. In practice, absorbance measurements are performed using a 1:1000 dilution of the RNA sample prepared in DEPC-treated H_2O. If such a dilution yields an A_{260} of < 0.01 then a lower dilution of the RNA should be prepared to give a reliable concentration.

3. Analysis of RNA

3.1 Choice of analysis method

Once RNA has been isolated it can be analysed by a number of methods. Northern blotting after electrophoresis through an agarose gel allows both the quality (i.e. integrity) and the quantity of the RNA to be analysed at the same time and is the most widely used analysis method. Slot and dot blotting analyses allow the accurate quantification of RNA levels using samples immobilized on to membranes under vacuum. These methods are useful if many samples are to be analysed and the integrity of the RNA has already been confirmed by electrophoresis. In certain circumstances when the levels of transcripts under study are low, more sensitive techniques such as RNase protection may be used. RNase protection assays allow the analysis of mul-tiple transcripts in the same sample but are technically more difficult to establish and are probably outside the skill level of the beginner molecular biologist. In practice probing Northern blots with complementary RNA (cRNA probes), which are directly complementary to the mRNA transcript of interest (Section 3.2), will allow the ready detection of the majority of RNA transcipts.

3.1.1 Northern blotting

Northern blotting involves the electrophoretic resolution of denatured RNA species through agarose gels followed by capillary transfer of the RNA to membranes and the subsequent hybridization of the membranes with specific complementary DNA (cDNA) or cRNA probes. This technique allows analysis of specific RNA species in a complex RNA population and allows an estima-tion of both the abundance of a specific mRNA species and its molecular size in the same sample. By varying the concentration of agarose in the gel resolution of large and small RNA transcripts can be achieved. Increasing the concen-tration of agarose allows resolution of smaller (< 1 kb) RNA transcripts whereas larger (> 5 kb) transcripts can be resolved on lower concentration gels.

Protocol 2. Northern blotting

Equipment and reagents.

- Electrophoresis tank and power supply
- Capillary transfer system
- Glycerol
- Hybridization membrane[a]
- Heat block
- Bromophenol blue
- Agarose[b]
- Ethidium bromide
- Whatman 3MM paper
- UV light source
- Deionized formamide
- Formaldehyde solution (37% v/v)
- 10× Mops Northern running buffer (0.2 M Mops (3-(N-Morpholino)propane-sulphonic acid, 0.5 M sodium acetate, pH 7.0, 0.01 M EDTA; filter sterilized
- 20× SSC (3 M sodium chloride, 0.3 M sodium citrate)

Method

1. Dissolve 0.8–2 g of agarose in 73 ml H_2O by heating in a microwave. Allow to cool to 60°C and add (in a fume hood) 17 ml formaldehyde and 10 ml 10× Mops buffer. Mix and pour into gel tray.

2. Whilst gel is setting prepare RNA samples. Use 5–20 μg of total RNA or 0.5–2 μg poly(A)$^+$ RNA per lane. Combine RNA in total volume of 6 μl, with 12.5 μl formamide, 2.5 μl 10× Mops and 4 μl formaldehyde. Heat to 65°C for 5 min, place on ice and add 2.5 μl of 50% glycerol/ 0.1 mg/ml bromophenol blue. The RNA loading buffer can be pre-mixed in advance and stored in aliquots at 20°C for more than 1 month.

3. Remove comb from gel and place gel in electrophoresis tank in 1× Mops. Wash wells to remove debris, load samples, and run gel until the blue dye has migrated two-thirds of the way down the gel. Power settings should be determined for each gel system. A midi-gel will run in 3–4 hours at 65 mA and can be run overnight at 10–12 mA

4. Once the gel has run visualize the RNA by soaking the gel in 1 μg/ml ethidium bromide for 30 min. Then wash for 30 min in water. The gel is ready for transfer. Intact RNA will contain clearly visible large (28S) and small (18S) ribosomal RNA bands. Degraded RNA will run as a low-molecular-weight smear.

5. Fill blotting tray with 20× SSC, make a platform to support the gel, and cover with a wick of pre-soaked Whatman 3MM. Place upturned gel on the wick so that no air bubbles are formed between the paper and the gel. Overlay the gel with Hybond-N membrane cut to the exact size of the gel, remove air bubbles, and surround gel with Parafilm to prevent buffer bypassing the gel during transfer. Place two layers of 3MM paper on top of the membrane and place a stack of cut paper towels on the top weighted down with a weight of > 500 g. Allow transfer to proceed for at least 4 h. Remove membrane and fix

Protocol 2. *Continued*

> either by baking for 2 h at 80°C or by using a transilluminator at 312 nm for 2–5 min. For orientation of the blot mark the position of the 28S and 18S ribosomal RNA bands. The exact time of fixation should be determined for each transilluminator. Store blots at room temperature until required.

> [a] I generally use Hybond-N nylon membranes from Amersham. Nylon membranes are stronger than nitrocellulose and can be rehybridized at least five times.
> [b] Seakem ME® (FMC) is good for Northern blotting work.

3.1.2 Slot and dot blotting

In these techniques a known amount of RNA is immobilized on to an inert support and hybridized with a radioactive probe. The RNA can be applied as elongated 'slots' or circular 'dots'. Slots give a more even distribution of RNA on the membrane and are therefore better for densitometric quantification. These blots can be quantified either by densitometry or by direct counting of the radioactivity associated with each excised slot/dot.

Protocol 3. Slot and dot blotting

Equipment and reagents
- Slot blotting or dot blotting manifold[a]
- Other reagents as in *Protocol 2*

Method

1. Make up the denaturing buffer by mixing 5 ml formamide, 1.62 ml formaldehyde, and 1 ml of 10× Mops buffer.

2. Add 200 μl of denaturing buffer to RNA sample in total volume of 20 μl. Use 1–5 μg of RNA and prepare either a serial dilution of the sample or samples prepared in duplicate. Heat samples at 65°C for 5 min and place on ice. Add equal volume of 20× SSC.

3. Assemble blotting apparatus according to the manufacturers' instructions using membrane pre-wetted in 10× SSC, and load samples into wells. Allow the samples to be pulled on to the membrane, wash wells with 2 × 400 μl of 10× SSC, and allow the vacuum to pull for at least 5 min.

4. Disassemble the apparatus, remove and fix filter, and store as for Northern blot (*Protocol 2*).

[a] Manifolds are produced by a number of producers; I use a Minifold II (Schleicher & Schuell) slot blot system.

3.2 Choice and labelling of probes

The exact type of probe to be used for hybridization of RNA blots depends on a number of factors. Best results are achieved with the use of antisense c RNA probes transcribed using specific RNA polymerases in the presence of radiolabelled ribonucleoside triphosphates (rNTPs) (8). The production of such probes requires that the cDNA of interest is cloned into a vector flanked by suitable polymerase recognition sites. Such vectors include the pBluescript (Stratagene) and the p-GEM (Promega) series of vectors. Plasmids to be used for transcription reactions should be digested to completion with a restriction enzyme which cuts beyond the end of the required transcript. Restriction enzymes which leave a 3′ overhang have been found to yield probes of variable quality and should therefore be avoided. Using well-prepared templates, RNA probes exactly complementary to the RNA of interest are generated with very high specific radio activities ($>10^8$ c.p.m./μg RNA). Blots hybridized with such probes give a strong signal requiring short exposure times. Labelled sense probes can also be generated from such vectors for use as negative controls in hybridization experiments.

Protocol 4. Transcription of cRNA probes

Equipment and reagents

- Linearized DNA template, 0.2–1 μg/μl
- [α-^{32}P]Cytidine triphosphate (CTP), 800 Ci/mol
- 5× Transcription buffer[a]: 200 mM Tris–HCl, pH 7.5, 30 mM MgCl$_2$, 10 mM spermidine, 50 mM NaCl
- RNase inhibitor
- RQ1 RNase-free DNase
- RNA polymerase (T3, T7, or SP6)
- 100 mM dithiothreitol (DTT)
- Ribonucleoside triphosphates (ATP/CTP/GTP/UTP), 25 mM
- 37 °C heat block

Method

1. Assemble the following components at room temperature in an RNase-free reaction tube: 4 μl 5× transcription buffer, 2 μl 100 mM DTT, 20 units RNase inhibitor, 4 μl 2.5 mM ATP/GTP/UTP mix, 2.4 μl 100 μM CTP, 1 μl linearized DNA template, 5 μl [α-^{32}P]CTP, 1 μl RNA polymerase, and H$_2$O to a total volume of 20 μl.

2. Incubate the reaction for 1 h at 37 °C.

3. Digest the DNA tempate with 1 unit of RQ1 RNase-free DNase at 37 °C for 15 min.

4. Remove unincorporated nucleotides from the transcribed probe (see below).

[a] I use Promega reagents for RNA transcription.

Double-stranded cDNA probes are easier to work with and are probably the best type of probe to use in initial experiments. When labelled by random hexamer priming with labelled dNTPs, probes with high specific radioactivity can be rapidly produced. In addition to using templates isolated from plasmids, products from polymerase chain reaction (PCR) amplifications can also be used with a minimum of manipulation. The signals generated from cDNA probes are less than those produced with cRNA probes but in general a 24–72 h exposure will produce a sufficiently strong signal.

Protocol 5. Random primer labelling of cDNA probes

Equipment and reagents

- DNA template, 25–50 ng/μl[a]
- 5× labelling buffer,[c] 250 mM Tris–HCl, pH 7.5, 25 mM MgCl$_2$, 10 mM DTT, 1 mM Hepes, pH 6.6, 20 A_{260} units/ml random primers

- [α-^{32}P] dCTP, 3000 Ci/mmol[b]
- Deoxynucleoside triphosphates (dATP/dCTP/dGTP/TTP), 25 mM
- Klenow DNA polymerase

Method

1. Add template DNA to 10 μl of H$_2$O and denature by heating at 95 °C for 2 min; rapidly cool on ice.

2. Assemble the following components on ice: 10 μl 5× buffer, 2 μl 20 μm dNTP (A/G/T) mix, denatured template DNA (25–50 ng), 2 μl 10 mg/ml bovine serum albumin, 5 μl, [α-^{32}P]dCTP, 5 units Klenow polymerase, and H$_2$O to 50 μl.

3. Mix and incubate at room temperature for 15–60 min.

4. Heat the tube to 95 °C for 2 min to stop the reaction.

5. Remove the unincorporated dNTPs and store at −20 °C.

[a] Digested insert or PCR product.
[b] Other [^{32}P] dNTPs can be used if required in addition to dCTP.
[c] I use Promega reagents for random priming.

When cloned DNA sequences are not available, complementary synthetic oligonucleotides, labelled by phosphorylation using [γ-^{32}P]ATP can be used as specific probes. The major advantage of using oligonucleotides is that probes can be produced from any published sequence. Additionally regions of the cDNA of interest can be chosen such that there is reduced likelihood of cross-hybridization with homologous sequences. Synthetic oligonucleotides are produced without 5′ phosphate groups and can be labelled by transfer of the γ-^{32}P from [γ-^{32}P]ATP using bacteriophage T4 polynucleotide kinase (PNK). The resultant probes are labelled to very high specific radioactivity, with every individual oligonucleotide being labelled with a single γ-^{32}P.

Protocol 6. End labelling of synthetic oligonucleotides using T4
polynucleotide kinase

Equipment and reagents

- T4 polynucleotide kinase (5–10 units/ml)[a]
- 10× PNK buffer (500 mM Tris–HCl, pH 7.5, 100 mM MgCl$_2$, 50 mM DTT, 1 mM spermidine)
- [γ-^{32}P]ATP, 3000–6000 Ci/mmol
- Oligonucleotide (10 pmol/μl)
- 0.5 M EDTA, pH 8
- 37°C heatblock

Method

1. At room temperature assemble the following: 5 μl 10 × PNK buffer, 15 μl [γ-^{32}P]ATP, 1 μl oligonucleotide (10 pmol/μl), 1 μl T4 polynucleotide kinase and H$_2$O to 50 μl.

2. Incubate at 37°C for 10–30 min.

3. Stop reaction by adding 2 μl of 0.5 M EDTA.

4. Remove unincorporated [γ-^{32}P]ATP and store probe at −20°C.

[a]I use Promega reagents for kinase reactions.

3.3 Removal of unincorporated NTPs

Following successful labelling reactions the majority of the labelled NTPs should be incorporated into the probe. If any remaining labelled NTPs are left in the probe and subsequently added to the hybridization reactions they can add significantly to the background signals seen following hybridization of northern and slot blots. For this reason it is best to remove free [^{32}P] NTPs from probes either by a technique that involves selective precipitation of the labelled probe using ammonium acetate (5) or by use of a chromatography-based system which removes contaminating NTPs on the basis of size. Techniques of selective precipitation can often lead to a poor recovery of the labelled probe; a problem which is much more pronounced with small oligonucleotide probes and for this reason it is recommended that a chromatography-based system is used. Such columns can be produced in the lab (6) or are available from a number of suppliers. I use the push column system from Stratagene which is based on size-exclusion chromatography.

4. Hybridization and washing conditions

Many methods have been developed which allow the hybridization of radioactive probes to immobilized RNA. These methods differ in a number of ways including, for example, the solvent used in the hybridization solution, the temperature of hybridization, the volume of solution used, the length of hybridization, the use of agents which block non-specific binding of the probe and the use of compounds which cause macromolecular crowding thereby increasing the rate of reassociation of the nucleic acids (5, 9). The exact choice of hybridization strategy is largely based on personal preference, but the method chosen should allow the formation of stable probe/RNA hybrids which will withstand stringent washing and allow highly specific detection of the transcript of interest. Due to the formation of highly stable RNA/RNA hybrids, when using cRNA probes, higher hybridization and wash temperatures can be used than when using cDNA and oligonucleotide probes. Both cRNA and cDNA probes can be considered to form essentially stable hybrids which will withstand highly stringent washing. This is not the case when using oligonucleotide probes and when using such probes a balance must be achieved which allows a high degree of specificity to the hybrid formation but also allows the hybrids to form at an acceptable rate. Due to the short length of synthetic oligonucleotide probes the formation of oligonucleotide/RNA hybrids can be considered to be fully reversible which in practice means that lower hybridization and washing temperatures should be employed. *Protocols 7, 8,* and *9* have been used successfully to hybridize cRNA, cDNA, and oligonucleotide probes, respectively. These outline protocols should be amended empirically as required.

Protocol 7. Hybridization of RNA blots with cRNA probes

Equipment and reagents

- Formamide
- 50× Denhardt's: 10 g Ficoll, 10 g polyvinylpyrrolidone, 10 g albumin per litre
- 10 mg/ml salmon sperm DNA, heat denatured, sonicated
- 20× SSC
- Hybridization oven or water bath and hybridization bags or bottles[a]
- 20% sodium dodecyl sulphate (SDS)
- Hybridization solution (5× SSC, 5× Denhardt's, 0.5% SDS, 50% formamide, 100 µg/ml salmon sperm DNA)

Methods

1. Pre-wet filters in 2× SSC and prehybridize for 1 h at 58.5°C in a volume of hybridization solution sufficient to keep the filters covered at all times.

2. Add cRNA probe to the pre-hybridized filters and allow hybridization to continue at 58.5°C overnight.

3. Wash filters as follows:

   (a) 2 × 20 min in 2× SSC at room temperature.
   (b) 1 × 20 min in 0.2× SSC/0.1 SDS at room temperature.
   (c) 1 × 20 min in 0.2× SSC/0.1% SDS at 65°C.

[a] The use of a hybridization oven is recommended as they are easier to maintain at a constant temperature than a water-bath. Additionally the hybridization tubes provide the researcher with a higher degree of protection from the radioactive probes.

Protocol 8. Hybridization of RNA blots with cDNA probes

Equipment and reagents
- 95°C heat block
- Additional reagents as for *Protocol 7*

Method

1. Pre-wet and pre-hybridize the filters as for cRNA probes but use a temperature of 42°C.

2. Heat denature the cDNA probe at 95°C for 5 min,[a] rapidly place on ice and add the chilled probe to the pre-hybridized filters.

3. Allow the hybridization to proceed overnight at 42°C.

4. Wash filters as follows:
   - (a) 2 × 20 min in 2× SSC at room temperature
   - (b) 1 × 20 min in 0.2× SSC/0.1% SDS at room temperature
   - (c) 1 × 20 min in 0.2× SSC/0.1% SDS at 50–55°C.

[a] cDNA probes must be heat denatured to produce single-stranded DNA molecules prior to being added to the hybridization reaction. If probes are not correctly denatured hybridization will not occur efficiently.

4.1 Hybridization of RNA blots with synthetic oligonucleotide probes

As mentioned above the determination of the hybridization temperature to be used when employing oligonucleotides as probes is critical. In practice, hybridization temperatures are chosen which are 10–15°C below the determined melting temperature (T_m) of the sequence being used as the probe. The melting temperature of oligonucleotides greater than 18 bases in length can be determined by the following calculation, $T_m = 86.35 + 0.41 (\%G + C) - (600/N)$, where N = the length of the oligonucleotide. When using an oligonucleotide probe for the first time it is often worth attempting to determine the optimal hybridization temperature by empirical means. When optimal hybridization conditions have been elucidated it should be possible to produce differential hybridization of oligonucleotide probes with a single base mismatch.

By altering the washing times and conditions an additional level of stringency can be added to the hybridization procedure. Low stringency washes, which use higher salt concentrations and lower temperatures, remove only loosely associated probe from the filters. By reducing salt concentration, adding a detergent (i.e. SDS) and increasing wash temperature it is possible to remove all but the most specific probe–RNA interactions. When using a new probe care should be taken not to use washing conditions which remove all the specific interactions. If in doubt, underwash the blot and expose to X-ray film as it is always possible to wash the blot further if required. With experience it should be possible to determine wash conditions by the use of a hand-held radioactivity monitor.

Protocol 9. Hybridization of RNA blots using oligonucleotide probes

Equipment and reagents

- Hybridization solution (5× SSC, 10 mM sodium phosphate, pH 6.8, 1 mM EDTA, pH 8.0, 0.5% SDS, 100 μg/ml salmon sperm DNA)
- 1 M sodium phosphate, pH 6.8
- 0.5 M EDTA
- Other requirements as for *Protocol 7*

Methods

1. Pre-wet and pre-hybridize the filters as in *Protocol 7*, using the pre-determined hybridization temperature (see above).
2. Add the labelled oligonucleotide[a] and hybridize overnight at the same temperature.
3. Wash filter at low stringency.
   (a) 1 × 10 min in 2× SSC at room temperature.
   (b) 1 × 10 min in 0.2× SSC at room temperature.
   (c) 1 × 10 min in 0.2× SSC at 37°C.

[a] Oligonucleotide probes do not need to be heat denatured.

---

5. Analysis of results

The exact period of time that a blot needs to be exposed to X-ray film to generate a successful autoradiographic image must be determined for every application. In general, with cRNA probes and transcripts of medium abundance an exposure time of 4–12 h will be sufficient. When cDNA or oligonucleotide probes are used exposure time should be longer, possibly overnight to a few days. In practice, exposure time will be determined empirically for every experiment,

When the X-ray images have been produced they will need to be analysed. Northern blots give information on both the size and relative abundance of a particular transcript and a good Northern blot should give a single clear hybridization signal in each lane where expression occurs. A smear of low-molecular-weight transcripts indicates the degradation of the RNA prior to Northern blotting. If Northern blots are to be used to provide a quantitative estimate of transcript abundance, then the equality of loading in each lane of the blot is very important. All other things being equal, if the amount of RNA in each lane is identical then direct comparisons can be drawn with little further manipulation of the data. Commonly, estimation of loading is made based on the abundance of the ribosomal RNA bands on the blot following transfer. Often where loading of the lane is not exactly equal it is possible to rehybridize the blot with a probe to an additional steady-state RNA (for example actin), subject the blots to densitometry and use a ratio of the abundance of the two transcripts. Efforts must be made to ensure that films used for densitometry are within the linear range for exposure. It is also very important to determine if the steady-state probe used truly reflects loading. This is particularly the case when different tissues are being analysed. I have found that the use of an oligonucleotide probe to the 28S ribosomal RNA (10) provides the best estimation of loading of a gel, as this transcript makes up >50% of total RNA.

Though Northern blots are often used to quantify transcript abundance, if quantitative data are required it is best to use slot or dot blots. These blots can be scanned by densitometry in the same way as Northern blots but by using duplicates (or triplicates) of each sample a better quantitative estimation of transcript abundance can be produced. It is an absolute requirement for successful slot blotting experiments to know that the hybridization and washing of the blots leaves only specific probe/RNA hybrids on the final blot. More quantitative data can be generated by cutting out and counting the individual slots of RNA, but the blot can not then be hybridized with a control steady-state probe.

References

1. Clerch, L. B. and Massaro, D. (1992). *Am. J. Physiol.*, **263**, L466.
2. Fleming, R. E., Whitman, I. P., and Gitlin, J. D. (1991). *Am. J. Physiol.*, **260**, L68.
3. Rahman, I., Clerch, L. B., and Massaro, D. (1991). *Am. J. Physiol.*, **260**, L412.
4. Visner, G. A., Dougall, W. C., Wilson, J. M., Burr, I. A., and Nick, H. S. (1990). *J. Biol. Chem.*, **265**, 2856.
5. Sambrook, J., Fritsch, E. F., and Maniatis, T. (1989). *Molecular cloning, a laboratory manual* (2nd edn). Cold Spring Harbor Laboratory Press, Cold Spring Harbor, NY.
6. Mason, P. J., Enver, T., Wilkinson, D., and Williams, J. G. (1993). In *Gene transcription, a practical approach* (ed. B. D. Hames and S. J. Higgins). IRL Press, Oxford.
7. Chirgwin, J. M., Przybyld, A. E., MacDonald, R. J., and Rutter, W. J. (1979). *Biochemistry*, **18**, 5294.
8. Melton, D. A., Kreig, P. A., Rebagliati, M. R., Maniatis, T., Zinn, K., and Green, M. R. (1984). *Nucleic Acids Res.*, **12**, 7035.
9. Britten, R. J. and Davidson, E. H. (1987). In *Nucleic acid hybridization, a practical approach* (ed. B. D. Hames and S. J. Higgins). IRL Press, Oxford.
10. Chan, Y.-L., Olvera, J., and Wool, I. G. (1983). *Nucleic Acids Res.*, **11**, 7819.

Colin D. Bingle

Further reading

Brown, T. A. (ed.) (1993), *Essential molecular biology: a practical approuch*, Vols I and II. IRL Press, Oxford.

A1

Addresses of suppliers

Activon Scientific Products Co. Pty Ltd, 2A Pioneer Avenue, Thornleigh, NSW 2120, Australia.

Aldrich Chemical Company, The Old Brickyard, New Road, Gillingham, Dorset SP8 4BR, UK; *Aldrich Chemical Co.*, 1001 West St Paul Ave, Milwaukee, WI 53233, USA.

Alltech, Carnforth, Lancashire, UK; *Alltech Associates Inc.*, 2051 Waukegan Road, Deerfield, IL 60015, USA; *Alltech Associates (Aust) Pty Ltd*, 27 Brookhollow Avenue, Baulkham Hills, Sydney, NSW 2153, Australia.

Amersham International, Amersham Place, Little Chalfont, Bucks, UK; *Amersham International*, 2636 S. Clearbrook Drive, Arlington Heights, IL, USA.

Amicon Ltd, Upper Mill, Stonehouse, Glos. GL10 2BJ, UK.

Anachem Ltd, Charles Street, Luton, Beds. LU2 0EB, UK.

Baxter Diagnostics Inc., 1430 Waukegan Road, McGaw Park, IL 60085, USA.

Beckman Instruments, Progress Road, High Wycombe, Bucks, UK; *Beckman Instruments Inc.*, 2500 Harbor Boulevard, Fullerton, CA 92634, USA; *Beckman Instruments (Australia) Pty Ltd*, 24 College Street, Gladesville, NSW 2111, Australia.

BDH: *BDH Merck Ltd*, Merck House, Poole, Dorset BH15 1TD, UK; *BDH Pty Ltd*, 242 Beecroft Road, Epping, Epping, NSW 2121, Australia; *BDH/E. Merck*, 350 Evans Avenue, Toronto, Ontario, Canada M8Z 1K5.

Bio-Rad: *Bio-Rad Laboratories Ltd*, Hemel Hempstead, Herts, UK; *Bio-Rad Laboratories*, 2000 Alfred Nobel Drive, Hercules, CA 94547, USA.

BOC Gases Australia Ltd, PO Box 247, Paramatta, NSW 2124, Australia.

Boehringer Mannheim Corporation, 9115 Hague Road, PO Box 50415, Indianapolis, IN 46250-0414, USA.

Borland International, 1800 Green Hills Road, PO Box 660001, Scotts Valley, CA 95067-0001, USA.

Brinkmann Instruments, One Cantiague Road, Westbury, NY 11590, USA.

Büchi Laboratoriums-Technik AG, PO Box, CH-9230 Flawil, Switzerland.

Calbiochem, PO Box 12087, La Jolla, CA 92039-2087, USA.

Cascade Biochem Ltd, The Innovation Centre, University of Reading, Whiteknights Campus, PO Box 68, Reading RG6 2BX, UK.

Cayman Chemical Co., 690 KMS Place, Ann Arbor, MI 48104, USA.

Continental Water Systems Pty Ltd, Unit 2/5 Abvil Road, Seven Hills, NSW 2147, Australia.

Crown Scientific Pty Ltd, 144 Moorebank Avenue, Moorebank, Sydney, NSW 2170, Australia.

Dako, Carpenteria, CA, USA.

Denley Instruments Ltd, Billingshurst, Surrey, UK.

Dionex, Sunnyvale, CA, USA.

EDT Instruments Ltd, Lorne Road, Dover, Kent CT16 2AA, UK.

Finnigan-MAT, 355 River Oaks Parkway, San Jose, CA 95134, USA.

Fisons, Loughborough, Leics., UK.

FMC Bioproducts: *Flowgen Instruments*, Broad Oak Road, Sittingbourne, Kent. *Flowgen Instruments*, 191 Thomaston Street, Rockland, ME, USA.

Gelman Sciences, 600 S. Wagner Road, Ann Arbor, MI 48106 1440, USA.

Gibco BRL, Unit 18, Thame Park Business Centre, Wenman Road, Thame, Oxon, UK.

Gilson Medical Electronics, SA, BP 45, 95400 Villiers-le-Bel, France.

Hewlett-Packard: *Hewlett-Packard Co.*, 3495 Deer Creek Road, Palo Alto, CA 94304, USA; *Hewlett-Packard (Canada) Ltd*, 6877 Goreway Drive, Mississauga, Ontario, Canada L4V 1M8.

HPLC Technology Ltd, Wellington House, Waterloo Street West, Macclesfield, Cheshire SK11 6PS, UK.

ICN-Biomedicals, Costa Mesa, CA 92626, USA.

ISCO Inc., Separation Instruments Division, PO Box 5347, Lincoln, NE 68505, USA.

J and W Scientific Inc., 91 Blue Ravine Road, Folston, CA 95630, USA.

Jones Chromatography Ltd, Tir-y-Berth Industrial Estate, New Road, Hengoed, Mid Glamorgan CF8 8AU, UK.

Lindbrook International, PO Box 522, Pennant Hills, NSW 2120, Australia.

LKB-Pharmacia, Pharmacia Pty Ltd, Biotechnology Division, 4 Byfield Street, North Ryde, NSW 2113, Australia.

Lotus Development Corporation, 55 Cambridge Parkway, Cambridge, MA 02141, USA.

Mac-Mod Analytical, Chads Ford, PA, USA.

Meeco Holdings Pty Ltd, 2–106 Grose Street, Parramatta, Sydney, NSW 2150, Australia.

E. Merck Pty Ltd, 207 Colchester Road, Kilsyth, VIC 3137, Australia.

Millipore Corporation, 43 Maple Street, Milford, MA 01757, USA.

Millipore Waters, Waters Australia Pty Ltd, Private Bag 18, Lane Cove, Sydney, NSW 2066, Australia.

Mitsubishi Gas Chemicals, Neville J. Webb Pty Ltd, 33 Ryde Road, Pymble. NSW 2073, Australia.

New England Nuclear, Boston, MA 02118, USA.

Novell, 1555 N. Technology Way, Orem, UT 84057, USA.

Organomation Associates Inc., PO Box 159, South Berlin, MA 01549, USA.

Perkin Elmer Ltd, Beaconsfield, Bucks, UK.

Pharmacia Biotech Inc., Piscataway, NJ 08854, USA.

Pierce, PO Box 117, Rockford, IL 61105, USA.

Polymer Laboratories Ltd, Essex Road, Church Stretton, Shropshire SY6 6AX, UK.

Porphyrin Products Inc., PO Box 31, 195 South 700, West Logan, UT 84321, USA.

Promega: Delta House, Enterprise Road, Southampton, Hants, UK; 2800 Woods Hollow Road, Madison, WI, USA.

Rainin Instrument Co. Inc., Mack Road, Box 4026, Woburn, MA 01888, USA.

Ratek Instruments Pty Ltd, Unit 1/3, Wadhurst Drive, Boronia, VIC 3155, Australia.

Rathburn Chemicals Ltd, Caberston Road, Walkerburn, Pebbleshire, EH43 6AU, UK.

Romel Chemicals Ltd, 63 Ashby Road, Shepshed, Loughborough, Leics LE12 9BS, UK.

Selby Scientific and Medical, 32 Birnie Avenue, Lidcombe, Sydney, NSW 2141, Australia.

Shimadzu Oceania Pty Ltd, PO Box 477, Rydalmere, Sydney, NSW 2116, Australia.

Sigma: *Sigma Chemical Co.*, Fancy Road, Poole, Dorset BH1 7NH, UK; *Sigma Chemical Co.*, PO Box 14508, St Louis, MO 63178, USA; *Sigma Aldrich Pty Ltd*, PO Box 970, Castle Hill, Sydney, NSW 2154, Australia.

Southern Biotechnology, Birmingham, Alabama, USA.

Stratagene: 140 Cambridge Innovation Center, Milton Road, Cambridge, Cambs., UK; 11011 North Torrey Pines Road, La Jolla, CA, USA.

Supelco Inc., Supelco Park, Bellefonte, PA 16823, USA.

Thermolyne, 2555 Kerper Boulevard, PO Box 797, Dubuque, IA 52004–0797.

Tropix, Bedford, MA, USA.

Upchurch Scientific, 660 West Oak Street, Oak Harbour, Washington 98277-1529, USA.

Varian Analytical Instruments, 3045 Hanover Street, M/S H-110, Palo Alto, CA 94304, USA.

Varian Instrument Group, 220 Humboldt Crescent, Sunnyvale, CA 94089, USA.

Waters–Millipore, The Boulevard, Blackmoor Lane, Watford, Herts WD1 8YW, UK.

Whatman Chemical Separation Inc., 9 Bridewell Place, Clifton, NJ 07014, USA.

Index

Index

Index

expression
  analyses 235
  detection 231–5
  various types 232
  measurement of activity 228–31
  precautions needed in assay 238
  protective role 227
  various types 227–8
glycoproteins 194
GSH 213
  assay using plate reader 219–20
  determination
    from biological samples 214, 217–18
    column methods 221
    by HPLC 223–4
  intracellular synthesis 213
  ion-exchange chromatography 221–2
  methods available to measure 217
  monobromobimane derivatization 223
  preparation of samples for assay 214–16
GSSG
  assay 220–1
  ion-exchange chromatography 222
  methods available to measure 217
Gray (unit) 52
G values, units for expressing 52

Haber-Weiss reaction 3
haemodynamics 84
heparin degradation, investigated by NMR 28
high performance liquid chromatography 6
  in carbonyl analysis 160, 163–4
  for determination of 4HNE 137–9
  measurement of lipid hydroperoxides
    120–5
  neutral sugar analysis 190–1
  purification of lipid extracts 263
  quantification of lipid hydroperoxides
    123–5
  usefulness 271
  vulnerability of equipment to corrosion by
    guanidine 164–5
  with electrochemical detection 271, 274
    passivation of system 282–3
4HNE
  biological activity 134
  choice of protocol 143–4
  determination by GC–MS 139–43
  determination by HPLC 137–9
  determination by UV spectroscopy 134–7
  quantification 134–5
  quantitative analysis by GC–MS 141–3
  uses of quantitative analysis 133–4
  see also aldehydes; carbonyls
hyaluronic acid, OFR degradation of 197
hydrogen peroxide 1

hydroxyl radical 1, 101–2
  detection by aromatic hydroxylation 104–5
  HPLC recommended for quantitative
    assessment 105
  methods of detection 102–3
  production 102
hydroxyvaline, *see* β-hydroxyvaline
hyperoxic lung injury 120
hypochlorous acid 1

immunoglobulin G, possible role for OFRs in
  denaturation 191
inflammation 67
  free radicals, cause or effect 7
  OFRs in 3
  role of microvascular blood vessels in 4
inflammatory conditions 4; *see also*
  rheumatoid arthritis
iron, role in generation of ROS 3; *see also*
  transition metals; Fenton reaction
iron-catalysed production of OFRs,
  carbohydrate damage by 187
ischaemia 4
  role of superoxide 241
ischaemia–reperfusion injury 120
isoluminol 174
  in determination of lipid hydroperoxides
    120–1
  problems 121

ketones, *see* carbonyls

lactate, as hydroxyl scavenger 32
lectin ELISA for complex sugar analysis 193
lectins 192
leucocytes, and tissue damage 4
lipid hydroperoxides
  chemiluminescent detection 121–2
  HPLC determination 120–5
  as mediators of oxidative stress 130
  in plasma, in diseased states 125, 128
lipid peroxidation
  chain reaction 2, 120
  and disease states 120
  as mechanism of oxidative stress 83
  methods of determination 120
  problems of *in vivo* determination 119
  see also cellular membrane; F_2-isoprostanes;
    polyunsaturated fatty acids, autoxidation
lipid peroxide 1
lipids
  hydrolysis 152
  methods of extraction 149, 150, 151
  oxidation of 119
liver injury 133
lycopene, *see* carotenoids

Index

malonaldehyde 189
malondialdehyde 133
  determination as measure of lipid
    peroxidation 125–6
  measurement by HPLC 126, 128–30
methanol, as solvent for pulse radiolysis 50
methionine, as target molecule for ROS 36
mitogenic rays, *see* chemiluminescence
mRNA, analysis by Northern hybridization
  236–8; *see also* RNA
multifunctional digital microfluorography
  83–4
  in assessment of oxidative stress 88–95
  calibration procedures 92
  choice of fluorescent probe 85
  intravital equipment 84
  problem of quantitative measurement 98
  technical considerations 88

nitric oxide, implication in rheumatoid
  arthritis and in Parkinson's disease 115
nitric oxide radical, as scavenger of superoxide
  83
nitrogen-centred free radical species 1
nitrous oxide, and leucocyte adhesion 93–4
non-essential PUFAs 6
Northern analysis 235, 238
Northern blotting 290, 291–2
Northern hybridization 289
nuclear magnetic resonance 6
  analysis of biofluids 27–8, 28–32
  analysis of cell cultures 36–9
  analysis of perchloric acid extracts 38
  detection of lipid peroxidation products 35
  investigation of effects of myocardial injury
    39
NMR spectroscopy 25, 26
  improvement of sensitivity 26
  range of information available 25–6
nucleic acids
  extraction 205–6
  oxidation products 201
  see also DNA, mRNA, RNA

8–OHDG, as marker for oxidative stress 208
oligonucleotides, end labelling 295
optical spectra of radicals
  found by pulse radiolysis 53
  table of examples 54
osteoarthritis 186
oxidants as mediators of cell injury 83
oxidation of lipids 119
  see also lipid peroxidation
oxidative stress
  and aldehydes 133
  as medical parameter 209

oxygen free radicals 1
  generated by ionizing radiation 4
  involvement in inflammatory conditions 4
  physiological regulation of levels 5
  toxicity 2, 133
  see also iron-catalysed production of OFRs
oxygen, vital importance of proportion in
  atmosphere 3
ozone layer 4

paraquat, protection against toxicity of 39
Parkinson's disease 115
peroxynitrate 1
peroxynitrite 1
phenylalanine
  detection of hydroxylation products 111–12
  disadvantages of use 112
  as trap for hydroxyl radical 109–15
ping-pong mechanism 241–2
plasma, *see* blood plasma
polyunsaturated fats, susceptibility to lipid
  peroxidation 119–20; *see also*
  polyunsaturated fatty acids
polyunsaturated fatty acids 6
  autoxidation 133
  free radical attack on, as marker of
    oxidative injury 119
  peroxidation products, detected by NMR
    27
  thermally induced deterioration, NMR
    investigation 40–4
  thermal stressing 27
  toxicological hazards 44
  see also polyunsaturated fats
post-transplantation graft injury 98
prawns, radiolytic damage to 39–41
prostaglandin, *see* F_2-isoprostanes
protein
  hydrolysis 172–3
  preparation of samples for hydrolysis 173
protein-bound DOPA 171, 179
  HPLC determination 180–1
  as initiator of secondary reactions 171–2
  methods for determining 172
protein hydroperoxides 171
  methods for determining 172, 174
  stabilization by reduction 174–5
protein hydroxides, methods for determining
  172
pulse radiolysis 6, 47–8
  calculation of absorbance 51–2
  design of experiments 53–5
  distinction between solute excited states and
    transient species 48–9
  need for specialized equipment 48

Index

vitamin C, *see* ascorbate
vitamin E, *see* tocopherols

Western blotting 168–9
  use in determination of carbonyl groups
  167

xanthine oxidase
  OFR generation by 197
  preparation 250
  as source of superoxide 250